Sewage and Biomass from Wastewater to Energy

Scrivener Publishing
100 Cummings Center, Suite 541J
Beverly, MA 01915-6106

Publishers at Scrivener
Martin Scrivener (martin@scrivenerpublishing.com)
Phillip Carmical (pcarmical@scrivenerpublishing.com)

Sewage and Biomass from Wastewater to Energy

Edited by
Inamuddin
Tariq Altalhi
Mohammad Luqman
and
Joseph Kapuku

Scrivener
Publishing

WILEY

This edition first published 2024 by John Wiley & Sons, Inc., 111 River Street, Hoboken, NJ 07030, USA and Scrivener Publishing LLC, 100 Cummings Center, Suite 541J, Beverly, MA 01915, USA

For more information about Scrivener publications please visit www.scrivenerpublishing.com.

Wiley Global Headquarters
111 River Street, Hoboken, NJ 07030, USA

For details of our global editorial offices, customer services, and more information about Wiley products visit us at www.wiley.com.

Library of Congress Cataloging-in-Publication Data

ISBN 978-139420-431-1

Front cover images supplied by Adobe Firefly
Cover design by Russell Richardson

Set in size of 11pt and Minion Pro by Manila Typesetting Company, Makati, Philippines

Printed in the USA

10 9 8 7 6 5 4 3 2 1

Contents

1

Thermal/Photocatalytic Conversion of Sewage Sludge and Biomass to Energy

Maria Siddique[1], Sumia Akram[2], Zainab Liaqat[1] and Muhammad Mushtaq[1]*

[1]Department of Chemistry, Government College University, Lahore, Pakistan
[2]Division of Science and Technology, University of Education, Lahore, Pakistan

Abstract

Recently, a great deal of research has been devoted to exploring renewable energy sources like sunlight, geothermal, hydropower, and bioenergy. Many developed countries like the United States of America are getting more than 20% of their energy needs from renewable sources and biomass can work as a fascinating renewable energy source. Biomass and sewage are considered viable alternatives to fossil fuels for sustainable energy production. Thermal and photocatalytic conversion technologies can generate energy from these materials, contributing to cost-effective and sustainable energy systems. Utilizing these technologies could reduce the environmental impact of energy production while creating new opportunities for economic growth and job creation. Additionally, utilizing biomass and sewage for energy production provides benefits such as addressing waste management concerns and mitigating the negative environmental impacts of waste disposal. Challenges and opportunities of biomass energy: The potential of thermal and photocatalytic conversion technologies to generate energy from biomass and sewage has been comprehensively highlighted in this chapter. A thorough examination of the various types of biomass and sewage and their energy content paved the way for scrutinizing the challenges and opportunities associated with using them as energy sources. Further investigation into thermal conversion technologies including pyrolysis, gasification, and combustion, in addition to the photocatalytic conversion process, was carried out. Scrutiny of each technology's respective advantages and disadvantages was done with great detail and analysis.

Keywords: Biomass, pyrolysis, photocatalysis, biofuel

**Corresponding author*: muhammad.mushtaq@gcu.edu.pk

Inamuddin, Tariq Altalhi, Mohammad Luqman and Joseph Kapuku (eds.) Sewage and Biomass from Wastewater to Energy, (1–42) © 2024 Scrivener Publishing LLC

1.1 Introduction

In recent years, policymakers have been very excited to switch to clean and green energy sources as it not only can help them combat environmental pollution but also provide more renewable energy sources [1]. Meanwhile, rapid industrialization and urbanization have led to a significant increase in global energy demands and a shortfall in fossil fuels. The growing consumption of fossil fuel-based resources for energy has caused several substantial rises in greenhouse gas emissions. These consequences hold immense importance not just for the economy but also for human health [2]. An increase in awareness about future energy and environmental concerns has spurred the utilization of available renewable viable energy sources. Biomass is now regarded as a green renewable energy source owing to its eco-sensitive nature, transformability, and, cost-effectiveness [3–5].

As cities continue to grow and become more heavily populated, per capita energy consumption increases and this may lead to high levels of carbon emissions [6]. The scenario becomes more alarming in countries such as China, Pakistan, India, Indonesia, and other certain parts of the world which totally rely on fossil fuels for their energy demands. With dwindling global oil reserves, treating and recycling copious amounts of wastewater produced by these so-called 'megacities' offered new horizons of challenges. Many countries have successfully switched toward solar energy but this technology remained limited to regions of intense and longer sunlight which is not the case with many countries like Russia and other European nations [7, 8]. Consequently, a single energy source may not work fine for all nations and we need a hybrid energy system. For example, countries situated beside oceans like Brazil can exploit algal biomass as a viable source of energy. It has been estimated that algae can produce 30 to 100 times more biomass than that is produced by other photosynthetic sources on Earth. Meanwhile, researchers claim large-scale production of algal biomass is associated with more carbon conversion/capturing present in the air [9, 10].

Although the exact definition and composition of biomass vary from country to country and source to source (Figure 1.1), we, in this monograph, are going to treat the underutilized agricultural residues, seaweeds, organic pollutants, and livestock residues as biomass. In a broader sense, biomass refers to a wide array of living entities that are obtained from the process of photosynthesis [11], including organisms of the plant, animal, microbe, and some byproduct classifications [12]. Various types of biomass

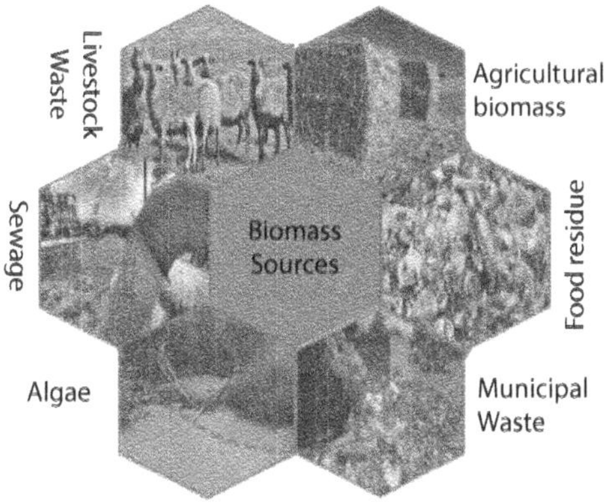

Figure 1.1 The key biomass/waste residues that can be exploited as alternative energy sources.

waste may comprise dissimilarities, for instance, agricultural waste forms encompass straw, crops, wood chips, and animal byproducts [13, 14]. Furthermore, industrial organic wastes include textile and food processing wastes, leftovers, and organic wastewater [15, 16], whereas municipal solid wastes cover food and kitchen waste, as well as domestic waste.

In countries undergoing rapid urbanization and an increase in population, the conversion of municipal waste into energy/electricity has been an essential area of research. The global annual municipal solid waste production is anticipated to be reached around 2200 million tons by 2025 as per the World Bank report [17]. The utilization of solid waste as an alternative energy source especially biomass and bio-solid wastes can help us to resolve not only waste management issues but also improve associated social and environmental standards [18, 19]. In another report, the yearly amount of worldwide solid waste production is estimated at 2400 million tons and is projected to reach above 2600 million tons in the next 6 years [20]. Furthermore, wastewater treatment plants produce around 1000 tons of solid waste each day [21]. With the proper technology, loads of energy required for the process can be minimized and also utilized for the production of sustainable energy sources.

1.2 Biomass as Energy Sources

Sewage and biomass are two promising renewable energy sources that have gained the interest of researchers, lately. Sewage or wastewater is an abundant source of organic matter along with other potentials; it can be useful in producing renewable energy such as biogas. Because sewage contains numerous complex components, it cannot be processed directly for energy production, rather it is required to be converted into a suitable form called sewage sludge [22, 23]. Therefore, the residue obtained from wastewater treatment is known as sewage sludge [24]. Sewage sludge can be distinguished both quantitatively and qualitatively, depending on the specific characteristics of the wastewater being processed and the approaches used for treatment. This process is relatively energy-intensive and definitively requires resources. This residue has long been classified as non-useful and often discarded in landfills [25]. A recent perspective regarding wastewater treatment approves of its conversion into energy and other useful products. It has been estimated that sewage sludge can potentially fulfill about 10% of our power demands as an alternative energy source [26, 27]. However, implementation of these perspective technologies is associated with high capital investment and needs considerable land area for sludge disposal. On the other side, sewage sludge production will keep growing with the population, and industrialization and its eco-friendly and sustainable management may cost much more particularly in developing countries [28].

Wastewater treatment plants are responsible for the sanitation of sewage water by separating its components, i.e., solids and minerals. It is also worth knowing that sewage sludge is an unavoidable material residue created at WWTPs, with significant amounts being created on a worldwide basis. According to the statistics, the annual dried sewage production of some countries was as follows: for the United States, it was 6.5 million tons in 2004 [29]; for China, it was almost 6.25 million tons in 2013 [30]; and for the European Union, it was 13 million tons in 2020 [31, 32].

Biomass is a popular phrase that relates to plant-derived and biological waste materials. However, the concept can include any sort of organic material. The total estimated production of carbon is 105 billion metric tons annually, evenly divided between terrestrial sources and oceans (algal biomass). Despite the fact that algal biomass constitutes a major part of world biomass volume, *wood* remains the primary biomass source used in various applications. Wood sources (cover forests) are often utilized for power generation and biopower, while agricultural remains like sugarcane bagasse, rice straw, cotton stalks, and wheat straw are classified as biomass [33–35].

Biomass can be employed as a powerful energy source to mitigate environmental risks associated with waste disposal and establish a sustainable and renewable energy alternative [36]. Bioenergy is another term used to refer to energy that is derived from biomass resources. It takes a series of stages, which include harvesting, drying, storage, moving, conversion, etc., to convert biomass into an energy source that can be utilized.

The direct combustion of wood and biomass has been used as an energy and heat source since prehistoric times. However, the conventional burning/incineration of agricultural residues causes a variety of environmental issues and their transformation into biofuels via a variety of processes (chemical, biological, and physical) renders them a renewable energy source. In many countries, the open burning/combustion of agricultural residues like sugarcane bagasse, rice straw, cotton stalks, and wheat straw is a punishable act [37]. Likewise, planned deforestation or accidental forest fires cause a variety of environmental issues. On the flip side, rural areas of many underdeveloped countries use agricultural and livestock biomass as energy/heat sources in houses, factories, and mills. The only thing that needs to be pointed out here is that direct conversion of agricultural, livestock, domestic waste, or industrial waste (paper and plastics) causes a variety of known and concealed environmental issues and health issues. For similar reasons, agriculture biomass, livestock waste, and municipal waste are subjected to controlled pyrolysis, anaerobic digestion (AD), gasification, and fermentation for the cogeneration of heat and electricity [38]. The algal biomass can work as a fascinating alternative energy source as its production involves the uptake of carbon dioxide and does not need valuable arable land and fresh water. The seaweeds have been successfully transformed into biofuel; however, another school of thought believes that algal species have the potential to be used as superfoods for human and animal consumption.

1.3 Biomass Types and Energy Content

The waste is a fusion of municipal and industrial materials, normally composed of sewage sludge, regular waste, food waste, shrubbery, paper items, latex, fabrics, firewood, debris, and soil. The biomass in contrast covers seaweed, food, and agricultural wastes, and a fraction of domestic or industrial refuse that contains carbon and hydrogen. The exact ratio of components in any waste collected from one location can differ from other sites [39]. Biomass often denotes a form of waste residue derived from sources such as firewood, timber, crops, aquatic plants, shrubs, municipal waste, and human waste. The exact composition of biomass depends upon its source, i.e., agricultural residues mainly consist of cellulose, hemicellulose, lignin, and trace amounts of minerals, vitamins, and extractives [40].

Besides, legal regulations, lifestyle and seasonal elements, preparation, and recycling procedures cause waste streams to vary with time. Underdeveloped nations often have waste that predominantly consists of decomposable material in comparison to plastics. Likewise, municipal waste may be rich in pigments, plastics, glass, and other nonbiodegradable materials. For the production of energy, chemical manufacturing, and biofuel production, it offers more benefits than traditional fuels [41]. According to the International Renewable Energy Agency (IRENA), biomass is expected to become one of the leading renewable energy sources by 2030. According to their roadmap, biomass is capable of meeting up to 60% of global energy consumption [42]. In many developed countries like the United States, biomass contributes more than 2% of renewable energy sources; however, for countries like Brazil which produces more than 20,000 tons (algae year^{-1}), the contribution of biomass as an energy source may exceed.

The use of biomass and other organic waste materials as renewable energy sources has become a fascinating choice for the global energy supply and it has already accounted for about 20% of the global energy production, securing its position as the fourth-largest energy source [27]. Biomass-based sources have been extensively explored to meet our energy demands for the past 30 years, and they can be speculated to be the top energy source fulfilling over half (~56%) of the demands as compared to renewable sources [43]. It has been also anticipated that clean energy sources like biofuel will cover 2 to 27% of our energy source in the immediate future [44]. Organic contents of sludge are typically made up of proteins (24 to 42%), carbohydrates (7 to 18%), and lipids (1 to 14%). The chief functional groups of such organic compounds comprise carbonyl moiety, carboxylic acids, different amines, amides, some aromatics, and methyl. In underdeveloped countries of Asia and Africa, biomass can account for around 33% of the total energy needs [45]. The widespread accessibility of biomass as an energy source makes it advantageous compared to other conventional energy sources [46].

1.4 Conversion of Biomass to Energy

Incineration has been applied for years to destroy hazardous materials or contaminants. For many years, the incineration of organic material has been applied to produce a stream of gasses or heat to run turbines and produce electricity. Nowadays, thermal and photocatalytic methods have been applied to initiate a chemical reaction for the conversion of biomass into energy. Thermal processes use heat to change organic matter into energy, whereas photocatalytic processes rely on light to initiate chemical reactions that generate energy afterward. The power generation system nowadays utilizes waste as fuel and produces energy from the abundant materials present in trash gathered from both residential and industrial sources. As the exact composition of waste or biomass varies with location, the exact incineration, thermal, or photolytic conversion process varies widely [39].

1.4.1 Thermal Conversion of Biomass to Energy

Figure 1.2 illustrates the general way out for the thermal conversion of biomass into a set of products. Despite having an effective system of recycling and reconversion of wastes, many biomass or waste ingredients are hard to break, it is important to have an exact composition of biomass for effective conversion [47]. A variety of approaches are used to convert solid waste into something more beneficial to the planet, including hydrothermal liquefaction [48], pyrolysis [49], and gasification [50]. The key objectives have to apply biological, chemical, or thermal approaches to generate heat and electricity. Using fuel produced from waste materials can have a smaller environmental impact than burning fossil fuels [37]. The energy production from the waste can solve the solid waste disposal issue and produce electricity. For example, during thermal conversion, the wastewater is heated to high temperatures that then generate gasses, especially methane,

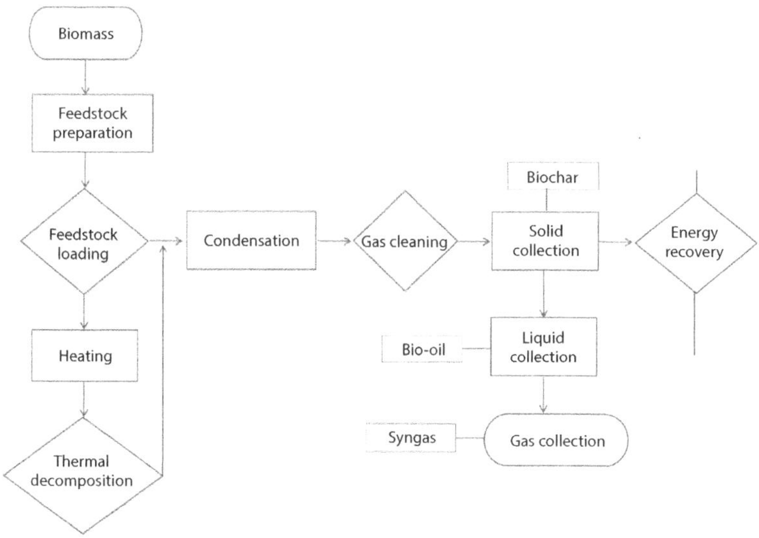

Figure 1.2 Process identification diagram for traditional thermal conversion of biomass.

and hydrogen as byproducts that further act as fuel to generate electricity and heat. However, the method uses a substantial amount of energy in getting high temperatures, making it generate harmful air pollutants.

In contrast, the photocatalytic conversion uses less energy to decompose the organic material in the wastewater. This method produces hydrogen gas and CO_2 which can serve as fuel. Photocatalysts (such as titanium oxide) offer several advantages over traditional thermal conversion such as minimal energy input, reduced pollutants, and higher conversion efficiency. The wastewater even has the potential to be transformed into energy using diverse methods mainly in the presence of catalysts, heat, and/or light. However, it is still in the experimental stage, and further research is necessary to improve its efficiency and scalability.

1.4.1.1 Combustion/Incineration

The incineration process involves burning and converting feed materials into beneficial products. This generates heat that can be used in vapor-based turbines and heat exchangers to generate power. The temperature range is in the vicinity of 800°C to 1000°C. Practically, the biomass rich in organic (carbon) that has been dried and holds a water level of less than 50% can be burned in the process [51]. The biomass with higher water levels is usually transformed into energy through biological means. Incineration can decrease the volume of solid waste by 80 to 85% [52]. However, this technique has been around for a long time and is associated with toxic airborne discharges, including dioxins and heavy metals. Also, the inorganic contents like arsenic, mercury, cadmium, and uranium in the ashes which cannot be burned can be dangerous to one's health [53].

1.4.1.2 Hydrothermal Carbonization

This technique involves temperature-based conversion and has captured the interest of many researchers, particularly attractive to scientists working with waste materials that have a moisture content higher than 80 to 90%. This process is considered a "wet" process as it transforms different kinds of waste into a usable energy source subjected to consistent pressures and relatively low-temperature conditions between 180°C and 350°C [54, 55]. This process has multiple advantages such as larger waste handling capacity, removal of odors along with contaminants, and shorter incineration period. Furthermore, an increase in the temperature of the procedure can decontaminate various hazardous materials. The byproducts and certain nutrients can be used for the production of gasses and fertilizers.

The hydrothermal carbonization process simultaneously causes hydrolysis, condensation, dehydration, and decarboxylation in the damped raw material [56, 57].

1.4.1.3 Biomass Pyrolysis

Pyrolysis is the thermal decomposition of material in an oxygen-free environment by using an external heat source to keep the temperature between 300°C and 850°C. The expression pyrolysis is derived from the Greek words 'pyro' signifying fire and 'lysis' implying disassembly or breaking into separate parts. The result of this process is typically char, solid residue, and a blend of carbon and ash. It is also possible to condense the product to yield tar, crude wax, lubricants, and oil. On average, the syngas created from pyrolysis contains a net calorific value of 10 to 20 MJ/Nm^3 [58].

Pyrolysis has grown significantly popular due to its adjustable nature, aptitude for using many different feedstocks, and possibility for varied outputs. During this process, which does not require oxygen, heat is applied from outside sources, and biomass components are broken down into gasses and vapors which usually have subsequent reactions, thus providing a broad scope of results. The circumstances that can have an immense influence on both the products and the process performance consist of feedstock, technology, reaction temperature, accompaniments, accelerants, vaporization stay time, solid stay time, and pressure. For thousands of years, pyrolysis has been utilized to create charcoal, yet only within the past 35 years have efforts been made to invent rapid pyrolysis for the production of liquids. Such pyrolysis needs to occur at approximately 500°C for a very brief period of time with the vapor's stay in this hot area lasting less than 2 s. This kind of quick pyrolysis is of particular fascination because it can produce liquids in high yields of up to 75 wt% [59]. The slow and intermediate pyrolysis process centers on the production of dry char and biomass, which can be directly utilized in a wide range of applications, offering energy density and/or serving as a biofuel source. More attention is now being paid to maximizing the resulting products. This method of burning has been practiced for more than seven decades in order to reduce the amount of disposed waste while making the remainder less ecologically damaging.

Thermochemical conversion of biomass takes place through pyrolysis, gasification, liquefaction, and combustion. Pyrolysis is the origin of all of these conversions, which breaks a compound down into solid, liquid, and gas components with no oxygen. Particular importance is placed on the technologies that can create either gaseous or liquid products which can be improved to become useful energy sources such as electricity and transportation fuel.

When placed inside a reactor and heated, the organic elements in the wastewater sludge commence transforming at around 200°C, leading to volatile products and a nonvolatile solid byproduct known as char created by several breaking and forming processes [60]. Primary pyrolysis and the vaporization of moisture contained in a material are both essential initial steps for all of the processes involved in transforming something through chemical reactions with heat, including burning, generating gas, forming a liquid, and changing the composition of carbon.

1.4.1.4 Noncatalytic Pyrolysis

Pyrolysis is an advantageous way to treat sewage sludge because it changes the solids in the sludge into biochar, bio-oil, and gasses (such as hydrogen, carbon monoxide, and some other frothy gasses) while emitting only a small amount of pollutants in a temperate to moderate temperature range, in the absence of oxygen. Specifically, pyrolysis involves thermally breaking down the sludge solids at high temperatures (400°C to 600°C) in an oxygen-free environment, to obtain the required output. Pyrolysis can be of various types including slow, fast, hydro, vacuum, and flash pyrolysis [61, 62]. Crucial elements that can affect the pyrolysis operation include heating rate, feedstock size, duration of residence of reactant, and temperature. Pyrolysis has been utilized to treat various feedstocks such as plastic waste, municipal solid waste, oily sludges, sewage sludges, and woody biomass.

1.4.1.5 Slow Pyrolysis (Carbonization/Torrefaction)

Slow pyrolysis (carbonization/torrefaction) typically comprises temperatures below than 400°C. The goal of this process is to condense moisture and densification, and also make the material more grindable. Generally, slow pyrolysis results in a greater volume of char and lower amounts of bio-oil and gas. The solid that is left behind is referred to as charcoal. The carbonization process eliminates the presence of NO_x, SO_x, and additional contaminants along with the densification of sludge and also raises the C/H ratio and calorific value [63, 64]. When sewage sludge is burnt at 900°C, it produces a noteworthy decrease in NH_3 (46%) [65]. Furthermore, sewage sludge or biomass can be mixed and burned to form a large biochar product. Factors of importance to assess include balances of energy and mass to identify the productivity of torrefying sewage sludge.

1.4.1.6 Rapid/Fast Pyrolysis

This type is employed for the conversion of sewage sludge into different bio-oil products. It differs from slow pyrolysis in its heating rate, stay time, and temperature range (400°C to 600°C). The fast pyrolysis of biomass produces more bio-oil [66]. Bio-oil produced via fast pyrolysis of sewage sludge mainly comprises hydrocarbons, organic acids, high molecular weight carbonyl compounds, phenols, ketones, a broad range of both aliphatic and aromatic compounds, acetic acid, different alcohols, nitrogen, and sulfur-containing compounds along with water [22, 67–69]. The bio-oil is found to be more eco-friendly than traditional heavy fuel oil, as it contains additional constituents and has higher moisture levels thanks to the presence of oxygenated compounds that divide it into two fractions, one lighter, and the other with a higher concentration of organic oil. The organic phase which is lightweight largely consists of water, with the heavier one being abundant in benzene, toluene, styrene, naphthalene, phenol, and so forth. The mean heat energy value of bio-oil (30.1 MJ/kg) is less than fuel oil (40.2 MJ/kg). Due to pyrolysis, the heat energy of the bio-oil becomes low and the resultant product is typically used for combustion and heating purposes, although the small heat energy acts as a hindrance to this method. Research on how to improve the quality of bio-oil has been conducted to raise the C/H ratio and diminish oxygen concentration.

The gas stream released as the result of sewage sludge pyrolysis consists of lighter gasses such as hydrogen (H_2), carbon dioxide (CO_2), carbon monoxide (CO), methane (CH_4), ethylene (C_2H_4), propane (C_3H_8), and butane (C_4H_{10}) with energy values of 32.2 MJ/kg for bio-oil, 12.3 MJ/kg for char, while 25 MJ/kg for gas, with the process taking place at 500°C [70]. If a higher temperature is used during the pyrolysis, the quantity of gas rises; though at the same time, increasing the possibility of emissions such as nitrous oxide (NO_x), sulfur oxide (SO_x), tar, polycyclic aromatic hydrocarbons (PAHs), and particulate matter (PM). The percentage of PAHs in sludge products of pyrolysis at different gas residence times and temperatures of 1000°C can rise from 1.14 wt% up to 6.33 wt%, thanks to high flow rates of the gasses produced. These pyrolysis gasses may have reduced energy content when heated in a medium to low range, yet they still have the possibility of being used to generate heat [71, 72].

The char obtained from the rapid thermal decomposition procedure includes a combination of carbon and has been utilized as an adsorbent due to possessing a very large exterior area or for more utilization in the generation of syngas and chemicals in the process of gasification. The heating values of the chars mainly vary according to their basic material, and the readings generally range from 11 to 15 MJ/kg [72].

1.4.1.7 *Co-Pyrolysis*

Incorporating alternative feedstocks alongside sewage sludge through co-pyrolysis proves to be a beneficial approach for improving both the quantity and quality of obtained pyrolysis outcomes. Among the feed fuels, reported to be used with sludge in such pyrolysis are pinewood saw [73], microalgae [74], wood dust [75], paddy straw [76], wheat chaff [77], bamboo wood dust [78], compost [79], filbert shell [80], wool + stalk [81], cannel shale/fossil oil [82], lignite coal/brown coal [83], megass [84], and waste wood [85]. Such feed fuels raise the output of H_2, CO, and other gasses, boost the texture of biochars, and help take out oxygenated rainwater and different pollutions in the bio-oil. Organic sources like bagasse, wheat chaff, paddy straw, and filbert supplement additional mass reduction at 200°C to 600°C heat range rather than the sludge, so this can set off the dismantling of volatile matter and a rise in the supply of oil and gas.

1.4.1.8 *Gasification*

Gasification (Figure 1.3) lies between pyrolysis and combustion in terms of its processing temperature; it requires oxygen, air, steam, or a mixture of them and a heat of >650°C [86]. The net heat output of gasification is usually less than that of combustion or incineration, usually in the range of 4 to 14 MJ/Nm^3. The process is exothermal, meaning an external source

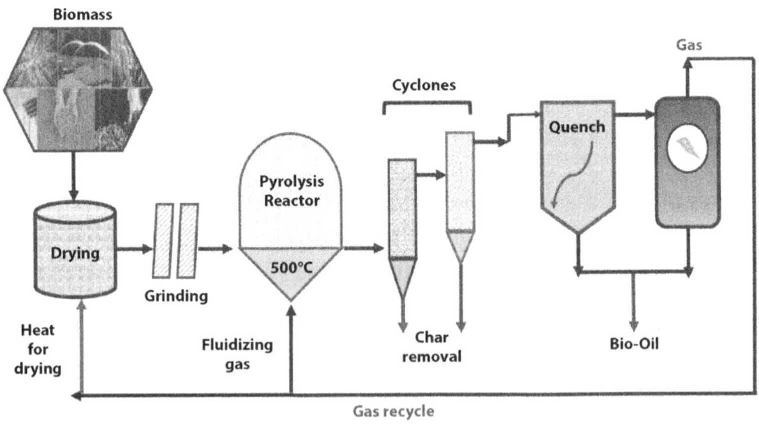

Figure 1.3 Layout for traditional thermal pyrolysis/gasification of biomass.

of energy must be supplied [87]. Gasification has its perks over conventional incineration or combustion systems, mostly in terms of cost and the flexibility offered via optimization of operating conditions like equivalence ratio or temperature and features of the apparatus like fixed-bed and fluidized-bed. Ed, an entrained-bed vertical-shaft furnace, rotary kiln, and plasma gasifier are used to get gas that can be employed for multiple purposes [88]. The syngas is formed of several gasses such as CO, CH_4, CO_2, and H_2 plus light C_XH_Y, for instance, C_2H_6 and C_3H_9, and denser hydrocarbons, like tars, that liquefy at temperatures of 250°C to 300°C. Unwanted fumes may also be present in the resulting gas, which depends on the treatment procedure and operational conditions [89, 90].

When sewage sludge is heated, the first step is a process of drying and devolatilizing, resulting in volatile substances being discharged and the remainder becoming solid char. Within a gasifier, additional gas-to-gas processes and gas-to-solid transformations, both involving the gasifying agent and the products of drying, pyrolysis, and cracking, occur in order to generate a larger total of combustible gasses. Gasification can be thus looked upon as an additional part of the pyrolysis process. The gasification of sewage sludge is often administered with air as the gasifying agent [91–96]. However, other gasifying media have also been utilized, including steam [97–101], air-steam [102–104], steam, CO_2 [105], steam-O_2 [106], and CO_2 [107].

Generally, a gasification system has four separate parts which are referred to as zones or stages. This includes the drying stage where sewage sludge is brought to a temperature of between 70°C and 200°C to evaporate the moisture content. Several factors like the surface area of the feed materials, the speed of recirculation, and the relative humidity regulate the drying. The humidity of the gas used for drying, the rate of moisture diffusion within the sewage sludge, and the variance between the feed and hot gasses all contribute to the dehydration process [91, 108]. For complete dehydration, the sludge should be below 15% damped [91, 108]. The next stage is the pyrolysis zone undergoes the thermal decomposition of the dried sewage sludge (between 350°C and 600°C). The next stage is the oxidation zone where products of pyrolysis (volatiles and char residue) get oxidized, making a swift rise in temperature up to 1100°C [91, 108]. The energy generated from the exothermic chemical reactions powers the drying and thermal cracking of sewage sludge and the gasification of char. The char is changed into a gaseous blend principally comprised of CO and H_2 (known as syngas) mainly through incomplete ignition, the Boudouard response, steam gasification, and hydrogasification. As well, a consistent gas-phase reforming of volatiles takes place, influencing the makeup of the resulting gas. The main issue with transforming sewage sludge into gas is the gasification of char, as it is a much slower process than the devolatilization or reforming of volatiles [109]. The speed of the char's gasification is dependent on the type of gasifying agent, the operating conditions, and the char's reactivity, and thus, this is a key factor to contemplate when creating a gasifier, as a reduced conversion of char decreases the overall success of the gasification process destroyed "during the char gasification in fluidized beds" [110, 111].

The effect of certain elements on the reactivity of char is accounted for, like the temperature and heating rate of pyrolysis, the gasification medium [105, 110, 112], the pore structure [113], the catalytic capability of ash [112, 113], and the activity of the char's carbon [113]. Moreover, the gasification process in fluidized beds is not able to completely break down all of the tars generated in the devolatilization phase, similar to that of biomass gasification. The geometry and size of the gasifier as well as the rates and nature of the processes that occur can greatly affect the completion of the gasification procedure. Subsequently, tar accumulation can diminish the effectiveness of gasification, so minimizing tar production should be taken into account when planning and using the gasifier [114].

1.4.1.8.1 Reactors

Reactors play a pivotal role in the gasification process. Scientists have identified several types of reactors that can be employed during sewage sludge gasification: fixed bed [97], fluidized bed [115, 116], circulating fluidized bed [117–119], rotary kiln [120], auger, and plasma [121, 122]. The gasification type, either updraft aka co-current or downdraft aka cross draft, can be evaluated by the nature of the method used for the injection of the gasifying agent into the gasifier. The composition of the feedstock—including its moisture and ash content—plays an important role in what the final product from the reactor looks like. The updraft gasifier uses a top opening for the introduction of the sewage sludge while gasses like oxygen, air, steam, or their combination are added from the bottom. The resulting gas then exits from the upper side. Lower efficiencies have been reported in this regard mainly due to the contamination in syngas. In the downdraft gasifier, the feedstock is fed in through the upper opening besides the gasifying gas occupies the zone of oxidation. The resulting syngas exists from the lower portion of the gasifier, and one advantage is that it has a low risk of contamination by tar. However, there are some issues with slag accumulation and a decrease in pressure within the fixed-bed reactor.

Fluidized bed reactors require a high velocity of air to break down sewage sludge into smaller particles, resulting in a large amount of produced gas. The transfer of heat is effective, and the gas is clean after the particles have been filtered out using cyclones and electrostatic precipitators. However, the high level of ash left from the sewage sludge poses a problem, as it reduces bed height and produces clinker and lumps. Fluidized bed and circulating fluidized bed reactors are effective with regard to generating solid/gas (sublimation) reactions, removing asphalt, along generating high-grade syngas. Furthermore, these reactor types provide an opportunity for char burning and catalyst regeneration. Unfortunately, high gas flow rates increase the risk of fuel particles being broken down, which could lead to lower quality outputs. Because these high flow rates also cause additional handling and maintenance costs, it is important to ensure that functioning is optimized. Regarding this situation, a two-stage reactor setup is used to create a high-grade synthesis gas along with ash removal. It has been documented that there is a three-stage configuration accessible with an additional two-stage setup reactor. An additional third stage might contain an auger or plasma reactor, fixed or fluidized bed, or rotary kiln in order to take out tar and enhance synthesis gas quality. However, this method has the constraint of additional costs.

Auger reactors have been noted for their better asphalt amputation and improved syngas grade. Meanwhile, the rotary kilns may be fitted along foils reserved in cylinders placed at a specific perspective in order to eliminate ash buildup as well as clumping. However, due to their rotation at certain speeds, rotary kilns typically have a slower rate of heat transfer and may suffer from inefficient seals. On the other hand, plasma reactors provide an option with several advantages. Promising outcomes from moisty sewage sludge treatment and elimination of contaminated gasses by heated processes are challenging to realize on a large business scale as they need a large amount of energy input. Several factors must be taken into account when choosing a certain reactor technology for sludge gasification, among which temperature, pressure, material nature, water content, feed particle size and rate, the relation between sewage and air sludge, and the ratio between steam and sewage sludge. An additional criterion aimed at tar (asphalt) along with syngas remodeling is the correct choice of a catalytic agent and where it is located (internally or externally) in relation to the reactor, as well as its overall effects on the end product. The consequences of each reactor technology when it comes to these elements and the given conditions differ. It may be advantageous to combine gasification with steam and the catalyst in the same system, as this could help reduce time and cost. It is strongly suggested that this setup be used.

1.4.1.8.2 Biomass Gasification

Gasification is a complex process wherein several reactions take place in an unpredictable way (Figure 1.4). The essential steps of this process are drying, reduction, and oxidation with the addition of tar breakdown in recent times for the production of a lighter C_XH_Y due to the high tar content. To execute all these reactions, a few different types of reactors with varying feed and operational requirements are used. Gasifiers are the reactors being used to process all sorts of wastes or biomass [123, 124]. The manner in which the raw material and agent interact can be co-current, counter-current, or crossflow and gasifiers may be powered by an external or internal heat source. The time spent in the reaction zone may range from a few hours to just a few minutes. The most common reactors used in the gasification of biomasses are entrained flow, fixed-bed, fluidized-bed, rotary-kiln, and plasma reactor [125]. Biomass gasification can be described as follows:

$$\text{biomass} + O_2\ (\text{or } H_2O) \rightarrow CO, CO_2, H_2O, H_2, CH_4 + \text{other hydrocarbons}$$

$$\rightarrow \text{tar} + \text{char} + \text{ash}$$

$$\rightarrow HCN + NH_3 + HCl + H_2S + \text{other sulfur gase}$$

1.4.1.8.3 Technology Advantages and Disadvantages

There are three main thermal conversion technologies for biomass and sewage: pyrolysis, gasification, and combustion with their distinguished pros and cons.

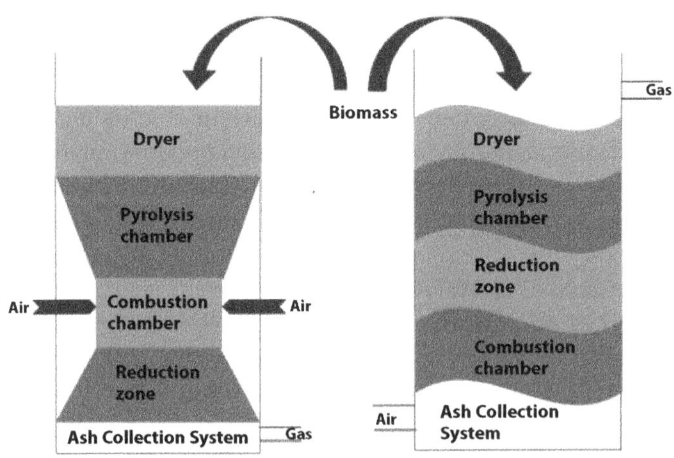

Figure 1.4 (a) Downdraft gasifier and (b) updraft gasifier [42].

Pyrolysis involves heating biomass or sewage without oxygen, producing biochar, gasses, and liquids that can be used as fuels. Biochar is useful for soil improvement. The advantages of pyrolysis are low emissions, biochar production, and high energy efficiency. However, the process requires high capital costs, unstable operating conditions, and extensive feedstock processing. Gasification is a process that involves heating biomass or sewage with limited oxygen to form a gas mixture that can be used as fuel for electricity and heat generation. The advantages of gasification include high energy efficiency, fuel flexibility, and low emissions. The disadvantages are high capital costs, ash disposal challenges, and the need for preprocessing feedstocks.

Combustion is a process that involves burning biomass or sewage in the presence of oxygen to produce heat and steam that can generate electricity. Combustion has the advantages of low capital costs, high energy efficiency, and fuel flexibility. However, the process results in emissions of PM, sulfur dioxide, and nitrogen oxides and requires air pollution control equipment. The choice of thermal conversion technology depends on factors such as feedstock type, scale of operation, and local regulations. Successful implementation of these technologies requires careful planning and management to ensure optimal performance and thermal conversion technologies are advantageous in their capacity to convert diverse feedstocks and their efficient conversion rates. They have broad applications in energy production, yielding electricity, heat, and biofuels. Thermal conversion technologies' negative attributes include high costs in operation and capital, intricate process design, and the creation of greenhouse gasses, such as carbon dioxide, nitrogen oxides, and sulfur dioxide. Additionally, the handling of residues such as ash requires specific disposal methods.

1.5 Biochemical Conversion

In the biochemical conversion of biomass into energy or fuel, the agricultural or algal feedstock is treated with a variety of microorganisms like *Lipomyces starkeyi* which hydrolyze biomass into lipids for their subsequent conversion into blends of hydrocarbons (gasoline, diesel, or jet fuel). The process mainly comprises a variety of steps like biomass pre-treatment, hydrolysis, biotransformation, chemical transformation, product recovery, refining, and energy production. The most commonly adopted biochemical process involves the fermentation of agricultural sources like wheat straw, corncob, beat sugar, and sugar cane bagasse into ethanol. There are two main processes that can be used: AD, for producing biogas—a mixture of mostly methane and carbon dioxide—and fermentation, which produces ethanol. The biochemical changes normally involve two processes: digestion which produces biogas and fermentation which produces ethanol. Biochemical conversion can be useful for feeds that feature a higher amount of feedstock, while thermochemical transformation is more suitable for feeds with less water content as it occurs at elevated temperatures and high reaction rates [126].

Sewage sludge contains a large amount of recyclable energy, but with high moisture, ash, metals, and organic pollutants, it causes massive environmental and health issues. The energy recovery pathways can be either chemical or biological. AD, a biological process, is financially beneficial; however, it might have a negative environmental effect brought about by pollutants released into the air. Hence, when it comes to picking the best way of regaining energy, factors such as environmental and economic aspects of resource and energy salvaging ought to be taken into account [127]. Nevertheless, the majority of thermochemical processes, which employ catalysts, are still not in full use for energy production from wastewater sludge. While certain approaches have reached the trial period, few have been truly successful [128] as evaluated in Table 1.1. Thermochemical processing is desirable due to its capacity to reduce sludge amounts, kill off hazardous germs, manufacture an expansive variety of products, and notably gain substantial energy restitution. In order to improve the grade and variety of thermochemical processes, the usage of activators and add-ons becoming noticeably valued [129].

1.5.1 Photocatalytic Conversion of Biomass

Diverse methods exist for transforming biomass into energy, including thermal, chemical, biological, aerobic, anaerobic, and photocatalytic processes. Thermal conversion includes combustion, incineration, pyrolysis, and gasification, leading to the production of heat, biochar, and syngas, respectively. The thermal conversion of biomass often generates electricity, gas, fuel, and heat. Alternatively, photocatalytic conversion uses photocatalysts and solar radiation to transform biomass into hydrogen or other valuable products, providing greater energy efficiency and product selectivity than thermal methods.

The IUPAC states that photocatalysis is a transformation in the pace of a chemical phenomenon or its starting upon interaction with ultraviolet, visible, or infrared light in the presence of a component—the photocatalyst—that takes in light and is involved in the alteration of the reaction ingredients [130]. On the other hand, the term photochemistry

Table 1.1 A comparison of efficiency and working conditions for biomass conversion technologies.

Year	Sample	Technology	Conversion process	Conversion efficiency/ product yield	Reaction conditions T	Ref.
2021	Sewage sludge	Pyrolysis	Carbonization-activation cascade waste heat recovery system	Under the optimal condition (T2 = 700 C, T3 = 200 C), the waste heat recovery ratio and the exergy recovery ratio reach 90.8 and 94.9%, respectively, which were improved by this method compared with the traditional water quenched method	Reactor: 300°C–800°C Activation reactor: 800°C–1100°C	[157]
2022	Paper and pulp industry wastewater	Photocatalysis as an advanced oxidation process	Photocatalytic degradation of organic pollutants in the wastewater leads to hydrogen production	An estimated amount of 0.45% of the total world's energy could be produced from the paper and pulp industry alone	Different composite photocatalysts at different loadings, pollutants, concentrations, and time	[158]

(Continued)

Table 1.1 A comparison of efficiency and working conditions for biomass conversion technologies. (*Continued*)

Year	Sample	Technology	Conversion process	Conversion efficiency/ product yield	Reaction conditions T	Ref.
2019	Sewage sludge	Gasification	Gasification	NA	298K	[159]
2019	Sewage sludge	Hydrothermal carbonization (HTC)	Hydrothermal carbonization (HTC)	NA	180°C and 300°C	[160]
2014	Biomass	Pyrolysis	Microwave-assisted pyrolysis	40–51% (bio-oil)	220°C–500°C	[161]
2017	Biodiesel wastewater	Photocatalysis	Photocatalytic oxidation	H_2 production rate 67.4 μmol/hour	30°C	[162]
2019	Biochar derived from sewage sludge	Photocatalysis	Metal-based photocatalysts	3845 μmol/g H_2 production	700°C for the pyrolysis of sewage sludge	[163]
2017	Sewage sludge mixed with sawdust	Hydrothermal carbonization (HTC)	Thermochemical conversion	47.06–71.60%	>260°C	[164]
2015	Biomass (algae-*chlorella* residue)	Pyrolysis	Slow pyrolysis	56.3, 66.2, and 65%	300°C, 500°C, and 700°C	[165]

(*Continued*)

Table 1.1 A comparison of efficiency and working conditions for biomass conversion technologies. *(Continued)*

Year	Sample	Technology	Conversion process	Conversion efficiency/ product yield	Reaction conditions T	Ref.
2009	Sewage sludge	Pyrolysis	Thermochemical conversion	39.52% biochar	350°C and 950°C	[166]
2013	Biomass (straw)	Pyrolysis	Microwave-assisted pyrolysis	33.7 wt% biochar	200°C	[167]
2019	Biomass (canola residue)	Pyrolysis	Slow pyrolysis	conversions of up to 70 wt% of volatile matter	<450°C	[168]
2022	Lignocellulose biomass	Pyrolysis		**Bio-oil**: 35–45% yield **Biochar**: 40–50% yield	500°C	[169]

is also often used to describe photoactivated or heated chemical processes. In a photochemical reaction, rather than having a photocatalyst, the reactant is put in an excited state. This same excitation process is used in all photo-sensitized reactions when, as an example, molecules in a dye absorb light and the associated excited state transfers an electron into the conduction band of a semiconductor that is suitable. The use of photocatalytic conversion technology holds great promise for the conversion of sewage and biomass to energy. This innovative approach involves the use of photocatalysts that can create free electrons and holes when exposed to light. These electrons and holes then engage in chemical reactions to break down organic pollutants in sewage and biomass. One of the key benefits of this technology is that it can operate under gentle conditions, such as at room temperature and atmospheric pressure, and does not generate harmful byproducts. Moreover, this technology offers the possibility of treating sewage and biomass while producing renewable energy, making it an attractive option for sustainable waste management.

Figure 1.5 illustrates steps taking place during the photocatalytic conversion to effectively treat sewage and biomass. This includes carefully selecting the proper photocatalyst, with TiO_2, ZnO, and CdS being the most commonly utilized. Designing the proper reactor is also crucial, with batch, continuous flow, and immobilized reactors all being employed. To ensure a successful conversion of sewage and biomass to energy, the process conditions must be optimized, such as accurately controlling the pH, temperature, and light intensity. Although this technology offers numerous benefits, like low operating costs, simultaneous treatment, and the ability to degrade various organic pollutants, there are still a few disadvantages involved.

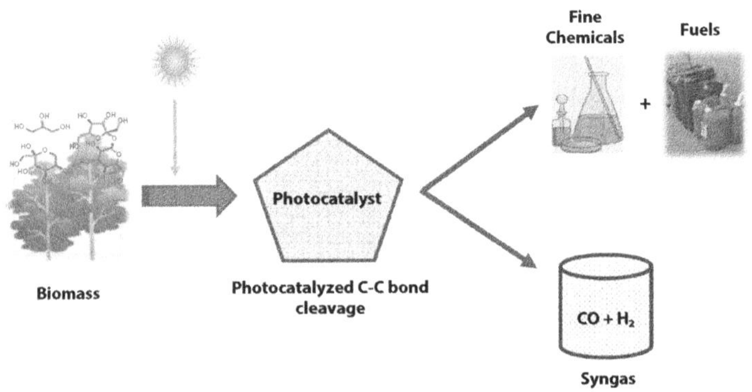

Figure 1.5 The generic layout for the photocatalytic conversion of biomass into various products.

For instance, high-intensity light sources are required, which can lead to longer treatment times and the potential for photocatalyst deactivation. Precise control of the reaction conditions may also pose a challenge.

The generation of heat and biofuels from biomass and waste materials can be accomplished through the utilization of three major processes: physicochemical, biochemical, and thermochemical. These methods are based on the number of inputs available, either in carbon or lignocellulosic form of biomass [131]. The result of the treatment relying on each of these three processes is distinct. The process of physicochemical conversion involves transforming feedstock into biodiesel and other products through various physicochemical combinations. This typically includes the conversion of raw or used vegetable oils, tallows (animal fat), and other suitable materials into liquid oils or biodiesel [132]. Common thermal conversion techniques involve burning (incineration), hydrothermal carbonization, pyrolysis, and gasification. As the world seeks to transition to renewable energy, the conversion of sewage and biomass to energy has become an area of increasing interest. In this regard, photocatalytic conversion technology has emerged as a promising approach for the conversion of sewage and biomass to energy.

1.5.2 Photocatalytic Hydrogen Production

The utilization of photocatalysts to split water molecules into hydrogen and oxygen via photocatalytic conversion technology is a promising approach to biomass-based hydrogen production [133]. Despite its potential to deliver a sustainable and environmentally friendly energy source, this technique is not without limitations related to technical complexities and economic viability. One potential application for photocatalytic conversion technology is the oxidation of wastewater, which involves using a photocatalyst to decompose organic compounds into carbon dioxide and water [134]. Despite its potential to simultaneously treat wastewater and generate energy, this technology faces multiple technical and economic challenges.

Solar energy can be converted to chemical/hydrogen energy by using environmentally friendly photocatalytic technology. The route of photocatalytic hydrogen production can be written as follows:

$$Photocalyst \xrightarrow{h\upsilon \geq E} h^+ + e^-$$

$$2H^+ + 2e^- \rightarrow H_2$$

$$H_2O + 2H^+ \rightarrow 0.5O_2 + 2H^+$$

While this application is sustainable, it requires more work before it can be commercialized for several reasons. These include its limitation to UV region only due to the wider bandgap, as opposed to the other 95% of the solar spectrum, and a rapid recombination rate. Even though photocatalysis is a sustainable practice, it still has a ways to go before it can be commercialized because the catalyst has a wider bandgap, which merely produces a reaction in the UV region— less than 5% of the solar spectrum. Additionally, there is an extraordinary recombinant rate and poor charge splitting, leading to lower viability in the visible region as well as photon-hydrogen conversion.

The use of wastewater for photocatalytic production would be significantly advantageous due to its organic pollutants acting as hole scavengers and decreasing the rate of recombination. However, certain issues such as mass transfer and solution stability could potentially hold up this process. Photocatalysis brings about organic compound mineralization which is expressed by stoichiometry as below:

$$C_xH_yO_z + (x + 0.25y - 0.5z)O_2 \rightarrow xCO_2 + 0.5yH_2O$$

The thermodynamic stability of this process is clear, as the end products are CO_2 and H_2O. The equation for mineralizing organic compounds using photocatalyst illustrates this as follows:

$$C_aH_bO_c + (2a - c)H_2O \rightarrow (2a + 0.5b - c)H_2 + aCO_2$$

The wastewater cannot be partially degraded with ease as the OH· are indiscriminate, having the capacity to completely break down substances. To effectively create hydrogen, the oxygen-deficient decomposition technique and anaerobic decomposition tactic should be employed to only partially break down the wastewater.

The use of photocatalysis is applicable to many fields, including the production of hydrogen, medical science, biology, antibacterial activity, and chemical synthesis. In 1972, Professor Kenichi Honda and Akira Fujishima emulated photocatalysis [135]. In photocatalysis, redox reactions are triggered by light and sensitized to produce charge carriers. Semiconductors are exploited when light has energy equal to or higher than their bandgap.

When electrons are excited, holes in the valence band (VB) are left behind. The reaction pathway is quite unclear without proper study since electrons participate in reduction while holes participate in oxidation. As a result, holes either oxidize organic pollutants to intermediates or react with adsorbed water molecules to form hydroxyl radicals. Alternatively, adsorbed oxygen reacts with electrons (e.g., H_2O_2, O_3, and air) to produce superoxide ions. Hydrogen peroxide is formed when the electrons and hydrogen ions react with these radicals [136–139] as depicted in Figure 1.6.

$$Photocatalyst + hv \rightarrow e^- + h^+$$

$$h^+ + H_2O \rightarrow OH\cdot + H^+$$

$$h^+ + OH^- \rightarrow OH\cdot$$

$$e^- + O_2 \rightarrow O_2^{\cdot -}$$

$$O_{2\cdot}^{\ -} + e^- + 2H^+ \rightarrow H_2O_2$$

$$H_2O_2 + e^- \rightarrow OH\cdot + OH^-$$

$$e^- + h+ \rightarrow heat$$

The homogenous photocatalytic processes, like ozonation, Fenton, and transition metal oxides, are effective but pose problems during the

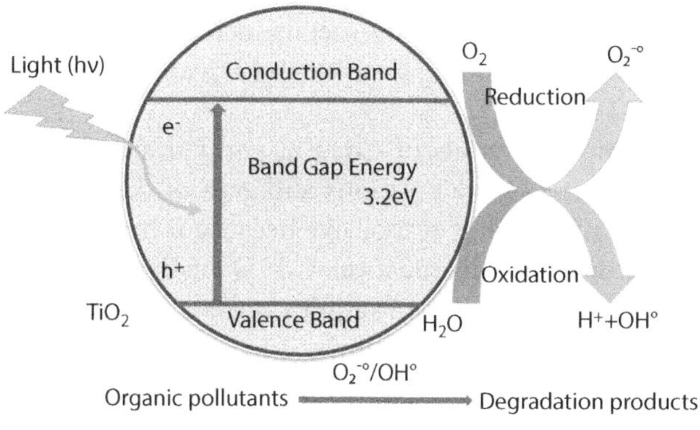

Figure 1.6 Photocatalytic conversion of pollutant/wastewater into hydrogen and other products.

separation and regeneration of catalysts. In heterogeneous photocatalytic processes, organic compounds are mineralized across various stages. The heterogeneous photocatalytic process is based on five stages including the addition of pollutants to catalyst surfaces, adsorption on the catalyst surface, photocatalysis, intermediates/products desorption, and allocation of removed substances to the main fluid.

1.5.2.1 Photocatalytic Materials

Such a route is not yet widely implemented because of its low quantum efficiency. Combining semiconductor photocatalysts greatly improves quantum efficiency [139]. Besides being active under UV light, most semiconductors are not photocatalysts; UV light contributes only 5% to solar light. To ensure efficiency, photocatalysts have to be photoactive, nontoxic, chemically and biologically inert, and stable. There are numerous combinations of noble metals used as heterogeneous photocatalysts, including Ru, Rh, Pd, Pt, V, Bi, Cu, Zn, and Ti oxides. In addition to increasing the pore size, stabilizing catalysts, lowering bandgaps, reducing recombination rates, and tuning surface properties, composite catalysts enhance photocatalytic properties [140].

Various metal oxides, including ZnO, TiO_2, CeO_2, $SrTiO_3$, SnO_2, $BiVO_4$, WO_3, and In_2O_3, are effective in removing organic and inorganic pollutants [141, 142]. In water, Groups I and II form oxides like K_2O, MgO, BaO, and SrO because of their high oxidation potential [143]. Although ZnO is photocatalytic, it is inactive and corrosion-resistant in visible light. A combination of CeO_2 and TiO_2 can yield excellent results when combined. With a high bandgap, SnO_2 and WO_3 are used as photocatalysts to enhance the activity of TiO_2 [144, 145]. Nevertheless, using the combination of SnO_2 and WO_3 along rare-earth elements metals such as Ce, Tb, and Ho, as well as transitional metals including Cr, Mn, and Ir, can provide extraordinary photoactivity [144].

The three sulfide materials of Cd, Zn, and Cu are all reactive in the visible region because of the positions and gaps of their bands [144, 146, 147]. However, ZnS is not as responsive because it has a wide bandgap of 3.35 eV. Aside from photodegradation, CdS is relatively responsive to visible light. During the recharging of sulfide catalysts, the electron is excitable enough to break its bond, which allows it to move between bonding and antibonding orbitals. Additionally, influence intensifies when the water is used, resulting in corrosion on the surface. Researchers are trying to explore sulfide-based catalysts with long-term endurance instead of relying on high efficiency.

Metallates are combinations of anions/oxides accompanied by metals along with small atom moieties. Certain metallates consist of built-in layered arrangements that are capable of responding to visible light. Examples of common metallates with visible light responsiveness are $BiVO_4$, Bi_2WO_6, Bi_2O_3, $CdWO_4$, $ZnWO_4$, $PbWO_4$, and Bi_2MoO_6BiOBr. Much of the focus has been on metallates due to their durability in the face of corrosion, larger surface area, as well as the decrease of bandwidth when accompanied by further fundamental catalysts. Studies have been done that have confirmed the quick movement and charge division by linking metallates with TiO_2.

Carbon-based photocatalysts on their own are not highly efficient; however, after combining with oxide or metallates, photostability and photoactivity drastically improve. To enhance the overall performance, the base catalyst is combined with carbon nanotubes (CNT), fullerenes (C60 and C70), and graphene oxide to create a synergistic effect [148]. As a result, conductivity and surface area increase which, in turn, curbs the recombinant rate. The generation of an electron is accompanied by carbon-based compounds that can effectively facilitate the movement of electrons and optimize their separation capabilities. Graphene stands out among all of the carbon compounds, displaying incredible conductivity [149, 150], electron mobility [140, 151], and large surface area.

Mesoporous materials contain organized pores and pathways that offer a great deal of surface area. While these materials are not responsive to light, they can still hasten the breakdown of other substances by virtue of their increased surface area, improved absorption, and facility for attachment. Zeolites, silicas, and alumina are some examples of this [152–154].

Organic materials can be converted into energy by using different technologies. Thermal conversion technologies, specifically pyrolysis, gasification, and combustion, are applied to produce gas, liquid, or solid products. Pyrolysis utilizes high temperatures without oxygen to decompose organic materials. Gasification partially oxidizes organic materials to generate syngas, which is a combustible gas mixture. Combustion, on the other hand, burns organic materials in the presence of oxygen resulting in heat and energy.

Another technology called photocatalytic conversion uses photocatalysts to initiate chemical reactions between organic materials and light energy resulting in energy-rich products like methane and hydrogen gas. This process can be performed under mild conditions at atmospheric pressure and room temperature. Nanoparticles or thin films can be produced as photocatalysts from materials such as zinc oxide, titanium dioxide, or tungsten oxide. The main difficulties associated with both photocatalytic hydrogen production and photocatalytic degradation are lack of efficient solar light capture, susceptibility to photo corrosion, inadequate charge separation, negligible hydrophobic contaminants adsorption, and uneven catalyst distribution.

Photocatalytic conversion technologies offer several benefits when compared to thermal conversion technologies—functioning under milder conditions, producing high-energy byproducts like hydrogen gas and methane, and emitting fewer greenhouse gasses. Also, they leave a lower environmental impact. However, their implementation requires specialized and costly equipment, and the low conversion efficiency, dependence on light energy, and necessity of extra energy sources for powering the reaction are some of the significant limitations to consider with the photocatalytic conversion process.

1.6 Conclusion and Future Prospects

The utilization of biomass and sewage as sources of energy through thermal and photocatalytic conversion technologies is highly encouraged for sustainable and renewable energy. However, significant challenges still need to be resolved to enhance the performance and minimize the ecological impact of these technologies. Addressing these difficulties requires future research, particularly in thermal conversion technologies, including the development of more economical and efficient equipment and optimization of operating conditions to reduce pollutants and recover more energy. Furthermore, other possible applications of waste heat from thermal conversion processes should be explored. For photocatalytic conversion technologies, research should concentrate on developing more efficient and durable photocatalytic materials and optimizing the design of photocatalytic reactors to maximize energy conversion efficiency. In addition, coupling photocatalytic conversion technology with renewable energy sources such as wind and hydropower could enhance its overall efficiency and reliability, and further research and development of both technologies are essential for achieving a sustainable and renewable future energy system.

Biomass and sewage are considered viable alternatives to fossil fuels for sustainable energy production. Thermal and photocatalytic conversion technologies can generate energy from these materials, contributing to cost-effective and sustainable energy systems. Utilizing these technologies could reduce the environmental impact of energy production while creating new opportunities for economic growth and job creation. Additionally, utilizing biomass and sewage for energy production provides benefits such as addressing waste management concerns and mitigating the negative environmental impacts of waste disposal. Challenges and opportunities of biomass energy:

Every day, different types of waste are being created all over the world. Some of these can be recycled or used to produce energy or chemicals, while certain kinds of them are just thrown away or prohibited, which then can taint the air. The incorrect disposal of garbage, such as municipal solid waste, has the potential to cause pollution [155]. It has been determined that increased amounts of aromatic hydrocarbons in sewage and municipal runoffs occur due to excessive amounts of grease. The removal of this type of waste can be achieved through landfilling and incineration, but the downside of these methods is that they can cause contamination of the environment [156]. To address this issue, research is being done to find a better solution that meets both environmental and economic needs by transforming unutilized resources into alternate energy sources using thermal transformation techniques.

Utilizing biomass as an energy source can pose a multitude of obstacles, encompassing a combination of burdensome logistical fees, variable feedstock quality, and competing agriculture needs. Nevertheless, we can conquer these impediments through the establishment of efficient biomass supply chains, technological innovations in biomass conversion, and the promotion of sustainability in biomass production. Despite these challenges, biomass presents a collection of prospects as an energy source. It is a viable option for power generation, heat, and biofuel production and can also be transformed into value-added commodities via biorefinery processes, featuring chemicals and materials. Furthermore, the implementation of biomass use can stimulate economic growth and propagate rural development, rendering it an alluring alternative to sustainable energy.

References

1. Dutta, S., Review on solar hydrogen: Its prospects and limitations. *Energy Fuel*, 35, 11613, 2021.
2. Hansen, S., Mirkouei, A., Diaz, L.A., A comprehensive state-of-technology review for upgrading bio-oil to renewable or blended hydrocarbon fuels. *Renew. Sustain. Energy Rev.*, 118, 109548, 2020.
3. Küçük, M.M. and Demirbaş, A., Biomass conversion processes. *Energy Convers. Manage.*, 38, 151, 1997.
4. Demirbaş, A., Biomass resource facilities and biomass conversion processing for fuels and chemicals. *Energy Convers. Manage.*, 42, 1357, 2001.

5. Srisasiwimon, N., Chuangchote, S., Laosiripojana, N., Sagawa, T., TiO2/lignin-based carbon composited photocatalysts for enhanced photocatalytic conversion of lignin to high value chemicals. *ACS Sustain. Chem. Eng.*, 6, 13968, 2018.
6. Rezk, H., Nassef, A.M., Inayat, A., Sayed, E.T., Shahbaz, M., Olabi, A.G., Improving the environmental impact of palm kernel shell through maximizing its production of hydrogen and syngas using advanced artificial intelligence. *Sci. Total Environ.*, 658, 1150, 2019.
7. Gill, S.S., Mehmood, M.A., Ahmad, N., Ibrahim, M., Rashid, U., Ali, S., Nehdi, I.A., Strain selection, growth productivity and biomass characterization of novel microalgae isolated from fresh and wastewaters of upper Punjab, Pakistan. *Front. Life Sci.*, 9, 190, 2016.
8. Wu, Z., Li, Y., Zhang, B., Yang, W., Yang, B., Co-pyrolysis behavior of microalgae biomass and low-rank coal: Kinetic analysis of the main volatile products. *Bioresour. Technol.*, 271, 202, 2019.
9. Afzal, I., Shahid, A., Ibrahim, M., Liu, T., Nawaz, M., Mehmood, M.A., Microalgae: A promising feedstock for energy and high-value products, in: *Algae Based Polymers, Blends, and Composites*, K.M. Zia, M. Zuber, M. Ali (Eds.), pp. 55–75, Elsevier, 2017.
10. Shahid, A., Ishfaq, M., Ahmad, M.S., Malik, S., Farooq, M., Hui, Z., Batawi, A.H., Shafi, M.E., Aloqbi, A.A., Gull, M., Mehmood, M.A., Bioenergy potential of the residual microalgal biomass produced in city wastewater assessed through pyrolysis, kinetics and thermodynamics study to design algal biorefinery. *Bioresour. Technol.*, 289, 121701, 2019.
11. Chiang, H.L., Huang, C.P., Chiang, P.C., The surface characteristics of activated carbon as affected by ozone and alkaline treatment. *Chemosphere*, 47, 257, 2002.
12. Akhtar, J. and Amin, N.A.S., A review on process conditions for optimum bio-oil yield in hydrothermal liquefaction of biomass. *Renew. Sustain. Energy Rev.*, 15, 1615, 2011.
13. Long, H., Li, X., Wang, H., Jia, J., Biomass resources and their bioenergy potential estimation: A review. *Renew. Sustain. Energy Rev.*, 26, 344, 2013.
14. Gallezot, P., Conversion of biomass to selected chemical products. *Chem. Soc. Rev.*, 41, 1538, 2012.
15. Tekin, K., Karagöz, S., Bektaş, S., A review of hydrothermal biomass processing. *Renew. Sustain. Energy Rev.*, 40, 673, 2014.
16. Li, D.C. and Jiang, H., The thermochemical conversion of non-lignocellulosic biomass to form biochar: A review on characterizations and mechanism elucidation. *Bioresour. Technol.*, 246, 57, 2017.
17. Campuzano, F., Brown, R.C., Martínez, J.D., Auger reactors for pyrolysis of biomass and wastes. *Renew. Sustain. Energy Rev.*, 102, 372, 2019.
18. Gollakota, A.R.K., Kishore, N., Gu, S., A review on hydrothermal liquefaction of biomass. *Renew. Sustain. Energy Rev.*, 81, 1378, 2018.

19. Pang, S., Fuel flexible gas production: Biomass, coal and bio-solid wastes, in: *Fuel Flexible Energy Generation*, J. Oakey (Ed.), pp. 241–269, Woodhead Publishing, Boston, 2016.
20. Ouda, O.K.M., Raza, S.A., Nizami, A.S., Rehan, M., Al-Waked, R., Korres, N.E., Waste to energy potential: A case study of Saudi Arabia. *Renew. Sustain. Energy Rev.*, 61, 328, 2016.
21. Rulkens, W., Sewage sludge as a biomass resource for the production of energy: Overview and assessment of the various options. *Energy Fuel*, 22, 9, 2008.
22. Gao, N., Quan, C., Liu, B., Li, Z., Wu, C., Li, A., Continuous pyrolysis of sewage sludge in a screw-feeding reactor: Products characterization and ecological risk assessment of heavy metals. *Energy Fuel*, 31, 5063, 2017.
23. Moraes, E.S., Reis, G.F., Cruz, J., Cozzolino, K., Neto, A.F.G., Andrade-Filho, T., Neto, A.M.J.C., Thermodynamics study of biokerosene from coconut and palm kernel oils and JP-8 aircraft fuels in the gas phase by the DFT method. *J. Mol. Model.*, 26, 79, 2020.
24. Buonocore, E., Mellino, S., De Angelis, G., Liu, G., Ulgiati, S., Life cycle assessment indicators of urban wastewater and sewage sludge treatment. *Ecol. Indic.*, 94, 13, 2018.
25. Morgano, M., Leibold, H., Richter, F., Stapf, D., Seifert, H., Screw pyrolysis technology for sewage sludge treatment. *Waste Manage.*, 73, 487, 2018.
26. Wilk, M., Magdziarz, A., Jayaraman, K., Chargot, M., Gökalp, I., Hydrothermal carbonization characteristics of sewage sludge and lignocellulosic biomass. A comparative study. *Biomass Bioenergy*, 120, 166, 2019.
27. Nagtode, V.S., Cardoza, C., Yasin, H.K.A., Mali, S.N., Tambe, S.M., Roy, P., Singh, K., Goel, A., Amin, P.D., Thorat, B.R., Cruz, J.N., Pratap, A.P., Green Surfactants (Biosurfactants): A Petroleum-Free Substitute for Sustainability-Comparison, Applications, Market, and Future Prospects. *ACS Omega.*, 8, 11674, 2023.
28. Fytili, D. and Zabaniotou, A., Utilization of sewage sludge in EU application of old and new methods—A review. *Renew. Sustain. Energy Rev.*, 12, 116, 2008.
29. Yu, X., Xue, J., Yao, H., Wu, Q., Venkatesan, A.K., Halden, R.U., Kannan, K., Occurrence and estrogenic potency of eight bisphenol analogs in sewage sludge from the U.S. EPA targeted national sewage sludge survey. *J. Hazard. Mater.*, 299, 733, 2015.
30. Demirbas, A., Edris, G., Alalayah, W.M., Sludge production from municipal wastewater treatment in sewage treatment plant. *Energy Sources Part A*, 39, 999, 2017.
31. Kelessidis, A. and Stasinakis, A.S., Comparative study of the methods used for treatment and final disposal of sewage sludge in European countries. *Waste Manage.*, 32, 1186, 2012.

32. Ramachandran, S., Yao, Z., You, S., Massier, T., Stimming, U., Wang, C.H., Wang, C.H., Life cycle assessment of a sewage sludge and woody biomass co-gasification system. *Energy*, 137, 369, 2017.
33. Jimenez, G., Monti, T., Titman, J.J., Montoya, V., Kingman, S.W., Binner, E.R., New insights into microwave pyrolysis of biomass: Preparation of carbon-based products from pecan nutshells and their application in wastewater treatment. *J. Anal. Appl. Pyrolysis*, 124, 113, 2017.
34. Mobarak, F., Fahmy, Y., Schweers, W., Production of phenols and charcoal from bagasse by a rapid continuous pyrolysis process. *Wood Sci. Technol.*, 16, 59, 1982.
35. Mobarak, F., Rapid continuous pyrolysis of cotton stalks for charcoal production. *Holzforschung*, 37, 251, 1983.
36. He, X., Zheng, N., Hu, R., Hu, Z., Yu, J.C., Hydrothermal and pyrolytic conversion of biomasses into catalysts for advanced oxidation treatments. *Adv. Funct. Mater.*, 31, 2006505, 2021.
37. Dixon, N. and Jones, D.R.V., Engineering properties of municipal solid waste. *Geotext. Geomembr.*, 23, 205, 2005.
38. Fahmy, T.Y.A., Fahmy, Y., Mobarak, F., El-Sakhawy, M., Zeid, R.E., Biomass pyrolysis: Past, present, and future. *Environ. Dev. Sustain.*, 22, 17, 2020.
39. Boer, E., Jędrczak, A., Kowalski, Z., Kulczycka, J., Szpadt, R., A review of municipal solid waste composition and quantities in Poland. *Waste Manage.*, 30, 369, 2010.
40. Sansaniwal, S.K., Pal, K., Rosen, M.A., Tyagi, S.K., Recent advances in the development of biomass gasification technology: A comprehensive review. *Renew. Sustain. Energy Rev.*, 72, 363, 2017.
41. Huber, G.W., Iborra, S., Corma, A., Synthesis of transportation fuels from biomass: Chemistry, catalysts, and engineering. *Chem. Rev.*, 106, 4044, 2006.
42. Hameed, Z., Aslam, M., Khan, Z., Maqsood, K., Atabani, A.E., Ghauri, M., Khurram, M.S., Rehan, M., Nizami, A.S., Gasification of municipal solid waste blends with biomass for energy production and resources recovery: Current status, hybrid technologies and innovative prospects. *Renew. Sustain. Energy Rev.*, 136, 110375, 2021.
43. Agugliaro, F., Alcayde, A., Montoya, F.G., Sierra, A., Gil, C., Scientific production of renewable energies worldwide: An overview. *Renew. Sustain. Energy Rev.*, 18, 134, 2013.
44. Hassan, H., Lim, J.K., Hameed, B.H., Recent progress on biomass co-pyrolysis conversion into high-quality bio-oil. *Bioresour. Technol.*, 221, 645, 2016.
45. Buragohain, B., Mahanta, P., Moholkar, V.S., Biomass gasification for decentralized power generation: The Indian perspective. *Renew. Sustain. Energy Rev.*, 14, 73, 2010.
46. Situmorang, Y.A., Zhao, Z., Yoshida, A., Abudula, A., Guan, G., Small-scale biomass gasification systems for power generation (<200 kW class): A review. *Renew. Sustain. Energy Rev.*, 117, 109486, 2020.

47. Burnley, S.J., A review of municipal solid waste composition in the United Kingdom. *Waste Manage.*, 27, 1274, 2007.
48. Conti, F., Toor, S.S., Pedersen, T.H., Seehar, T.H., Nielsen, A.H., Rosendahl, L.A., Valorization of animal and human wastes through hydrothermal liquefaction for biocrude production and simultaneous recovery of nutrients. *Energy Convers. Manage.*, 216, 112925, 2020.
49. Sotoudehnia, F., Rabiu, A., Alayat, A., McDonald, A.G., Characterization of bio-oil and biochar from pyrolysis of waste corrugated cardboard. *J. Anal. Appl. Pyrolysis.*, 145, 104722, 2020.
50. Zhang, Y., Cui, Y., Chen, P., Liu, S., Zhou, N., Ding, K., Fan, L., Peng, P., Min, M., Cheng, Y., Wang, Y., Wan, Y., Liu, Y., Li, B., Ruan, R., Gasification Technologies and Their Energy Potentials, in: *Sustainable Resource Recovery and Zero Waste Approaches*, M.J. Taherzadeh (Ed.), pp. 193–206, Elsevier, 2019.
51. Quispe, C.A.G., Coronado, C.J.R., Carvalho Jr., J.A., Glycerol: Production, consumption, prices, characterization and new trends in combustion. *Renew. Sustain. Energy Rev.*, 27, 475, 2013.
52. Autret, E., Berthier, F., Luszezanec, A., Nicolas, F., Incineration of municipal and assimilated wastes in France: Assessment of latest energy and material recovery performances. *J. Hazard. Mater.*, 139, 569, 2007.
53. Katami, T., Yasuhara, A., Shibamoto, T., Formation of dioxins from incierration of foods found in domestic garbage. *Environ. Sci. amp; Technol.*, 38, 1062, 2004.
54. Libra, J.A., Ro, K.S., Kammann, C., Funke, A., Berge, N.D., Neubauer, Y., Titirici, M.M., Fühner, C., Bens, O., Kern, J., Emmerich, K.H., Hydrothermal carbonization of biomass residuals: A comparative review of the chemistry, processes and applications of wet and dry pyrolysis. *Biofuels*, 2, 71, 2011.
55. Hoekman, S.K., Broch, A., Robbins, C., Hydrothermal carbonization (HTC) of lignocellulosic biomass. *Energy Fuels.*, 25, 1802, 2011.
56. Li, L., Diederick, R., Flora, J.R.V., Berge, N.D., Hydrothermal carbonization of food waste and associated packaging materials for energy source generation. *Waste Manage.*, 33, 2478, 2013.
57. Berge, N.D., Ro, K.S., Mao, J., Flora, J.R.V., Chappell, M.A., Bae, S., Hydrothermal carbonization of municipal waste streams. *Environ. Sci. Technol.*, 45, 5696, 2011.
58. Schmitt, E., Bura, R., Gustafson, R., Cooper, J., Vajzovic, A., Converting lignocellulosic solid waste into ethanol for the State of Washington: An investigation of treatment technologies and environmental impacts. *Bioresour. Technol.*, 104, 400, 2012.
59. Lu, Q., Li, W.Z., Zhu, X.F., Overview of fuel properties of biomass fast pyrolysis oils. *Energy Convers. Manage.*, 50, 1376, 2009.
60. Li, M., Xiao, B., Wang, X., Liu, J., Consequences of sludge composition on combustion performance derived from thermogravimetry analysis. *Waste Manage.*, 35, 141, 2015.

61. Hassan, S.S.A., Wang, Y., Hu, S., Su, S., Xiang, J., Thermochemical processing of sewage sludge to energy and fuel: Fundamentals, challenges and considerations. *Renew. Sustain. Energy Rev.*, 80, 888, 2017.
62. Samolada, M.C. and Zabaniotou, A.A., Comparative assessment of municipal sewage sludge incineration, gasification and pyrolysis for a sustainable sludge-to-energy management in Greece. *Waste Manage.*, 34, 411, 2014.
63. Martínez, M., Fonts, I., Lázaro, L., Ceamanos, J., Gea, G., Fast pyrolysis of torrefied sewage sludge in a fluidized bed reactor. *J. Chem. Eng.*, 259, 467, 2015.
64. Martínez, M., Fonts, I., Ábrego, J., Ceamanos, J., Gea, G., Sewage sludge torrefaction in a fluidized bed reactor. *J. Chem. Eng.*, 222, 534, 2013.
65. Hernández, A.B., Okonta, F., Freeman, N., Sewage sludge charcoal production by N2- and CO2-torrefaction. *J. Environ. Chem. Eng.*, 5, 4406, 2017.
66. Huang, F., Yu, Y., Huang, H., Temperature influence and distribution of bio-oil from pyrolysis of granular sewage sludge. *J. Anal. Appl. Pyrolysis*, 130, 36, 2018.
67. Fonts, I., Gea, G., Azuara, M., Ábrego, J., Arauzo, J., Sewage sludge pyrolysis for liquid production: A review. *Renewable Sustain. Energy Rev.*, 16, 2781, 2012.
68. Kan, T., Strezov, V., Evans, T., Effect of the heating rate on the thermochemical behavior and biofuel properties of sewage sludge pyrolysis. *Energy Fuels*, 30, 1564, 2016.
69. Trinh, T.N., Jensen, P.A., Johansen, K., Knudsen, N.O., Sørensen, H.R., Influence of the pyrolysis temperature on sewage sludge product distribution, bio-oil, and char properties. *Energy Fuels*, 27, 1419, 2013.
70. Barry, D., Barbiero, C., Briens, C., Berruti, F., Pyrolysis as an economical and ecological treatment option for municipal sewage sludge. *Biomass Bioenergy*, 122, 472, 2019.
71. Inguanzo, M., Domínguez, A., Menéndez, J.A., Blanco, C.G., Pis, J.J., On the pyrolysis of sewage sludge: The influence of pyrolysis conditions on solid, liquid and gas fractions. *J. Anal. Appl. Pyrolysis*, 63, 209, 2002.
72. Zhang, L., Xiao, B., Hu, Z., Liu, S., Cheng, G., He, P., Sun, L., Tar-free fuel gas production from high temperature pyrolysis of sewage sludge. *Waste Manage.*, 34, 180, 2014.
73. Alvarez, J., Amutio, M., Lopez, G., Bilbao, J., Olazar, M., Fast co-pyrolysis of sewage sludge and lignocellulosic biomass in a conical spouted bed reactor. *Fuel*, 159, 810, 2015.
74. Verdugo, A., Goos, E., Godino, A., Hernando, N., Riedel, U., Pyrolysis of biofuels of the future: Sewage sludge and microalgae – Thermogravimetric analysis and modelling of the pyrolysis under different temperature conditions. *Energy Convers. Manage.*, 138, 261, 2017.
75. Zhu, J., Yang, Y., Yang, L., Zhu, Y., High quality syngas produced from the co-pyrolysis of wet sewage sludge with sawdust. *Int. J. Hydrog. Energy*, 43, 5463, 2018.

76. Naqvi, S.R., Hameed, Z., Tariq, R., Taqvi, S.A., Ali, I., Niazi, M.B.K., Noor, T., Hussain, A., Iqbal, N., Shahbaz, M., Synergistic effect on co-pyrolysis of rice husk and sewage sludge by thermal behavior, kinetics, thermodynamic parameters and artificial neural network. *Waste Manage.*, 85, 131, 2019.
77. Wang, X., Deng, S., Tan, H., Adeosun, A., Vujanović, M., Yang, F., Duić, N., Synergetic effect of sewage sludge and biomass co-pyrolysis: A combined study in thermogravimetric analyzer and a fixed bed reactor. *Energy Convers. Manage.*, 118, 399, 2016.
78. Jin, J., Wang, M., Cao, Y., Wu, S., Liang, P., Li, Y., Zhang, J., Zhang, J., Wong, M.H., Shan, S., Christie, P., Cumulative effects of bamboo sawdust addition on pyrolysis of sewage sludge: Biochar properties and environmental risk from metals. *Bioresour. Technol.*, 228, 218, 2017.
79. Gómez, N., Quispe, V., Ábrego, J., Martínez, M., Murillo, M.B., Gea, G., Co-pyrolysis of sewage sludge and manure. *Waste Manage.*, 59, 211, 2017.
80. Zhao, B., Xu, X., Li, H., Chen, X., Zeng, F., Kinetics evaluation and thermal decomposition characteristics of co-pyrolysis of municipal sewage sludge and hazelnut shell. *Bioresour. Technol.*, 247, 21, 2018.
81. Wang, Z., Liu, K., Xie, L., Zhu, H., Ji, S., Shu, X., Effects of residence time on characteristics of biochars prepared via co-pyrolysis of sewage sludge and cotton stalks. *J. Anal. Appl. Pyrolysis*, 142, 104659, 2019.
82. Lin, Y., Liao, Y., Yu, Z., Fang, S., Lin, Y., Fan, Y., Peng, X., Ma, X., Co-pyrolysis kinetics of sewage sludge and oil shale thermal decomposition using TGA–FTIR analysis. *Energy Convers. Manage.*, 118, 345, 2016.
83. Yang, X., Yuan, C., Xu, J., Zhang, W., Potential method for gas production: High temperature co-pyrolysis of lignite and sewage sludge with vacuum reactor and long contact time. *Bioresour. Technol.*, 179, 602, 2015.
84. Lin, Y., Liao, Y., Yu, Z., Fang, S., Ma, X., A study on co-pyrolysis of bagasse and sewage sludge using TG-FTIR and Py-GC/MS. *Energy Convers. Manage.*, 151, 190, 2017.
85. Chen, Q., Liu, H., Ko, J., Wu, H., Xu, Q., Structure characteristics of bio-char generated from co-pyrolysis of wooden waste and wet municipal sewage sludge. *Fuel Process. Technol.*, 183, 48, 2019.
86. Zhang, Y., Xu, P., Liang, S., Liu, B., Shuai, Y., Li, B., Exergy analysis of hydrogen production from steam gasification of biomass: A review. *Int. J. Hydrog. Energy*, 44, 14290, 2019.
87. Mekhilef, S., Saidur, R., Kamalisarvestani, M., Effect of dust, humidity and air velocity on efficiency of photovoltaic cells. *Renew. Sustain. Energy Rev.*, 16, 2920, 2012.
88. McKendry, P., Energy production from biomass: Gasification technologies. *Bioresour. Technol.*, 83, 55, 2002.
89. Wu, Y., Yang, W., Blasiak, W., Energy and exergy analysis of high temperature agent gasification of biomass. *Energies*, 7, 2107, 2014.

90. Liu, B. and Ji, S., Comparative study of fluidized-bed and fixed-bed reactor for syngas methanation over Ni-W/TiO2-SiO2 catalyst. *J. Energy Chem.*, 22, 740, 2013.
91. Midilli, A., Dogru, M., Howarth, C.R., Ling, M.J., Ayhan, T., Combustible gas production from sewage sludge with a downdraft gasifier. *Energy Convers. Manage.*, 42, 157, 2001.
92. Manyà, J.J., Sánchez, J.L., Gonzalo, A., Arauzo, J., Air gasification of dried sewage sludge in a fluidized bed: Effect of the operating conditions and in-bed use of alumina. *Energy Fuels*, 19, 629, 2005.
93. Manyà, J.J., Aznar, M., Sánchez, J.L., Arauzo, J., Murillo, M.B., Further experiments on sewage sludge air gasification: Influence of the nonstationary period on the overall results. *Ind. Eng. Chem. Res.*, 45, 7313, 2006.
94. Aznar, M., Manyà, J.J., García, G., Sánchez, J.L., Murillo, M.B., Influence of freeboard temperature, fluidization velocity, and particle size on tar production and composition during the air gasification of sewage sludge. *Energy Fuels*, 22, 2840, 2008.
95. Mun, T.Y., Kang, B.S., Kim, J.S., Production of a producer gas with high heating values and less tar from dried sewage sludge through air gasification using a two-stage gasifier and activated carbon. *Energy Fuels*, 23, 3268, 2009.
96. Mun, T.Y., Kim, J.W., Kim, J.S., Air gasification of dried sewage sludge in a two-stage gasifier: Part 1. The effects and reusability of additives on the removal of tar and hydrogen production. *Int. J. Hydrog. Energy*, 38, 5226, 2013.
97. Nipattummakul, N., Ahmed, I.I., Kerdsuwan, S., Gupta, A.K., Hydrogen and syngas production from sewage sludge via steam gasification. *Int. J. Hydrog. Energy*, 35, 11738, 2010.
98. Cano, D., Barea, A., Nilsson, S., Ollero, P., The influence of temperature and steam on the yields of tar and light hydrocarbon compounds during devolatilization of dried sewage sludge in a fluidized bed. *Fuel*, 108, 341, 2013.
99. Chun, Y.N., Ji, D.W., Yoshikawa, K., Pyrolysis and gasification characterization of sewage sludge for high quality gas and char production. *J. Mech. Sci.*, 27, 263, 2013.
100. Li, H., Chen, Z., Huo, C., Hu, M., Guo, D., Xiao, B., Effect of bioleaching on hydrogen-rich gas production by steam gasification of sewage sludge. *Energy Convers. Manage.*, 106, 1212, 2015.
101. Lee, U., Dong, J., Chung, J.N., Production of useful energy from solid waste materials by steam gasification. *Int. J. Energy Res.*, 40, 1474, 2016.
102. Lalaguna, N., Sánchez, J.L., Murillo, M.B., Rodríguez, E., Gea, G., Air–steam gasification of sewage sludge in a fluidized bed. Influence of some operating conditions. *J. Chem. Eng.*, 248, 373, 2014.
103. de Andrés, J.M., Narros, A., Rodríguez, M.E., Air-steam gasification of sewage sludge in a bubbling bed reactor: Effect of alumina as a primary catalyst. *Fuel Process. Technol.*, 92, 433, 2011.

104. Roche, E., de Andrés, J.M., Narros, A., Rodríguez, M.E., Air and air-steam gasification of sewage sludge. The influence of dolomite and throughput in tar production and composition. *Fuel*, 115, 54, 2014.
105. Nilsson, S., Barea, A., Ollero, P., Gasification of char from dried sewage sludge in fluidized bed: Reaction rate in mixtures of CO2 and H2O. *Fuel*, 105, 764, 2013.
106. Choi, Y.K., Cho, M.H., Kim, J.S., Steam/oxygen gasification of dried sewage sludge in a two-stage gasifier: Effects of the steam to fuel ratio and ash of the activated carbon on the production of hydrogen and tar removal. *Energy*, 91, 160, 2015.
107. Kwon, E.E., Yi, H., Kwon, H.H., Thermo-chemical process with sewage sludge by using CO2. *J. Environ. Manage.*, 128, 435, 2013.
108. Dogru, M., Midilli, A., Howarth, C.R., Gasification of sewage sludge using a throated downdraft gasifier and uncertainty analysis. *Fuel Process. Technol.*, 75, 55, 2002.
109. Nilsson, S., Barea, A., Cano, D.F., Gasification reactivity of char from dried sewage sludge in a fluidized bed. *Fuel*, 92, 346, 2012.
110. Inguanzo, M., Menéndez, J.A., Fuente, E., Pis, J.J., Reactivity of pyrolyzed sewage sludge in air and CO2. *J. Anal. Appl. Pyrolysis*, 58-59, 943, 2001.
111. Urych, B. and Smoliński, A., Kinetics of sewage sludge pyrolysis and air gasification of its chars. *Energy Fuels*, 30, 4869, 2016.
112. Nowicki, L. and Markowski, M., Gasification of pyrolysis chars from sewage sludge. *Fuel*, 143, 476, 2015.
113. Scott, S.A., Davidson, J.F., Dennis, J.S., Fennell, P.S., Hayhurst, A.N., The rate of gasification by CO2 of chars from waste. *Proc. Combust. Inst.*, 30, 2151-2159, 2005.
114. Bridgwater, A.V., Renewable fuels and chemicals by thermal processing of biomass. *J. Chem. Eng.*, 91, 87, 2003.
115. Campoy, M., Barea, A., Ollero, P., Nilsson, S., Gasification of wastes in a pilot fluidized bed gasifier. *Fuel Process. Technol.*, 121, 63, 2014.
116. Zhu, J.G., Yao, Y., Lu, Q.G., Gao, M., Ouyang, Z.Q., Experimental investigation of gasification and incineration characteristics of dried sewage sludge in a circulating fluidized bed. *Fuel*, 150, 441, 2015.
117. Thomsen, T.P., Sárossy, Z., Gøbel, B., Stoholm, P., Ahrenfeldt, J., Frandsen, F.J., Henriksen, U.B., Low temperature circulating fluidized bed gasification and co-gasification of municipal sewage sludge. Part 1: Process performance and gas product characterization. *Waste Manage.*, 66, 123, 2017.
118. Petersen, I. and Werther, J., Experimental investigation and modeling of gasification of sewage sludge in the circulating fluidized bed. *Chem. Eng. Process.: Process Intensif.*, 44, 717, 2005.
119. Petersen, I. and Werther, J., Three-dimensional modeling of a circulating fluidized bed gasifier for sewage sludge. *Chem. Eng. Sci.*, 60, 4469, 2005.
120. Freda, C., Cornacchia, G., Romanelli, A., Valerio, V., Grieco, M., Sewage sludge gasification in a bench scale rotary kiln. *Fuel*, 212, 88, 2018.

121. Striūgas, N., Valinčius, V., Pedišius, N., Poškas, R., Zakarauskas, K., Investigation of sewage sludge treatment using air plasma assisted gasification. *Waste Manage.*, 64, 149, 2017.
122. Mountouris, A., Voutsas, E., Tassios, D., Plasma gasification of sewage sludge: Process development and energy optimization. *Energy Convers. Manage.*, 49, 2264, 2008.
123. Zhang, J., Weng, X., Han, Y., Li, W., Gan, Z., Gu, J., Effect of supercritical water on the stability and activity of alkaline carbonate catalysts in coal gasification. *J. Energy Chem.*, 22, 459, 2013.
124. Rauch, R., Hrbek, J., Hofbauer, H., Biomass gasification for synthesis gas production and applications of the syngas. *Wiley Interdiscip. Rev.: Energy Environ.*, 3, 343, 2014.
125. Pereira, E.G., da Silva, J.N., de Oliveira, J.L., Machado, C.S., Sustainable energy: A review of gasification technologies. *Renew. Sustain. Energy Rev.*, 16, 4753, 2012.
126. McKendry, P., Energy production from biomass (part 2): Conversion technologies. *Bioresour. Technol.*, 83, 47, 2002.
127. Mills, N., Pearce, P., Farrow, J., Thorpe, R.B., Kirkby, N.F., Environmental & economic life cycle assessment of current & future sewage sludge to energy technologies. *Waste Manage.*, 34, 185, 2014.
128. Piippo, S., Lauronen, M., Postila, H., Greenhouse gas emissions from different sewage sludge treatment methods in north. *J. Clean. Prod.*, 177, 483, 2018.
129. Werle, S. and Wilk, R.K., A review of methods for the thermal utilization of sewage sludge: The Polish perspective. *Renew. Energy*, 35, 1914, 2010.
130. Braslavsky, S.E., Glossary of terms used in photochemistry, (IUPAC Recommendations 2006). *Pure Appl. Chem.*, 79, 293, 2007.
131. Pham, T.P.T., Kaushik, R., Parshetti, G.K., Mahmood, R., Balasubramanian, R., Food waste-to-energy conversion technologies: Current status and future directions. *Waste Manage.*, 38, 399, 2015.
132. Dabe, S.J., Prasad, P.J., Vaidya, A.N., Purohit, H.J., Technological pathways for bioenergy generation from municipal solid waste: Renewable energy option. *Environ. Prog. Sustain.*, 38, 654, 2019.
133. Granone, L.I., Sieland, F., Zheng, N., Dillert, R., Bahnemann, D.W., Photocatalytic conversion of biomass into valuable products: A meaningful approach? *Green Chem.*, 20, 1169, 2018.
134. Hirakawa, T., Yawata, K., Nosaka, Y., Photocatalytic reactivity for O2− and OH radical formation in anatase and rutile TiO2 suspension as the effect of H2O2 addition. *Appl. Catal. A: Gen.*, 325, 105, 2007.
135. Li, X., Yu, J., Wageh, S., Al-Ghamdi, A.A., Xie, J., Graphene in photocatalysis: A review. *Small*, 12, 6640, 2016.
136. Pirkanniemi, K. and Sillanpää, M., Heterogeneous water phase catalysis as an environmental application: A review. *Chemosphere*, 48, 1047, 2002.

137. Turchi, C.S. and Ollis, D.F., Photocatalytic degradation of organic water contaminants: Mechanisms involving hydroxyl radical attack. *J. Catal.*, 122, 178, 1990.
138. Hoffmann, M.R., Martin, S.T., Choi, W., Bahnemann, D.W., Environmental applications of semiconductor photocatalysis. *Chem. Rev.*, 95, 69, 1995.
139. Braham, R.J. and Harris, A.T., Review of major design and scale-up considerations for solar photocatalytic reactors. *Ind. Eng. Chem. Res.*, 48, 8890, 2009.
140. Bolotin, K.I., Sikes, K.J., Jiang, Z., Klima, M., Fudenberg, G., Hone, J., Kim, P., Stormer, H.L., Ultrahigh electron mobility in suspended graphene. *Solid State Commun.*, 146, 351, 2008.
141. Khalil, L.B., Mourad, W.E., Rophael, M.W., Photocatalytic reduction of environmental pollutant Cr(VI) over some semiconductors under UV/visible light illumination. *Appl. Catal. B: Environ.*, 17, 267, 1998.
142. Khalil, L.B., Rophael, M.W., Mourad, W.E., The removal of the toxic Hg(II) salts from water by photocatalysis. *Appl. Catal. B: Environ.*, 36, 125, 2002.
143. Dahl, M., Liu, Y., Yin, Y., Composite titanium dioxide nanomaterials. *Chem. Rev.*, 114, 9853, 2014.
144. Alonso, M.D., Fresno, F., Suárez, S., Coronado, J.M., Development of alternative photocatalysts to TiO2: Challenges and opportunities. *Energy Environ. Sci.*, 2, 1231, 2009.
145. Hong, E., Kim, D., Kim, J.H., Heterostructured metal sulfide (ZnS–CuS–CdS) photocatalyst for high electron utilization in hydrogen production from solar water splitting. *J. Ind. Eng. Chem.*, 20, 3869, 2014.
146. Yue, M., Wang, R., Cheng, N., Cong, R., Gao, W., Yang, T., ZnCr2S4: Highly effective photocatalyst converting nitrate into N2 without over-reduction under both UV and pure visible light. *Sci. Rep.*, 6, 30992, 2016.
147. Chen, D., Huang, F., Ren, G., Li, D., Zheng, M., Wang, Y., Lin, Z., ZnS nanoarchitectures: Photocatalysis, deactivation and regeneration. *Nanoscale*, 2, 2062, 2010.
148. Han, C., Yang, M.Q., Weng, B., Xu, Y.J., Improving the photocatalytic activity and anti-photocorrosion of semiconductor ZnO by coupling with versatile carbon. *Phys. Chem. Chem. Phys.*, 16, 16891, 2014.
149. Balandin, A.A., Ghosh, S., Bao, W., Calizo, I., Teweldebrhan, D., Miao, F., Lau, C.N., Superior thermal conductivity of single-layer graphene. *Nano Lett.*, 8, 902, 2008.
150. Balandin, A.A., Thermal properties of graphene and nanostructured carbon materials. *Nat. Mater.*, 10, 569, 2011.
151. Du, X., Skachko, I., Barker, A., Andrei, E.Y., Approaching ballistic transport in suspended graphene. *Nat. Nanotechnol.*, 3, 491, 2008.
152. Yang, J., Zhang, J., Zhu, L., Chen, S., Zhang, Y., Tang, Y., Zhu, Y., Li, Y., Synthesis of nano titania particles embedded in mesoporous SBA-15: Characterization and photocatalytic activity. *J. Hazard. Mater.*, 137, 952, 2006.

153. Aronson, B.J., Blanford, C.F., Stein, A., Solution-phase grafting of titanium dioxide onto the pore surface of mesoporous silicates: Synthesis and structural characterization. *Chem. Mater.*, 9, 2842, 1997.
154. Wang, Z., Zhang, F., Yang, Y., Xue, B., Cui, J., Guan, N., Facile postsynthesis of visible-light-sensitive titanium dioxide/mesoporous SBA-15. *Chem. Mater.*, 19, 3286, 2007.
155. Latimer, J.S., Hoffman, E.J., Hoffman, G., Fasching, J.L., Quinn, J.G., Sources of petroleum hydrocarbons in urban runoff. *Water Air Soil Pollut.*, 52, 1, 1990.
156. Wang, K. and Nakakubo, T., Comparative assessment of waste disposal systems and technologies with regard to greenhouse gas emissions: A case study of municipal solid waste treatment options in China. *J. Clean. Prod.*, 260, 120827, 2020.
157. Zuo, Z., Feng, Y., Li, X., Luo, S., Ma, J., Sun, H., Bi, X., Yu, Q., Zhou, E., Zhang, J., Guo, J., Lin, H., Thermal-chemical conversion of sewage sludge based on waste heat cascade recovery of copper slag: Mass and energy analysis. *Energy*, 235, 121327, 2021.
158. Moses, A., Komandur, J., Maarisetty, D., Mohapatra, P., Baral, S.S., A review on photocatalytic hydrogen production potential from paper and pulp industry wastewater. *Biomass Convers. Biorefin.*, 2022.
159. Werle, S. and Sobek, S., Gasification of sewage sludge within a circular economy perspective: A Polish case study. *Environ. Sci. Pollut. Res.*, 26, 35422, 2019.
160. Tasca, A.L., Puccini, M., Gori, R., Corsi, I., Galletti, A.M.R., Vitolo, S., Hydrothermal carbonization of sewage sludge: A critical analysis of process severity, hydrochar properties and environmental implications. *Waste Manage.*, 93, 1, 2019.
161. Al Shra'ah, A. and Helleur, R., Microwave pyrolysis of cellulose at low temperature. *J. Anal. Appl. Pyrolysis*, 105, 91, 2014.
162. Ngat, P., Jedsukontorn, T., Hunsom, M., Simultaneous H2 production and pollutant removal from biodiesel wastewater by photocatalytic oxidation with different crystal structure TiO2 photocatalysts. *J. Taiwan Inst. Chem. Eng.*, 78, 386, 2017.
163. Norouzi, O., Kheradmand, A., Jiang, Y., Di Maria, F., Masek, O., Superior activity of metal oxide biochar composite in hydrogen evolution under artificial solar irradiation: A promising alternative to conventional metal-based photocatalysts. *Int. J. Hydrog. Energy*, 44, 28698, 2019.
164. Zhai, Y., Peng, C., Xu, B., Wang, T., Li, C., Zeng, G., Zhu, Y., Hydrothermal carbonisation of sewage sludge for char production with different waste biomass: Effects of reaction temperature and energy recycling. *Energy*, 127, 167, 2017.
165. Chang, Y.M., Tsai, W.T., Li, M.H., Chemical characterization of char derived from slow pyrolysis of microalgal residue. *J. Anal. Appl. Pyrolysis*, 111, 88, 2015.

166. Sánchez, M.E., Menéndez, J.A., Domínguez, A., Pis, J.J., Martínez, O., Calvo, L.F., Bernad, P.L., Effect of pyrolysis temperature on the composition of the oils obtained from sewage sludge. *Biomass Bioenergy*, 33, 933, 2009.
167. Mašek, O., Budarin, V., Gronnow, M., Crombie, K., Brownsort, P., Fitzpatrick, E., Hurst, P., Microwave and slow pyrolysis biochar—Comparison of physical and functional properties. *J. Anal. Appl. Pyrolysis*, 100, 41, 2013.
168. Tahir, M.H., Çakman, G., Goldfarb, J.L., Topcu, Y., Naqvi, S.R., Ceylan, S., Demonstrating the suitability of canola residue biomass to biofuel conversion via pyrolysis through reaction kinetics, thermodynamics and evolved gas analyses. *Bioresour. Technol.*, 279, 67, 2019.
169. Liu, W.J. and Yu, H.Q., Thermochemical conversion of lignocellulosic biomass into mass-producible fuels: Emerging technology progress and environmental sustainability evaluation. *ACS Environ. Au*, 2, 98, 2022.

2

Sewage Sludge Conversion to Sustainable Energy: Biogas, Methane, Hydrogen, and Biofuels

El Asri Ouahid[1]*, Ben EL Caid Mohamed[2], Yousfi Ikram[3] and BenKaddour Rachid[3]

[1]*Laboratory of Microbial Biotechnology and Vegetal Protection, Faculty of Sciences, Ibn Zohr University, Agadir, Morocco*
[2]*Laboratory of Biotechnology and Valorisation of Natural Resources, Faculty of Sciences, Ibn Zohr University, Agadir, Morocco*
[3]*Laboratory of Agricultural Production Improvement, Biotechnology, and Environment, Faculty of Sciences, Mohammed First University, Oujda, Morocco*

Abstract
Today, humanity is caught between two constraints, the first being that of water. The permanent generation of wastewater has become a daily challenge for municipalities worldwide. In addition, water has become scarce, and we are witnessing a lack of drinking water; the generated wastewater produces a significant amount of sludge sewage that is generally not valued. The second constraint is energy; we face a depletion of energy sources, increased prices, and energy dependence on imports. Thus, we owe the need to recover wastewater by producing energy to participate in resolving the environmental, economic, and social problems of our nations. This chapter presents an update on new technology for transforming sludge and biomass from wastewater into energy sources like biogas, methane, hydrogen, and biofuel. We have mainly dissected technology with a sustainable and protective aspect of the environment to participate in sustainable socio-economic development.

Keywords: Anaerobic digestion, biogas, treatment, sustainable development, water

**Corresponding author*: o.elasri@uiz.ac.ma; elasriouahid@yahoo.fr

Inamuddin, Tariq Altalhi, Mohammad Luqman and Joseph Kapuku (eds.) Sewage and Biomass from Wastewater to Energy, (43–72) © 2024 Scrivener Publishing LLC

2.1 Introduction

Since the industrial revolution, the flow of matter between the socioeconomic system and the ecosystems has been profoundly unbalanced to the detriment of the environment, which is in progressive degradation [1]. The world nowadays has experienced a substantial demographic development involving an increasing economic effort. Currently, human activities require a massive demand for energy resources which are rapidly depleting, which implies the influence and increase of pollution. Thus, the deposits of fossil energy are diminished under enormous exploitation. Shafiee and Topal (2009) have calculated that fossil fuel deposits (oil, coal, and gas) will be exhausted in 35, 107, and 37 years, respectively [2]. In addition, the energy price represents a challenge for developing countries in import, transport, and storage. The increase in the price of hydrocarbons by companies producing fossil energy and the monopolization of global marketing generate very high prices for developing countries [3].

The pollution of the environment by the production of fossil energies generates high ecological risks linked to the transport of crude oil along the oceans [4]; the enormous expenditure of points related to the primary petrochemical transformation in refineries is at risk by accidents at the time of industrial chemical production, disposal of production waste and unwanted by-products, environmental pollution by post-use decomposition products, especially carbon dioxide (CO_2), and other pollutants in the biosphere [5, 6]. Thus, the protection of the environment and the preservation of natural resources are an essential component of sustainable development.

In front of this modernization, we have witnessed a significant increase in drinking water consumption and wastewater generation. In France, wastewater treatment plants treat about 5 billion m^3 of water annually [7]. Wastewater is combined with 100 to 500 g of feces and between 1 and 1.3 L of urine each person can produce daily [8]. The total volume of wastewater generated worldwide has been estimated at around 380 billion m^3, i.e., 380 trillion liters per year [9]. This enormous quantity is generally rejected in developing countries' seas, oceans, and rivers without valorization. This spill, without treatment, can harm public health and the environment [10]. Additionally, 40% of deaths worldwide are linked to water quality [11]. The North African region treats 43% of the wastewater generated [12]. On the other hand, developed countries have developed several technologies that allow the treatment of these waters and their valorization by producing energy because these waters contain a significant fraction of organic matter that can be converted into energy resources. In addition to these environmental problems, the depletion of energy resources and the massive use of fossil fuels have prompted the desire to establish concepts of sustainable development. We have elaborated this chapter to highlight the different possibilities and technologies used in converting biomass and sludge from wastewater in energy production. This chapter is divided into three axes: firstly, we have discussed the typology of wastewater, the characteristics of each type, then the world situation in the generation of wastewater and the problems generated by this excessive production; and finally, we have discussed the biological, physicochemical composition of wastewater. We dissected the notion of sludge and biomass in the second axis, mainly the typical definition. Then, we discussed the reciprocal relationship between sustainable energy and wastewater. Finally, we detailed new technologies for converting wastewater sludge and its residual biomass into biogas, methane, hydrogen, and fuel. We have focused our technical presentation on sustainable technology, which presents a new trend in the sustainable socio-economic development of our nations.

2.2 Wastewater: Origins, Characteristics, Types, and Problems

Wastewater is mainly produced from four source types: domestic, industrial, agricultural, and runoff. The first type comes from sinks, toilets, baths, and dishwashers. Factories and industrial facilities produce the second type. Agricultural wastewater originates from farm activities, including livestock, crops, laboratories, and food processing plants. However, the runoff comes from rain, melted snow, roofs, streets, and impermeable soils [13]. Therefore, wastewater is water that has either already been used and is discharged after use, in homes, factories, farms, or any other facility, or from rain. Therefore, wastewater contains wastes, chemicals, and organic materials that must be treated before they can be reused or released into the environment [10].

There are several classifications of wastewater, and they differ according to the classification criteria, for example, composition, origin, impact, degree of contamination, type of recovery, and final destination. The most commonly used classification is based on the degree of contaminant and their implications for public health and the environment. We can distinguish five categories of wastewater (Figure 2.1).

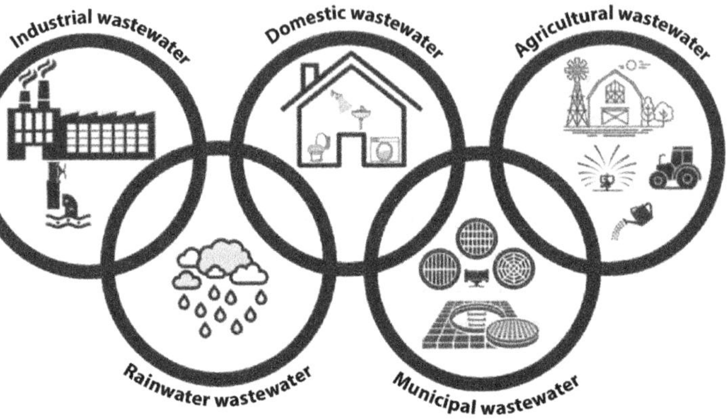

Figure 2.1 Different types of wastewater.

The global wastewater situation is very worrying. Millions of people worldwide do not have access to adequate sanitation systems and are at risk of disease and surface water contamination [14]. In addition, much-untreated sewage is discharged directly into rivers and oceans, with adverse environmental consequences [15]. In developing countries, urban wastewater includes five types of wastewater because the sewage pipeline is standard between the different sources. So, the urban sewage pipeline is a system of pipes and pipes that collects and transports wastewater from buildings and homes to the point of treatment or discharge into aquatic ecosystems. Moreover, the sewage pipes are generally constructed of concrete in these countries. They are installed at various depths according to the terrain and the local climatic conditions, from which their alterations and destruction are frequent because we witness a mixture of these waters. Moreover, the wastewater contains products that can harm human health and the environment, such as pathogenic bacteria (*Giardia*, *Salmonella typhimurium*, *Vibrio cholerae*, and *Cryptosporidium*), viruses, parasites, and chemicals [16]. These elements produce severe and even fatal illnesses and infections. Wastewater can also be a source of pollution for freshwater and soil, leading to water and soil quality degradation and decreased biodiversity [17]. Wastewater can also contain harmful chemicals, such as heavy metals (zinc, copper, nickel, and arsenic), which can cause long-term health problems and soil and groundwater contamination [18, 19]. Wastewater must therefore be treated before being discharged into public waters. Industrial plants and hospitals can install wastewater treatment systems to remove these toxic substances.

Therefore, wastewater is water used or not used for different purposes (domestic, agricultural, commercial, or industrial). However, the chemical and physical characteristics are highly variable depending on the type of wastewater, such as domestic, industrial, or agricultural wastewater. So, the composition of sewage can change over time and can be affected by weather, population, and treatment processes. Additionally, wastewater infrastructure is vulnerable to the hazards linked to a changing climate [20]. However, generally, the main components of wastewater typically include five materials (organic matter, nutrients, pathogens microbes, suspended solids, dissolved solids, and heavy metals) (Figure 2.2). So, the characteristics of wastewater can also vary depending on the source. Furthermore, wastewater contains much organic matter, constituting a vital biomass. Therefore, wastewater has a very complex biomass composition, with a wide variation in the concentrations of solid biomass (350 to 1200 mg/L), dissolved and particulate biomass (250 to 1000 mg/L), and microbial biomass (up to 109 microbes/ml) [21]. This biomass can be recovered and converted into solid, liquid, and gaseous fuels and electricity. Therefore, biomass from wastewater produces different types of energy. Thus, wastewater treatment is obligatory to reduce its dangerousness before being released into the environment or reused. Still, creating energy using new treatment technologies is necessary for a sustainable character. Therefore, in this chapter, we will discuss only biomass wastewater, focusing on its

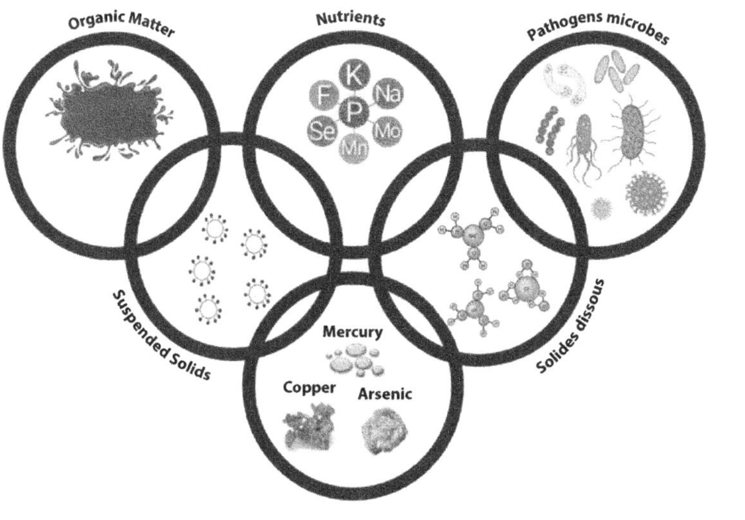

Figure 2.2 Main constituents of wastewater.

conversion to energy. The other constituents, such as heavy metals, are discussed in other forthcoming works.

2.3 World Situation in Wastewater and Energy

The exact amount of global wastewater produced each year is difficult to determine. According to the UN, around 80 to 90% of sewage in developing countries is discharged into the environment untreated, amounting to approximately 6.5 billion m^3 of wastewater annually [22]. For example, more than 150,000 European citizens living in urban agglomerations can generate a daily amount of 41.5 million m^3 of wastewater [23]. Therefore, it is challenging to give a precise quantity of wastewater generated worldwide because the volume of wastewater varies according to the strategies of supply and consumption of apoté by each nation.

The total volume of wastewater produced each year globally is estimated to be about 380 billion m^3 [9]. Qadir *et al.* (2010) estimated that global wastewater production will increase by 24% by 2030 and 51% by 2050 compared to the current level [12]. There are no precise data on the exact amount of wastewater in Africa. Still, it is estimated that more than 80% of wastewater in African and Asian countries is untreated and discharges directly into the environment [24]. Estimates of the quantity of wastewater in Africa vary depending on the source and the methods used to estimate it. Industriel activities in the Middle East and North Africa region generated 109 m^3 of wastewater [25]. According to El Moussaoui *et al.* (2022), Morroco produces 750 million m^3 of wastewater, but half of this quantity (48%) is discharged directly into aquatic ecosystems [26]. Therefore, we can say that we have a large amount of water that is not valorized and produces environmental and public health problems. Hence the need to declare that the waste of wastewater is a global concern, as it contributes to the loss of drinking water and inequalities in access to water. Water scarcity is a global problem affecting millions of people worldwide, which can be reduced by using this wastewater. Morocco has implemented many programs and projects to reuse 325 million m^3 in 2030 [27]. Wastewater is a valuable water source, and it is not used efficiently.

Energy security is a global concern, as it represents an essential source of stability and security for countries. It depends on access to affordable and reliable energy sources and the diversification of energy sources and technologies adopted. The global energy situation is complex and constantly changing. Oil and gas are the world's primary energy sources, accounting for around 60% of global energy consumption. Coal accounts for about 25%, hydropower around 10%, and renewables around 5% [28, 29]. However, the share of renewable energies is constantly growing and is expected to increase in the coming years.

In addition, there is another energy problem; it is energy dependence. It is the relations between the main producing and consuming countries of oil, gas, and other energy sources. This energy dependence has deepened since the Second World War and has been reinforced by the rise in oil prices since the 1970s [30]. Energy dependence is a global phenomenon that affects all countries, but some countries are more dependent than others. The major producing countries are the United States, Saudi Arabia, Russia, and Venezuela [31]. The primary consumer countries are Japan, China, Germany, and the United States. Producing and consuming countries depend on each other to ensure their energy supply and demand.

Morocco is a country that is highly dependent on imported energy for its consumption. This dependence is mainly due to the weakness of its energy resources, which do not cover the country's growing needs [32]. As a result, Morocco primarily imports oil, gas, and coal to meet its energy needs. The share of imports in total primary energy consumption is over 90% [33]. So, the majority of energy is imported from other countries.

Faced with this energy dependence on the import of energy resources and the enormous quantity of wastewater, we can combine wastewater treatment and energy production. Hence, the importance of producing this chapter emerges from the need to treat wastewater by delivering energy in sustainable development.

2.4 Composition of Wastewater

Wastewater is composed mainly of water. However, they also contain sediments such as solid particles, and organic and inorganic matter from domestic sewage and stormwater. Moreover, wastewater can contain chemicals, heavy metals, and drugs [18, 19]. In addition, bacteria, viruses, and parasites are also present in sewage [16]. So, the wastewater comprises a complex mixture of chemical, mineral, and biological materials.

The primary chemical constituents of wastewater are ammonia, carbon dioxide, nitrates, phosphates, salts, organic acids, oxygen, sulfur, and chlorine. Wastewater can also contain chemicals such as pesticides, hydrocarbons, heavy metals, and chemicals of industrial origin, which can be toxic to the environment [34, 35]. Wastewater can also contain biological matter, such as bacteria, viruses, and parasites, which can harm the environment.

The amount of ammonium present in wastewater depends on several factors, such as the origin and treatment of the wastewater, as well as the type of pollutants present in it. Ammonia is one of the principal components of wastewater; in domestic variety, the concentrations are between 10 and 50 mg industrial wastewater and can increase to 200 mg/L [36]. In untreated wastewater, the ammonium concentration can reach up to 45 mg/L; in treated wastewater by anaerobic digestion (AD), it is usually 600 to 1000 mg/L in the reject water [37]. The amount of phosphorus in wastewater depends on many factors, including water source and quality, agricultural and urban practices, wastewater treatment, and industrial processes. Phosphorus is present in domestic and industrial wastewater at concentrations below 10 mg/L [38]. However, phosphorus concentrations in wastewater vary considerably, from less than 0.1 mg/L to more than 100 mg/L [38]. Treated wastewater can contain a phosphorus concentration between 0.5 and 15 mg/L; this quantity does not produce eutrophication in the receiving ecosystems [39]. Regulatory standards for phosphorus in wastewater vary by country and region. The amount of organic matter in wastewater depends on the source of the wastewater and the activities that take place in it. Organic matter comes mainly from domestic wastewater, including food and industrial wastewater. Organic matter is a significant source of wastewater pollution and must be removed before the water is discharged into the environment. Domestic wastewater can contain up to 350 mg/L of organic matter [8]. In some industrial wastewater, the organic matter is to be in the range of 3000 to 5000 mg/L [40]. So, the amount of biomass in wastewater can vary greatly depending on the type of water source.

In general, wastewater can contain varying amounts of organic matter, including bacteria, algae, worms, fish, insects, and other aquatic organisms. Wastewater can also include different charges of chemicals that can alter water quality and harm the health of aquatic organisms. The biomass in wastewater can be measured using a test called Kjeldahl nitrogen, which calculates the total organic matter content in the water [41]. Wastewater contains various biological compounds such as bacteria, viruses, algae, protozoa, and parasites. They also have organic materials and chemicals from human activities, such as detergents, pharmaceuticals, gasoline, solvents, and agricultural chemicals. Bacteria, algae, and other living organisms consume these organic materials and produce wastes that are again dangerous for human health and the environment. So, the biological, organic, and inorganic fraction of wastewater constitutes the biomass of sewage; it is the most represented fraction in the general composition of wastewater. This biomass under sludge can be used to produce different types of energy. Therefore, we can conclude that all the suspended fraction is concentrated in the biomass of the sludge after treatment of the wastewater.

2.5 What is Sewage Sludge?

Sewage sludge is the solid material that remains after the treatment of wastewater. Wastewater contains a variety of organic and inorganic substances, including human and animal waste, food scraps, chemicals, and other contaminants. During the treatment process, the solids and liquids are separated, and the solids are further treated to produce sewage sludge [42, 43].

The quantity of sewage sludge produced from wastewater treatment varies worldwide, depending on population density, technology, and local regulations. However, it is estimated that globally, around 311 million tons of sewage sludge is produced annually from wastewater treatment plants. The typical wastewater treatment plant can have 2 to 3 L of sludge per person daily [44]. The daily dry sludge is 60 to 90 g per capita [45]. Advanced wastewater treatment technologies have made more sewage sludge in Europe and North America. Europe and the United States produced 10 million and 8 million tons of sewage sludge, respectively [46]. In other areas, such as Africa and Asia, the amount of sewage sludge produced is lower due to a lack of infrastructure for wastewater treatment.

The quantity of sewage sludge produced in Africa is not well-documented, as there is limited data available on the amount of wastewater treated and the efficiency of treatment plants in the region. However, the total amount of sewage produced in Africa is estimated to be around 50 billion m^3 annually. Due to a lack of adequate infrastructure for wastewater treatment in many African countries, a significant proportion of this wastewater is discharged into the environment without treatment, which can have serious environmental and health impacts.

In Europe, the quantity of sewage sludge produced from wastewater treatment is significant, as the region has a large population and a well-established wastewater treatment infrastructure. According to the latest available data from Eurostat, from 2003 to 2006, the dry biomass of sewage sludge generated in 27 European Union member countries (EU-27) was approximately 10 million tons; after 4 years (2010), it was 11.5 million tons; in 2020, it was expected to reach 13.5 million tons [47]. So, we see increasing production of sewage sludge in European countries. The quantity of sewage sludge produced in Europe can vary by country and by region within a country, depending on factors such as population density and the efficiency of wastewater treatment plants. Countries such as Germany, France, and the UK are among Europe's largest producers of sewage sludge due to their large populations and extensive wastewater treatment infrastructure [48].

The composition of sewage sludge in wastewater can vary depending on the source of the wastewater and the treatment process used. Generally, sewage sludge contains five material types:

a) Organic matter: This includes human and animal waste, food scraps, and other organic materials in wastewater. The dry sludge can contain (50 to 70%) organic matter [49]. Organic matter in sewage sludge can be a valuable source of energy.
b) Nutrients: Sewage sludge can contain high levels of nutrients such as nitrogen and phosphorus, which can be valuable for use as a fertilizer [50].
c) Pathogens: Wastewater can contain bacteria, viruses, and other pathogens that can threaten public health. Proper sewage sludge treatment is necessary to ensure these pathogens are destroyed or reduced to safe levels [51].
d) Heavy metals: Sewage sludge can contain heavy metals such as lead, cadmium, and mercury, which can be toxic to humans and the environment [52].
e) Organic chemicals: Wastewater can contain various organic substances, such as pharmaceuticals, personal care products, and industrial chemicals, which can be present in sewage sludge [53].

So, the composition of sewage sludge in wastewater is complex and can vary depending on various factors. For example, sewage sludge can contain a high concentration of organic matter converted to energy. However, it can also contain pathogens, heavy metals, and other contaminants, posing a risk to human health and the environment if not properly treated and managed. So, treatment and management of sewage sludge are necessary for protecting public health and the environment. Furthermore, the sustainable technologies of treating and managing wastewater sewage sludge can reduce the risk of disease transmission, prevent contamination of water and soil resources, reduce greenhouse gas emissions, and produce green energies. This last goal is the hinge of our chapter.

2.6 What Possibilities for the Conversion of Wastewater to Sustainable Energy?

There are many reasons why we need to convert wastewater into energy. The first reason is that it will help us reduce our carbon dioxide emissions. Wastewater is a significant carbon dioxide source contributing to the greenhouse effect and rising global temperatures. By converting this wastewater into energy, we can reduce our carbon footprint and help fight climate change. Second, converting biomass from wastewater to power will help us reduce our dependence on fossil fuels and air pollution. This substrate can be converted into energy through AD [54]. These methods produce energy from organic materials without having carbon dioxide or air pollutants. So, we participate in the establishment of the 17 sustainable development objectives. Finally, converting wastewater to energy can help reduce our energy costs. Wastewater is an abundant and inexpensive source of energy. If we can find efficient and economical ways to convert this wastewater into power, we can reduce our energy costs and improve our energy efficiency.

Converting wastewater to energy is a great way to promote sustainable development. Indeed, this technology makes it possible to recover the energy produced by waste and convert it into electricity and heat. This energy can then power buildings, factories, or municipal systems. In addition, this technology can help reduce greenhouse gas emissions and improve air quality. It can also help fight wastewater pollution, reducing the amount of harmful substances that are discharged into rivers and streams. Converting wastewater to energy is an effective way to promote sustainable development and improve the quality of the environment [55].

So, the conversion of sewage and biomass from wastewater to energy produces several benefits: Providing renewable energy, reducing our reliance on fossil fuels and other non-renewable sources of energy, conserving natural resources, saving the costs of wastewater treatment and disposal, potentially saving money for wastewater treatment facilities and their customers, reducing the environmental impact of waste management, reused for non-potable purposes such as irrigation and industrial processes, and reducing the demand for freshwater resources. Finally, converting wastewater to energy can provide a sustainable and cost-effective solution for managing wastewater and generating renewable energy.

2.7 Different Technologies, from Wastewater to Sustainable Energy

Several technologies can convert sewage and biomass wastewater into energy in scientific literature. Protecting the environment and preserving natural resources are essential to sustainable development. Additionally, several municipalities aim to adopt technologies that obey sustainable development [56]. So, we discuss in this part of the chapter alone that the technologies offer promising ways to convert sewage and biomass from wastewater into energy and valuable resources while reducing environmental pollution and improving public health. Overall, we bring together sustainable technologies.

2.7.1 Anaerobic Digestion: Production of Biogas and Methane

AD is a biological process that breaks down organic matter in wastewater in anaerobic conditions. The AD is carried out by several microorganisms, breaking the organic matter into biogas and digestate [56]. Microbes are the key players in this process, so AD is microbial biotechnology [54]. The microbial community in AD is complex and diverse. In AD, microorganisms, including bacteria, archaea, fungi, and protozoa, break down complex organic molecules into simpler compounds such as biogas and digestate [57]. The biogas produced by AD is a mixture of methane and carbon dioxide, and it can be used as a renewable energy source for heating or electricity generation [58]. The digestate, the remaining slurry after digestion, can be used as a nutrient-rich fertilizer. So, this process can be used to treat wastewater to produce electricity and fertilizer. The AD process typically occurs in a sealed vessel known as a digester, which provides an oxygen-free environment for the anaerobic bacteria to thrive [59]. Hanum *et al.* [60] have declared that the number of AD treatments based on sewage sludge exceeded 1300 digesters. The digester is usually heated to a temperature of around 35°C to 40°C to promote the growth of the bacteria and speed up the digestion process [54].

Sewage sludge is the solid material left over after wastewater treatment, and it typically contains a high concentration of organic matter. So, the biogas and digestate are produced by sewage sludge of wastewater. The AD process of sewage sludge involves the following steps:

a) Pre-treatment: Before AD, sewage sludge is usually pre-treated to break down large organic molecules and increase the surface area for the microorganisms to digest. Pinnekamp (1988) has confirmed that thermal pretreatment of sewage sludge before AD strongly affects energy production; it increases biogas production from 60 to 223 L/kgvs [61]. High-pressure homogenization pretreatment of sewage sludge has successfully improved biogas production to 240 ml/(gTCOD) [62]. Hydrothermal biomass treatments increase methane concentrations up to 79% in biogas [63]. Pre-treatment can involve physical, chemical, or thermal methods [64].

b) Digestion: The pre-treated sewage sludge is then introduced into an anaerobic digester, a closed vessel where AD occurs. Inside the digester, a complex microbial community, primarily anaerobic bacteria, breaks down the sewage sludge's organic matter and produces biogas.
c) Biogas production: Biogas is a mixture of methane and carbon dioxide, the primary product of the AD process. The biogas can be captured and used as a renewable source of energy. For example, it can be burned to generate heat and electricity or purified and compressed to be used as a vehicle fuel or injected into the natural gas grid. For example, Germany's 1258 AD plants can generate 3517 GWh/year of electricity. Malaysia has 35 large digesters, which produce 247 GWh/year of electricity [60].
d) Digestate production: The remaining material, after the digestion process, known as digestate, is a nutrient-rich fertilizer that can be used in agriculture. Applying anaerobic digestates from sewage sludge increases the biomass and height of tomato plants (*Solanum lycopersicum L.*) to 37.5- and 6-fold, respectively [65].

So, converting sewage sludge to biogas has several benefits, such as reducing greenhouse gas emissions, generating renewable energy, and providing a sustainable sewage treatment method [54, 60]. Additionally, the production of digestate can provide a valuable source of fertilizer for agriculture [65]. Finally, converting sewage sludge to biogas through AD is an effective and sustainable sewage treatment and renewable energy production method.

There are several benefits to using AD for wastewater treatment and energy production. First, it can help reduce the amount of organic wastewater, reducing greenhouse gas emissions. It can also provide a renewable energy source, which can help reduce dependence on fossil fuels.

Additionally, it can help reduce the environmental impact of wastewater discharge by producing high-quality effluent. So, AD is a promising technology for sustainable wastewater treatment and energy production, and it has the potential to play an essential role in the transition to a more sustainable and circular economy [66].

The conversion of sewage and biomass wastewater to methane through AD is an effective and sustainable method of wastewater treatment and renewable energy production. The methane produced during AD can be captured and used to generate heat and electricity, purified and compressed as a vehicle fuel, or injected into the natural gas grid as a renewable energy source. In addition to energy production, AD can reduce greenhouse gas emissions and help reduce the volume of wastewater that needs to be treated and the associated costs.

2.7.2 Microbial Fuel Cells: Hydrogen Production

Microbial fuel cells (MFCs) use bacteria to convert organic matter, such as sewage sludge, into electrical energy. MFCs operate on the principle of electroactive bacteria capable of transferring electrons from the organic matter in sewage sludge to an electrode, generating a flow of electrical current [67]. This generates an electrical current that can be harnessed for various applications, such as powering sensors or lighting. For example, the sludge added to chamber MFC chamber increased energy production from 0.30–0.32 V and 19.9–22.6 mW/m^2 (raw sludge) to 0.34–0.36 V and 25.5–28.6 mW/m^2 (sterilized sludge), and 0.41–0.43 V and 37.1–40.8 mW/m^2 (base pretreated sludge), respectively [68]. In addition, the double-chamber MFC reactor sludge prepared by graphite electrodes can produce more energy at 312.98 mW/m^2 and 39.07 $\mu A/cm^2$ [69].

The use of MFCs for sewage sludge treatment is still in the research and development stage, and there are several challenges to overcome, such as low power output and scalability. However, there is increasing interest in MFC technology as a potential solution for sustainable wastewater treatment and renewable energy production [67]. Additionally, due to its high organic content, sewage sludge can be a suitable substrate for MFCs. Therefore, using MFCs for sewage sludge treatment can offer several advantages over traditional treatment methods, including generating renewable energy and reducing the sludge volume [70]. Several types of MFCs have been developed for various applications (mediator-based MFCs, membrane-based MFCs, microbial desalination cells, MFCs, and sediment MFCs). Each type has its unique advantages and limitations. However, the MFCs that can use sewage sludge as a substrate are four principal types:

1. Microbial desalination cells use bacteria to remove salt from water while generating electricity. The sewage sludge as a substrate, the bacteria consume organic matter in the sludge, producing electrons and protons as waste products. The electrons are transferred to the anode electrode and then flow through an external circuit to the cathode electrode, generating an electrical current. In addition to generating electricity, this process using sewage sludge as a substrate can also remove salt from water [71]. The bacteria consume ions in the saltwater, driving them toward the cathode compartment, where an ion-exchange membrane can remove them [72]. This technology using dehydrated sludge can simultaneously generate electricity 1.97 ± 0.33 W/m^2 with an open circuit voltage of 0.998 ± 0.011 V for more than 50 days, desalination (6 to 31%), and remove biomass up to 32% [73]. Another study can produce maximum salt removal of more than 65%, a power density of 47.1 W/m^2, and 58% biomass removal when the initial COD is more than 6500 mg/liter [74]. So, the result is that the water that exits has a lower salt concentration than the water that enters treatment technology.
2. Sediment MFCs: The sewage sludge is placed in the anode compartment of the MFCs, where bacteria consume the organic matter and produce electrons and protons as waste products. The electrons are transferred to the anode electrode, which flows through an external circuit to the cathode electrode, generating an electrical current [75]. Generally, in these MFCs, the bacteria consume the organic matter in the sludge, generating an electrical current in the anode compartment. Wang *et al.* (2018) have used anaerobic and activated sludge to produce energy and biomass removal (10 mW/m^2, 62%) and (3.9 mW/m^2, 58.2%), respectively [76]. To ensure a good optimization of this process, we need to maintain a consistent flow of sewage sludge matter to the anode and the potential for bacterial fouling or biofilm formation on the electrodes to increase the efficiency of electricity generation.
3. Microbial electrolysis cells are a type of MFC that use an external voltage to drive the reduction of hydrogen ions to hydrogen gas. Bacteria consume organic matter in sewage sludge and produce hydrogen gas at the cathode. So, this

technology uses bacteria that generate hydrogen gas by breaking down sewage sludge's organic matter, producing electrons and protons as waste products. The electrons are transferred to the anode electrode and then flow through an external circuit to the cathode electrode, generating an electrical current. At the cathode, protons and electrons combine to form hydrogen gas. Combining ultrasonic pretreatment and microbial electrolysis cells of activated sludge improves the biogas and methane production to 3.68-fold and 2.56-fold, respectively [77]. So, this technology can produce more energy because hydrogen (141 MJ/kg) has 2.6 times methane (55 MJ/kg). Furthermore, the microbial electrolysis cells combined with AD of activated sludge with an applied 0.6 to 1.2 V can increase energy production to 30 to 200% [78]. So, Microbial electrolysis cells using sewage sludge as a substrate have several potential advantages, such as generating hydrogen gas from organic waste and operating in environments where other forms of hydrogen generation may not be possible.

4. Benthic MFCs are designed to generate electricity using microorganisms living on the bottom of aquatic environments such as oceans, rivers, or lakes [79]. This process uses sewage sludge as a substrate utilizing the bacteria to oxidize organic matter to produce electrons and protons. So, this process converts the sewage sludge to electricity by consuming organic matter in the sludge. The electrons are transferred to the anode electrode, which can flow through an external circuit to the cathode electrode, generating an electrical current [80]. This technology can produce minimal power from 5 to 50 mW, with 42% of this power storage in the battery [81]. Meanwhile, the protons migrate to the cathode compartment, combining with oxygen to produce water. The benthic MFCs, called bioelectrochemical fuel cells, can generate electricity and wastewater treatment.

2.7.3 Thermal Hydrolysis

Thermal hydrolysis (TH) is a process that uses heat and pressure to break down biomass in wastewater. This process involves heating the sludge in a pressurized vessel (from 6 to 9 bars) to a high temperature (typically 140°C to 170°C) for a short period (typically 30 min to 1 hour) absence of oxidants [82]. TH of the sewage sludge at 220°C can produce important caramelization of starch, an increasing amount of dextrose, maltose, and breakdown of unsaturated lipids. TH at 150°C can reduce the ammonification of sludge, reducing the risk of methanogenic inhibition [83]. TH mainly improved the liberation of low molecular weight proteins and polysaccharides [84]. So, the heat and pressure cause the sludge to break down and release more organic material, making it more readily available for subsequent treatment. Generally, TH of sewage sludge is a process used for treating sewage sludge before AD. This process can increase the efficiency of AD by making the organic matter more accessible to the bacteria that break it down. The TH of the sewage sludge positively affects the anaerobic biodegradability; it reduces volatile solids by 62% greater than that of the control test. It can also improve this waste's viscosity and dehydration because it could be operated at a higher matter loading rate (2.3 times than usual) [85]. TH at 172 °C processing showed better biomass solubilization and higher energy generation [84]. TH technology is powerfully combined with AD in large tanks in more than 120 installations on five continents [86]. So, TH improves the efficiency and effectiveness of AD.

TH of sewage sludge has several benefits: it increased biogas production during AD and improved dewatering of the sludge and the production of a more stable, pathogen-free biosolid that can be used as a fertilizer. Additionally, TH can reduce the volume of sludge by up to 50%, reducing the costs associated with transportation and disposal. However, some challenges are related to the TH of sewage sludge, such as the high energy costs associated with heating the sludge and the potential for the process to release volatile organic compounds or odors.

2.7.4 Gasification: Biofuel Production

Generally, gasification technology is a process that converts organic matter into a gas that can be burned to generate electricity or heat. For example, in wastewater treatment, gasification can convert sludge into a gas that can be burned to generate electricity. In Italy, the annual gasification of sewage sludge generated 714 and 1142 GWh of electrical and thermal energy, respectively [87]. The gasification of sewage sludge is a thermal process that involves converting the organic matter into a gas mixture, commonly known as "syngas or synthesis gas," by exposing the sludge to high temperatures in the presence of oxygen or steam. During the gasification step, the sludge is heated to temperatures between 700°C and 1000°C in the gasification reactor named the gasifier [88, 89]. As a result, the organic matter of sewage sludge breaks down and releases gasses such as hydrogen, carbon monoxide, nitrogen, and methane. The syngas produced is directly proportional to the amount of organic matter in the sewage sludge.

Gasification of sewage sludge involves several steps: a) drying the sludge by TH; b) the pyrolysis stage converts the organic matter in the wastewater to solid char and volatilized gas; c) the gasification stage involves heating the volatilized gas in the presence of oxygen or steam to produce syngas; and d) cleaning and conditioning the syngas: this product is cleaned and conditioned to remove impurities such as tars, sulfur, and other contaminants. The syngas produced can be burned directly to generate electricity or used as a heating fuel or feedstock for producing chemicals and other value-added products.

Gasification of sewage sludge produces syngas with a lower heating value (LHV) is up to 5 MJ/m^3, which can be intensely utilized in an internal combustion engine [90]. Elbl *et al.* (2023) have confirmed that converting sewage sludge to syngas with air as the gasifying agent can produce an LHV of 4.11 MJ/m^3 [91]. The combined heat and power production from sewage sludge at 900°C of gasification température is 2.73 kWh/kg dry solid [87]. In Croatia, the LHV of syngas produced by sewage sludge demonstrated an essential range of 10 to 20 MJ/m^3 [92]. So, the syngas composition depends on various factors such as the gasification temperature, gasification agent, and the properties of the sewage sludge. This process requires passage of the sludge in the process described above (thermal hydrolysis) to remove water, reduce the moisture content, and increase the calorific value of the sludge.

Additionally, the gasification of sewage sludge can reduce the volume of sludge by up to 90%. This process produces an ash residue that can be used as a soil amendment or construction material. The amount of phosphate produced by the gasification of sewage sludge was equal to 22.06%; this quality of P_2O_5 is similar to natural phosphate rocks [90]. However, some challenges are associated with the gasification of sewage sludge, such as the high capital and operating costs of gasification equipment, the need for high-quality feedstock with consistent properties, and the potential for emissions of air pollutants during the gasification process. So, gasification of sewage sludge is a promising technology for converting a waste product into a valuable resource.

2.7.5 Membrane Bioreactors: Improving Conversion Processes

Membrane bioreactor (MBRs) is a trending technology that offers promising ways to convert wastewater into energy and valuable resources while reducing environmental pollution and improving public health. Currently, the global market of water treatment membranes exceeds $10.4 billion [93]. Gao *et al.* (2022) have demonstrated that the MBR process is more profitable than conventional activated sludge [94]. It is a type of wastewater treatment technology that combines biological treatment with membrane filtration [95]. Generally, this process can produce high-quality effluent that can be reused or discharged to the environment.

MBRs technology is resumed with six essential steps:

1. Pre-treatment: The influent wastewater is screened to remove large solids, such as rags, plastics, and debris, that could clog the membrane pores. The pre-treatment improves the behavior of membranes [96].
2. Biological treatment: The pre-treated wastewater is then introduced into the bioreactor tank, where microorganisms, such as bacteria and fungi, break down the organic matter in the sewage sludge through activated sludge. The bacteria type is the central part of microorganisms (more than 90%) in the activated sludge [97]. The microorganisms consume the sewage sludge's organic matter, nitrogen, and phosphorus, converting them into biomass and other byproducts.
3. Membrane filtration: The treated effluent is then separated from the sewage sludge using membrane filtration, which uses either microfiltration, ultrafiltration, or nanofiltration membranes [98]. The membrane pores are small enough to retain the suspended solids and microorganisms in the sewage sludge while allowing clean water to pass through. Microfiltration and ultrafiltration are significant categories of filtration used in this technology with size pores 0.1 to 0.2 μm and 0.005 μm, respectively. Nanofiltration has fewer size pores than 10 nm, which stops any virus that may be present in wastewater [99].
4. Sludge handling: The concentrated sewage sludge, known as mixed liquor suspended solids, is withdrawn from the bottom of the bioreactor tank and sent to a sludge holding tank for further processing.
5. Disinfection: The treated effluent is then disinfected to remove any remaining pathogenic microorganisms, such as viruses and bacteria. This is typically achieved through chemical disinfectants, such as chlorine or ozone, or UV disinfection [100].

6. Effluent discharge or reuse: The treated effluent can then be discharged into a nearby water body or reused for non-potable applications, such as irrigation, fish processing industry, aquaculture, or industrial processes [101]. So, the different types of sewage sludge are treated by microorganisms in different bioreactor tanks, and the treated effluent is then separated from the sludge using a membrane filtration system. The membrane acts as a physical barrier, allowing only clean water to pass through while retaining the suspended solids and bacteria in the sludge.

We can distingue two main types of MBRs that can be used for sewage sludge treatment: submerged MBR and external MBR. Submerged are usually made of polymeric materials and have a pore size of around 0.1 to 0.4 μm. The mixed liquor is continuously circulated through the membranes, allowing the microorganisms to degrade the organic matter in the sewage sludge while retaining the suspended solids and bacteria [102]. On the other hand, external MBRs are typically used in larger-scale applications and consist of membrane modules placed outside the bioreactor tank. The membranes are usually made of ceramic or stainless steel and have a pore size of around 0.01 to 0.1 μm [93]. The mixed liquor is circulated through the bioreactor tank and pumped through the membranes to produce the permeate. External membrane types are generally more expensive than submerged types. Still, they can handle higher volumes of wastewater and produce higher-quality permeate. Martínez *et al.* [103] have confirmed that submerged MBRs present lower capital costs and energy consumption (0.038 to 5.68 kWh/m^3) against external MBRs that give high capital costs and increased consumption (3 to 7.3 kWh/m^3).

MBRs of sewage sludge do not directly produce energy, but they are often combined with other technologies to recover energy from the sewage sludge. So, this technology can be combined with previous technologies because it can also produce high-quality sewage sludge with a low water content, which can be further treated by AD, gasification, or other thermal processes to generate energy or biofuels. In addition, the low water content of sludge reduces the energy required for dewatering and drying, making it a more cost-effective feedstock for energy recovery. In large anaerobic with MBR produces an excellent effluent quality with 90 to 95% COD and BOD_5 removal efficiency and significant biogas production from 0.09 to 0.10 L/L of raw wastewater. This biogas presents a good amount of methane (75 to 81%) [104]. The energy electricity produced by AD with MBR varies from 0.5 to 12.0 kWh/m^3; this electricity can cover the installation energy demand [105].

The amount of energy produced by the MBR of sewage sludge depends on different parameters like the volume and characteristics of the wastewater and sewage sludge generated, the design and efficiency of the membrane, and the method used to recover energy from the sludge. Generally, this process produces biogas through AD of the concentrated sludge. So, MBRs have several advantages over conventional activated sludge systems, including high-quality effluent, low biomass production, compact footprints, and greater flexibility in plant design. In addition, producing high-quality effluent can be used for water reuse applications, such as irrigation, industrial processes, or even potable water production.

2.8 Conclusion

In this chapter, we can consider wastewater sludge as much as a renewable energy source. We noticed that they have a lot of sludge-to-energy technology and are different in many aspects. Combining the technologies to reach this objective is necessary to maximize the yield of electricity and reduce the resin treatment residues. Thus, the conversion of sludge and biomass from wastewater mainly into sustainable green electricity requires the contextual choice of one or more conversion technologies adapted to each country.

References

1. Chopra, K. and Dasgupta, P., Economic systems and ecosystems: Interlinkages, co-evolution or disparate movement?, in: *Ecosystem Functions and Management: Theory and Practice*, H. Sandhu, (Ed.), pp. 17–35, Springer International Publishing, Cham, 2017.
2. Shafiee, S. and Topal, E., When will fossil fuel reserves be diminished? *Energy Policy*, 37, 181, 2009.
3. Chow, E.C. and Rincon-Romero, J.-A., Energy and the Economy in North America, in: *The Palgrave Handbook of International Energy Economics*, M. Hafner, and G. Luciani, (Eds.), pp. 713–729, Springer International Publishing, Cham, 2022.
4. Wilberforce, T., Olabi, A.G., Sayed, E.T., Elsaid, K., Abdelkareem, M.A., Progress in carbon capture technologies. *Sci. Total Environ.*, 761, 143203, 2021.

5. Clews, R.J., Fundamentals of the Petroleum Industry, in: *Project Finance for the International Petroleum Industry*, pp. 83–99, Elsevier, 2016.
6. Nagtode, V.S., Cardoza, C., Yasin, H.K.A., Mali, S.N., Tambe, S.M., Roy, P., Singh, K., Goel, A., Amin, P.D., Thorat, B.R., Cruz, J.N., Pratap, A.P., Green Surfactants (Biosurfactants): A Petroleum-Free Substitute for Sustainability-Comparison, Applications, Market, and Future Prospects. *ACS Omega*, 8, 11674, 2023.
7. Brice, R.P., Anastasia, S., Somar, K., Corinne, L.G.L., Karine, W., Vincent, G., Gaël, P., Continuous degradation of micropollutants in real world treated wastewaters by photooxidation in dynamic conditions. *Water Res.*, 221, 118777, 2022.
8. Gerba, C.P. and Pepper, I.L., Wastewater Treatment and Biosolids Reuse, in: *Environmental Microbiology*, Second E., R.M. Maier, I.L. Pepper, C.P. Gerba, (Eds.), pp. 503–530, Academic Press, San Diego, 2009.
9. Kundu, D., Dutta, D., Samanta, P., Dey, S., Sherpa, K.C., Kumar, S., Dubey, B.K., Valorization of wastewater: A paradigm shift towards circular bioeconomy and sustainability. *Sci. Total Environ.*, 848, 157709, 2022.
10. Grassi, M., Rizzo, L., Farina, A., Endocrine disruptors compounds, pharmaceuticals and personal care products in urban wastewater: Implications for agricultural reuse and their removal by adsorption process. *Environ. Sci. Pollut. Res.*, 20, 3616, 2013.
11. Evans, A.E.V., Hanjra, M.A., Jiang, Y., Qadir, M., Drechsel, P., Water Quality: Assessment of the Current Situation in Asia. *Int. J. Water Resour. Dev.*, 28, 195, 2012.
12. Qadir, M., Bahri, A., Sato, T., Al-Karadsheh, E., Wastewater production, treatment, and irrigation in Middle East and North Africa. *Irrig. Drain. Syst.*, 24, 37, 2010.
13. Rashid, S.S. and Liu, Y.Q., Assessing environmental impacts of large centralized wastewater treatment plants with combined or separate sewer systems in dry/wet seasons by using LCA. *Environ. Sci. Pollut. Res.*, 27, 15674, 2020.
14. Meili, D., Schelbert, V., Alam, M.U., Antwi-Agyei, P., Simiyu, S., Adjei, K.A., Dwumfour-Asare, B., Rahman, M., Lüthi, C., Günther, I., Indicators for Sanitation Quality in Low-Income Urban Settlements: Evidence from Kenya, Ghana, and Bangladesh. *Soc Indic. Res.*, 162, 683, 2022.
15. Dodds, W.K. and Whiles, M.R., Aquatic Chemistry and Factors Controlling Nutrient Cycling, in: *Freshwater Ecology*, Second E., W.K. Dodds, and M.R. Whiles, (Eds.), pp. 289–321, Academic Press, London, 2010.
16. Zhuang, L.L., Yang, T., Zhang, J., Li, X., The configuration, purification effect and mechanism of intensified constructed wetland for wastewater treatment from the aspect of nitrogen removal: A review. *Bioresour. Technol.*, 293, 122086, 2019.
17. Sampei, N., Matsumoto, K., Mihara, C., Sakurai, S., Takeuchi, T., Ohkouchi, Y., Sugita, K., Inaba, K., Historical analysis of domestic wastewater pollution

and its measures in Lake Teganuma basin using surfactant concentrations recorded in sediment core samples. *SN Appl. Sci.*, 1, 2019.

18. Al-Enezi, G., Hamoda, M.F., Fawzi, N., Ion Exchange Extraction of Heavy Metals from Wastewater Sludges. *J. Environ. Sci. Health - Part A Toxic/ Hazard. Subst. Environ. Eng.*, 39, 455, 2004.
19. Liu, P.C., Vilando, A.C., Lu, M.C., Treatment of synthetic zinc and nickel wastewater and identification of its crystallization products by fluidized bed homogeneous crystallization technology. *Process Saf. Environ. Prot.*, 164, 154, 2022.
20. Langeveld, J.G., Schilperoort, R.P.S., Weijers, S.R., Climate change and urban wastewater infrastructure: There is more to explore. *J. Hydrol.*, 476, 112, 2013.
21. Warwick, C., Guerreiro, A., Soares, A., Sensing and analysis of soluble phosphates in environmental samples: A review. *Biosens. Bioelectron.*, 41, 1, 2013.
22. Bijekar, S., Padariya, H.D., Yadav, V.K., Gacem, A., Hasan, M.A., Awwad, N.S., Yadav, K.K., Islam, S., Park, S., Jeon, B.H., The State of the Art and Emerging Trends in the Wastewater Treatment in Developing Nations. *Water (Switzerland)*, 14, 2537, 2022.
23. Gkeka-Serpetsidaki, P. and Tsoutsos, T., Sustainable site selection of offshore wind farms using GIS-based multi-criteria decision analysis and analytical hierarchy process. Case study: Island of Crete (Greece), in: *Low Carbon Energy Technologies in Sustainable Energy Systems*,.G.L. Kyriakopoulos (Ed.), pp. 329–342, Academic Press, 2021.
24. Chen, Y.P., Fu, B.J., Zhao, Y., Wang, K.B., Zhao, M.M., Ma, J.F., Wu, J.H., Xu, C., Liu, W.G., Wang, H., Sustainable development in the Yellow River Basin: Issues and strategies. *J. Clean. Prod.*, 263, 121223, 2020.
25. Sakhel, S.R., Geissen, S.U., Vogelpohl, A., Virtual industrial water usage and wastewater generation in the Middle East and North Africa 2011–2015. *Euro-Mediterr. J. Environ. Integr.*, 2, 2017.
26. El Moussaoui, T., Belloulid, M.O., Elharbili, R., El Ass, K., Ouazzani, N., Simultaneous assessment of purification performances and wastewater byproducts management plans towards a circular economy: Case of Marrakesh WWTP. *Case Stud. Chem. Environ. Eng.*, 6, 100228, 2022.
27. Mansir, I., Oertlé, E., Choukr-Allah, R., Evaluation of the performance and quality ofwastewater treated by m'zar plant in Agadir, Morocco. *Water (Switzerland)*, 13, 954, 2021.
28. Khare, V., Khare, C., Nema, S., Baredar, P., Productivity and manufacturing economics of solar energy system, in: *Decision Science and Operations Management of Solar Energy Systems*, V. Khare, C. Khare, S. Nema, P. Baredar, (Eds.), pp. 111–145, Academic Press, 2023.
29. Amador, A.T., Neto, A.F.G., Cruz, J.N., Magno, F.N.B., Marques, F.C., Da Silva, C.A.B., Nero, J., Neto, A.M.J.C., Molecular Electronics Including Temperature Effects Based on Dyes Pigments. *J. Nanosci. Nanotechnol.*, 19, 3631, 2019.

30. Yeoman, I., High oil prices — Haven't we seen this all before! *J. Revenue Pricing Manage.*, 7, 239, 2008.
31. Bradshaw, M., Van de Graaf, T., Connolly, R., Preparing for the new oil order? Saudi Arabia and Russia. *Energy Strategy Rev.*, 26, 100374, 2019.
32. Elasri, O. and El amin Afilal, M., Potential for biogas production from the anaerobic digestion of chicken droppings in Morocco. *Int. J. Recycl. Org. Waste Agric.*, 5, 195, 2016.
33. Ainou, F.Z., Ali, M., Sadiq, M., Green energy security assessment in Morocco: green finance as a step toward sustainable energy transition. *Environ. Sci. Pollut. Res.*, 30, 61411, 2023.
34. Devault, D.A. and Karolak, S., Wastewater-based epidemiology approach to assess population exposure to pesticides: a review of a pesticide pharmacokinetic dataset. *Environ. Sci. Pollut. Res.*, 27, 4695, 2020.
35. Garba, Z.N., Abdullahi, A.K., Haruna, A., Gana, S.A., Risk assessment and the adsorptive removal of some pesticides from synthetic wastewater: a review. *Beni-Suef Univ. J. Basic Appl. Sci.*, 10, 2021.
36. Körner, S., Das, S.K., Veenstra, S., Vermaat, J.E., The effect of pH variation at the ammonium/ammonia equilibrium in wastewater and its toxicity to Lemna gibba. *Aquat. Bot.*, 71, 71, 2001.
37. Landreau, M., Byson, S.J., You, H.J., Stahl, D.A., Winkler, M.K.H., Effective nitrogen removal from ammonium-depleted wastewater by partial nitritation and anammox immobilized in granular and thin layer gel carriers. *Water Res.*, 183, 116078, 2020.
38. Yuan, Z., Pratt, S., Batstone, D.J., Phosphorus recovery from wastewater through microbial processes. *Curr. Opin. Biotechnol.*, 23, 878, 2012.
39. Omwene, P.I. and Kobya, M., Treatment of domestic wastewater phosphate by electrocoagulation using Fe and Al electrodes: A comparative study. *Process Saf. Environ. Prot.*, 116, 34, 2018.
40. Threrujirapapong, T., Khanitchaidecha, W., Nakaruk, A., Treatment of high organic carbon industrial wastewater using photocatalysis process. *Environ. Nanotechnol. Monit. Manage.*, 8, 163, 2017.
41. Manu, D.S. and Thalla, A.K., Artificial intelligence models for predicting the performance of biological wastewater treatment plant in the removal of Kjeldahl Nitrogen from wastewater. *Appl. Water Sci.*, 7, 3783, 2017.
42. Chen, H., Yan, S., Ye, Z., Meng, H., Zhu, Y., Utilization of urban sewage sludge: Chinese perspectives. *Environ. Sci. Pollut. Res.*, 19, 1454, 2012.
43. Scrinzi, D., Ferrentino, R., Baù, E., Fiori, L., Andreottola, G., Sewage Sludge Management at District Level: Reduction and Nutrients Recovery via Hydrothermal Carbonization. *Waste Biomass Valorization*, 2022.
44. Senfter, T., Fritsch, L., Berger, M., Kofler, T., Mayerl, C., Pillei, M., Kraxner, M., Sludge thickening in a wastewater treatment plant using a modified hydrocyclone. *Carbon Resour. Convers.*, 4, 132, 2021.

45. Ma, H., Guo, Y., Qin, Y., Li, Y.Y., Nutrient recovery technologies integrated with energy recovery by waste biomass anaerobic digestion. *Bioresour. Technol.*, 269, 520, 2018.
46. Enebe, N.L., Chigor, C.B., Obileke, K., Lawal, M.S., Enebe, M.C., Biogas and Syngas Production from Sewage Sludge: A Sustainable Source of Energy Generation. *Methane*, 2, 192, 2023.
47. Havukainen, J., Saud, A., Astrup, T.F., Peltola, P., Horttanainen, M., Environmental performance of dewatered sewage sludge digestate utilization based on life cycle assessment. *Waste Manage.*, 137, 210, 2022.
48. Mateo-Sagasta, J., Raschid-Sally, L., Thebo, A., Global wastewater and sludge production, treatment and use, in: *Wastewater: Economic Asset in an Urbanizing World*, P. Drechsel, M. Qadir, D. Wichelns, (Eds.), pp. 15–38, Springer Netherlands, Dordrecht, 2015.
49. Maw, M.M., Boontanon, N., Fujii, S., Boontanon, S.K., Rapid and efficient removal of organic matter from sewage sludge for extraction of microplastics. *Sci. Total Environ.*, 853, 158642, 2022.
50. Kirchmann, H., Börjesson, G., Kätterer, T., Cohen, Y., From agricultural use of sewage sludge to nutrient extraction: A soil science outlook. *Ambio.*, 46, 143, 2017.
51. Straub, T.M., Pepper, I.L., Gerba, C.P., Hazards from pathogenic microorganisms in land-disposed sewage sludge, in: *Reviews of Environmental Contamination and Toxicology*, G.W. Ware, (Ed.), pp. 55–91, Springer, New York, NY, 1993.
52. Geng, H., Xu, Y., Zheng, L., Gong, H., Dai, L., Dai, X., An overview of removing heavy metals from sewage sludge: Achievements and perspectives. *Environ. Pollut.*, 266, 115375, 2020.
53. Santana, J.M., Fraga, S.V.B., Zanatta, M.C.K., Martins, M.R., Pires, M.S.G., Characterization of organic compounds and drugs in sewage sludge aiming for agricultural recycling. *Heliyon*, 7, e06771, 2021.
54. Martins, M.G., da Silva Arouche, T., Neto, A.F.G., da Cruz, J.N., da Costa, F.L.P., Fernandes, L.L., de Carvalho Junior, R.N., da Silva Costa, J.F., de Jesus Chaves Neto, A.M., Density functional theory for the thermodynamic gas-phase investigation of butanol biofuel and its isomers mixed with gasoline and ethanol. *J. Mol. Model.*, 27, 80, 2021.
55. El Asri, O., Fadlaoui, S., Ramdani, M., Errochdi, S., Microbial Degradation of Biowaste for Hydrogen Production, in: *Recent Advances in Microbial Degradation*, Inamuddin, M.I. Ahamed, R. Prasad, (Eds.), pp. 431–447, Springer, Singapore, 2021.
56. El Asri, O., Fadlaoui, S., Afilal, M.E., Applications of Microbes in Municipal Solid Waste Treatment, in: *Application of Microbes in Environmental and Microbial Biotechnology*, Inamuddin, M.I. Ahamed, R. Prasad, (Eds.), pp. 587–607, Springer Nature, Singapore, 2022.

57. El Asri, O. and Afilal, M.E., Comparison of the experimental and theoretical production of biogas by monosaccharides, disaccharides, and amino acids. *Int. J. Environ. Sci. Technol.*, 15, 1957, 2018.
58. Asri, O., Fadlaoui, S., Ramdani, M., Anaerobic Digestion Technology of Solid and Liquid Forms of Olive Wastes in the Mediterranean Region, in: *Mediterranean Fruits Bio-wastes: Chemistry, Functionality and Technological Applications*, M.F. Ramadan, and M.A. Farag, (Eds.), pp. 67–89, Springer International Publishing, Cham, 2022.
59. El Asri, O., Ramdani, M., Latrach, L., Haloui, B., Mohamed, R., elaminAfilal, M., Energetic valorization of Nador lagoon algae and proposal to use it as a means of elimination of the eutrophication in this lagoon. *Ecol. Eng.*, 103, 236, 2017.
60. Hanum, F., Yuan, L.C., Kamahara, H., Aziz, H.A., Atsuta, Y., Yamada, T., Daimon, H., Treatment of sewage sludge using anaerobic digestion in Malaysia: Current state and challenges. *Front. Energy Res.*, 7, 2019.
61. Gahlot, P., Balasundaram, G., Tyagi, V.K., Atabani, A.E., Suthar, S., Kazmi, A.A., Štěpanec, L., Juchelková, D., Kumar, A., Principles and potential of thermal hydrolysis of sewage sludge to enhance anaerobic digestion. *Environ. Res.*, 214, 113856, 2022.
62. Nabi, M., Liang, J., Zhang, P., Wu, Y., Fu, C., Wang, S., Ye, J., Gao, D., Shah, F.A., Dai, J., Anaerobic digestion of sewage sludge pretreated by high pressure homogenization using expanded granular sludge blanket reactor: Feasibility, operation optimization and microbial community. *J. Environ. Chem. Eng.*, 9, 104720, 2021.
63. Aragon-Briceño, C.I., Ross, A.B., Camargo-Valero, M.A., Strategies for the Revalorization of Sewage Sludge in a Waste Water Treatment Plant Through the Integration of Hydrothermal Processing. *Waste Biomass Valorization*, 14, 105, 2023.
64. Neumann, P., Pesante, S., Venegas, M., Vidal, G., Developments in pre-treatment methods to improve anaerobic digestion of sewage sludge. *Rev. Environ. Sci. Biotechnol.*, 15, 173, 2016.
65. Cristina, G., Camelin, E., Tommasi, T., Fino, D., Pugliese, M., Anaerobic digestates from sewage sludge used as fertilizer on a poor alkaline sandy soil and on a peat substrate: Effects on tomato plants growth and on soil properties. *J. Environ. Manage.*, 269, 110767, 2020.
66. Xu, Y., Gong, H., Dai, X., High-solid anaerobic digestion of sewage sludge: achievements and perspectives. *Front. Environ. Sci. Eng.*, 15, 2021.
67. Nawaz, A., ul Haq, I., Qaisar, K., Gunes, B., Raja, S.I., Mohyuddin, K., Amin, H., Microbial fuel cells: Insight into simultaneous wastewater treatment and bioelectricity generation. *Process Saf. Environ. Prot.*, 161, 357, 2022.
68. Xiao, B., Yang, F., Liu, J., Enhancing simultaneous electricity production and reduction of sewage sludge in two-chamber MFC by aerobic sludge digestion and sludge pretreatments. *J. Hazard. Mater.*, 189, 444, 2011.

69. Ayol, A., Biryol, İ., Taşkan, E., Hasar, H., Enhanced sludge stabilization coupled with microbial fuel cells (MFCs). *Int. J. Hydrogen Energy*, 46, 29529, 2021.
70. Roy, H., Rahman, T.U., Tasnim, N., Arju, J., Rafid, M.M., Islam, M.R., Pervez, M.N., Cai, Y., Naddeo, V., Islam, M.S., Microbial Fuel Cell Construction Features and Application for Sustainable Wastewater Treatment. *Membranes (Basel)*, 13, 490, 2023.
71. Carmalin Sophia, A., Bhalambaal, V.M., Lima, E.C., Thirunavoukkarasu, M., Microbial desalination cell technology: Contribution to sustainable waste water treatment process, current status and future applications. *J. Environ. Chem. Eng.*, 4, 3468, 2016.
72. Ramírez-Moreno, M., Rodenas, P., Aliaguilla, M., Bosch-Jimenez, P., Borràs, E., Zamora, P., Monsalvo, V., Rogalla, F., Ortiz, J.M., Esteve-Núñez, A., Comparative Performance of Microbial Desalination Cells Using Air Diffusion and Liquid Cathode Reactions: Study of the Salt Removal and Desalination Efficiency. *Front. Energy Res.*, 7, 2019.
73. Meng, F., Zhao, Q., Zheng, Z., Wei, L., Wang, K., Jiang, J., Ding, J., Na, X., Simultaneous sludge degradation, desalination and bioelectricity generation in two-phase microbial desalination cells. *Chem. Eng. J.*, 361, 180, 2019.
74. Ebrahimi, A., Yousefi Kebria, D., Darzi, G.N., Improving bioelectricity generation and COD removal of sewage sludge in microbial desalination cell. *Environ. Technol. (United Kingdom)*, 39, 1188, 2018.
75. Xu, B., Ge, Z., He, Z., Sediment microbial fuel cells for wastewater treatment: Challenges and opportunities. *Environ. Sci. Water Res. Technol.*, 1, 279, 2015.
76. Wang, C.T., Sangeetha, T., Zhao, F., Garg, A., Chang, C.T., Wang, C.H., Sludge selection on the performance of sediment microbial fuel cells. *Int. J. Energy Res.*, 42, 4250, 2018.
77. Hu, K., Chen, W., Jia, S.Q., Wang, W., Han, F., Enhanced degradation of waste activated sludge in microbial electrolysis cell by ultrasonic treatment. *Front. Microbiol.*, 10, 2019.
78. Wang, X.T., Zhang, Y.F., Wang, B., Wang, S., Xing, X., Xu, X.J., Liu, W.Z., Ren, N.Q., Lee, D.J., Chen, C., Enhancement of methane production from waste activated sludge using hybrid microbial electrolysis cells-anaerobic digestion (MEC-AD) process – A review. *Bioresour. Technol.*, 346, 126641, 2022.
79. Prakash, O., Mungray, A., Kailasa, S.K., Chongdar, S., Mungray, A.K., Comparison of different electrode materials and modification for power enhancement in benthic microbial fuel cells (BMFCs). *Process Saf. Environ. Prot.*, 117, 11, 2018.
80. Umar, M.F., Rafatullah, M., Abbas, S.Z., Mohamad Ibrahim, M.N., Ismail, N., Advancement in benthic microbial fuel cells toward sustainable bioremediation and renewable energy production. *Int. J. Environ. Res. Public Health*, 18, 3811, 2021.
81. Reimers, C.E., Wolf, M., Alleau, Y., Li, C., Benthic microbial fuel cell systems for marine applications. *J. Power Sources*, 522, 231033, 2022.

82. Ngo, P.L., Udugama, I.A., Gernaey, K.V., Young, B.R., Baroutian, S., Mechanisms, status, and challenges of thermal hydrolysis and advanced thermal hydrolysis processes in sewage sludge treatment. *Chemosphere*, 281, 130890, 2021.
83. Wilson, C.A. and Novak, J.T., Hydrolysis of macromolecular components of primary and secondary wastewater sludge by thermal hydrolytic pretreatment. *Water Res.*, 43, 4489, 2009.
84. Lu, D., Sun, F., Zhou, Y., Insights into anaerobic transformation of key dissolved organic matters produced by thermal hydrolysis sludge pretreatment. *Bioresour. Technol.*, 266, 60, 2018.
85. Oosterhuis, M., Ringoot, D., Hendriks, A., Roeleveld, P., Thermal hydrolysis of waste activated sludge at Hengelo Wastewater Treatment Plant, the Netherlands. *Water Sci. Technol.*, 70, 1, 2014.
86. Svensson, K., Thermal hydrolysis sludge pretreatment, in: *Clean Energy and Resources Recovery: Biomass Waste Based Biorefineries*, vol. 1, V. Tyagi, and K. Aboudi, (Eds.), pp. 391–398, Elsevier, 2021.
87. Di Fraia, S., Massarotti, N., Uddin, M.R., Vanoli, L., Conversion of Sewage Sludge to combined heat and power: Modeling and optimization. *Smart Energy*, 5, 100061, 2022.
88. Korberg, A.D., Mathiesen, B.V., Clausen, L.R., Skov, I.R., The role of biomass gasification in low-carbon energy and transport systems. *Smart Energy*, 1, 100006, 2021.
89. Nipattummakul, N., Ahmed, I., Kerdsuwan, S., Gupta, A.K., High temperature steam gasification of wastewater sludge. *Appl. Energy*, 87, 3729, 2010.
90. Werle, S. and Sobek, S., Correction to: Gasification of sewage sludge within a circular economy perspective: a Polish case study. *Environ. Sci. Pollut. Res.*, 26, 35, 35422–35432, 2019, 10.1007/s11356-019- 05897-2.
91. Elbl, P., Baláš, M., Lisý, M., Lisá, H., Sewage sludge and digestate gasification in an atmospheric fluidized bed gasifier. *Biomass Convers. Biorefin.*, 2023.
92. Đurđević, D. and Blecich, P., Jurić, Ž., Energy recovery from sewage sludge: The case study of Croatia. *Energies*, 12, 1927, 2019.
93. Singh, R., Introduction to Membrane Technology, in: *Membrane Technology and Engineering for Water Purification*, Second E., R. Singh, (Ed.), pp. 1–80, Butterworth-Heinemann, Oxford, 2015.
94. Gao, T., Xiao, K., Zhang, J., Xue, W., Wei, C., Zhang, X., Liang, S., Wang, X., Huang, X., Techno-economic characteristics of wastewater treatment plants retrofitted from the conventional activated sludge process to the membrane bioreactor process. *Front. Environ. Sci. Eng.*, 16, 2022.
95. Le-Clech, P., Membrane bioreactors and their uses in wastewater treatments. *Appl. Microbiol. Biotechnol.*, 88, 1253, 2010.
96. Moustafa, M.A.E., Effect of the pre-treatment on the performance of MBR, Berghausen WWTP. Germany. *Alexandria Eng. J.*, 50, 197, 2011.
97. Al-Asheh, S., Bagheri, M., Aidan, A., Membrane bioreactor for wastewater treatment: A review. *Case Stud. Chem. Environ. Eng.*, 4, 100109, 2021.

98. Wang, Z., Ultrafiltration Membrane Bioreactor, in: *Encyclopedia of Membranes*, E. Drioli, and L. Giorno, (Eds.), pp. 1–2, Springer, Berlin Heidelberg, 2015.
99. Adelodun, B., Tiamiyu, A.G.O., Ajibade, F.O., Odey, G., Ibrahim, R.G., Goala, M., Bakare, H.O., Ajibade, T.F., Adeniran, J.A., Adeniran, K.A., Choi, K.S., Presence, detection, and persistence of SARS-CoV-2 in wastewater and the sustainable remedial measures, in: *Environmental and Health Management of Novel Coronavirus Disease (COVID-19)*, M.H. Dehghani, R.R. Karri, S. Roy, (Eds.), pp. 91–114, Academic Press, 2021.
100. Sun, H., Liu, H., Han, J., Zhang, X., Cheng, F., Liu, Y., Chemical cleaning-associated generation of dissolved organic matter and halogenated byproducts in ceramic MBR: Ozone versus hypochlorite. *Water Res.*, 140, 243, 2018.
101. Hoinkis, J., Gukelberger, E., Atiye, T., Galiano, F., Figoli, A., Gabriele, B., Mancuso, R., Mamo, J., Clough, S., Hoevenaars, K., Membrane Bioreactor (MBR) Treated Domestic Wastewater for Reuse in a Recirculating Aquaculture System (RAS), in: *Advances in Science, Technology and Innovation*, V. Naddeo, K.-H. Choo, M. Ksibi, (Eds.), pp. 153–155, Springer International Publishing, Cham, 2022.
102. Moraes, E.S., Reis, G.F., Cruz, J., Cozzolino, K., Neto, A.F.G., Andrade-Filho, T., Neto, A.M.J.C., Thermodynamics study of biokerosene from coconut and palm kernel oils and JP-8 aircraft fuels in the gas phase by the DFT method. *J. Mol. Model.*, 26, 79, 2020.
103. Martínez, R., Ruiz, M.O., Ramos, C., Cámara, J.M., Diez, V., Comparison of external and submerged membranes used in anaerobic membrane bioreactors: Fouling related issues and biological activity. *Biochem. Eng. J.*, 159, 107558, 2020.
104. Kong, Z., Wu, J., Rong, C., Wang, T., Li, L., Luo, Z., Ji, J., Hanaoka, T., Sakemi, S., Ito, M., Kobayashi, S., Kobayashi, M., Qin, Y., Li, Y.Y., Large pilot-scale submerged anaerobic membrane bioreactor for the treatment of municipal wastewater and biogas production at 25 °C. *Bioresour. Technol.*, 319, 124123, 2021.
105. Zielińska, M., Bernat, K., Mikucka, W., Membrane bioreactor technology: The effect of membrane filtration on biogas potential of the excess sludge. *Membranes (Basel)*, 10, 1, 2020.

3

Biodiesel from Sewage Sludge

Sonica Sondhi* **and Roopali Sharma**

Department of Biotechnology, Chandigarh Group of Colleges, Landran, Mohali, Punjab, India

Abstract

The usage of biofuels as an alternative to traditional mineral diesel generated from fossil fuels is gaining widespread acceptance across the world. In spite of this, the increased costs connected with the feedstock led to a lower level of competitiveness in comparison to commercial diesel. Research is now being conducted on a worldwide scale with the goal of developing fuels that are derived from renewable biomass as an alternative to conventional sources that are already in use. Due to the ease with which it can be obtained and the relatively cheap price at which it can be purchased, municipal sewage sludge (also known as MSS), which is rich in lipids, has the potential to be used as a profitable feedstock. The technique of transesterification may be used to turn lipids into biodiesel, while the pyrolysis method can be used to turn lipids into bio-oil. This article provides a comprehensive analysis of the many aspects of the process of producing biofuel from lipids that have been obtained from sludge. In addition, an in-depth discussion is provided on the critical aspects that have a significant impact on the process as well as the economic sustainability of the endeavor. This study highlights the need for future research to focus on the use of modern process intensification methodology to boost the effectiveness of manufacturing biodiesel from sustainable biomass. Specifically, this review highlights the requirement for future research to concentrate on the utilization of contemporary process intensification methodologies.

Keywords: Biodiesel, municipal sewage sludge, bio-sludge, bio-oil, transesterification, fossil fuels

**Corresponding author*: sonicasondhi@gmail.com

Inamuddin, Tariq Altalhi, Mohammad Luqman and Joseph Kapuku (eds.) Sewage and Biomass from Wastewater to Energy, (73–92) © 2024 Scrivener Publishing LLC

3.1 Introduction

According to projections, the global population is anticipated to increase to 9.8 billion by the year 2050 and subsequently to 11.2 billion by the year 2100. The rise in population may result in a notable surge in fuel consumption to meet the energy needs in the forthcoming years [1]. According to the most recent edition of the British Petroleum (BP) Statistical Review of World Energy, which was published in June 2017, the amount of oil that was used on a global scale was estimated to be 1.6 million barrels per day (Mb/d), while the increase in the amount of oil that was produced on a global scale was only 0.4 Mb/d. Furthermore, it has been reported that the worldwide confirmed oil reserves amounted to 1707 billion barrels in the year 2016, which would be sufficient to sustain oil production for a period of 50.6 years [2]. The reduction in energy generation necessitates the exploration of alternative sources that can effectively mitigate the emission of greenhouse gasses (GHG) into the atmosphere. Fossil fuel and industrial processes are major contributors to GHG emissions, and therefore, the adoption of new sources is crucial in reducing the earth's temperature [3].

In comparison to traditional fossil fuels, sustainable alternative fuels are characterized by a greater degree of renewability, cost-effectiveness, efficiency, and a lower impact on the surrounding environment [4]. In the course of the most recent several decades, a variety of different types of biofuels have been developed. Some examples of these biofuels include bio-ethanol, biomethanol, biodiesel, dimethyl ether, bio-oil, biogas, and biomethane, although the list is not exhaustive [5]. It has come to light that biodiesel and bioethanol are the ones that are employed the most often. The reason for their acceptance can be attributed to their specific characteristics, such as minimal or no modifications required for the current engine hardware and fuel delivery infrastructure, simplified production processes, and economic feasibility [6]. However, the use of bioethanol as an additive for enhancing the octane number of petrol and improving combustion quality and engine performance is restricted in its applicability. The fuel in question has thermal qualities that are comparable to those of diesel that is readily accessible on the market [7]. These attributes include the calorific value, the cetane number, and the volatility properties. However, there are significant differences in some attributes of fuels, such as density, viscosity, and the amount of oxygen they contain, in addition to the need for them to have a stoichiometric fuel-to-air ratio. Despite having a lower heating value of approximately 10% in comparison to mineral diesel, biodiesel's higher specific gravity ultimately results in a 5% lower energy content on a unit volume basis [8]. The utilization of biodiesel in an engine leads to a higher thermal efficiency compared to the utilization of conventional diesel [9].

In addition, the study of data from life cycle assessments reveals that the use of biodiesel leads to a reduction in greenhouse gas emissions, with a net energy ratio that is more than 1.3, in contrast to the standard mineral diesel, which has a ratio that is just 0.19 [10]. Biodiesel gives the impression of being a viable alternative fuel source that does away with the need to rely on nonrenewable fossil fuels and has the ability to reduce the amount of pollution that is caused all over the globe. In addition to this, it may provide economic possibilities in suburban and rural regions [11].

In everyday parlance, biodiesel is referred to as alkyl esters of fatty acid. It is possible to generate biodiesel by combining different lipid fractions with alcohol and catalyzing the reaction using an acid, a base, a solid, or an enzyme [12]. The high cost of biodiesel's feedstock, which is mostly composed of vegetable oils and accounts for more than 80% of the sector's overall production costs, is the key factor preventing the business from being commercially viable [13]. In addition, the use of vegetable oil as a source for the creation of biodiesel presents a difficulty because of the possibility that it will compete with the need for human consumption of the oil [14]. The use of used cooking oils as an alternative feedstock that may be produced at a lower cost for the manufacturing of biodiesel gives a possibility for a solution [15]. However, the variability in properties of the waste cooking oils poses a challenge to the consistency of the resulting biodiesel, necessitating additional costs for purification, as noted in previous studies. Researchers are now looking at the viability of using nonedible oil sources such as algae, jatropha, karanja, and neem as viable alternatives to the feedstock used in biodiesel production [16]. Nevertheless, the cultivation of these sources takes a significant amount of land, and the accessibility of these sources continues to be a source of concern. As a consequence of this, it is of the utmost importance to investigate several alternative sources of biodiesel feedstock that are not only simple to get but also economical [17].

Energy has evolved as a component that is essential for the survival of the human race. At this time, in the present, modern civilization uses significant amounts of energy in order to maintain a high quality of life, guarantee economic development and expansion, and guarantee social progress [18]. A natural and unavoidable yearly rise in the demand for energy is going to occur as a direct result of the continued growth of both the world population and industrialization. The predicted growth of the global economy, developments in cutting-edge technology, the creation of fresh entrepreneurial activities, and a rise in the population of the globe are all factors that are likely to contribute to an increase in the demand for

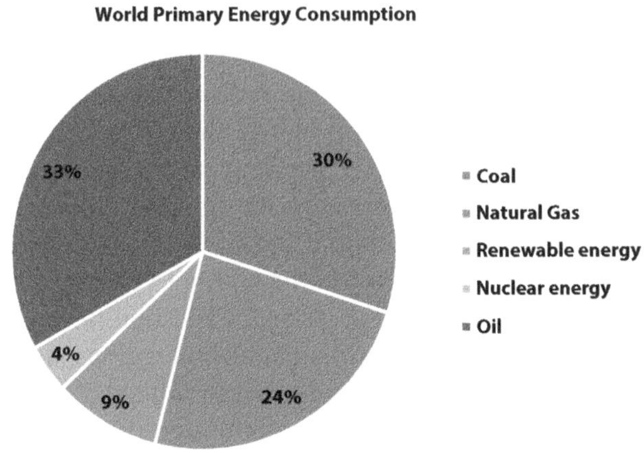

Figure 3.1 Global primary energy consumption [24].

energy worldwide [19]. According to research that was only recently made public, it is expected that the demand for energy on a worldwide scale will rise by 1.3% per year until the year 2040. The increasing economic boom of developing economies like China and India will lead to a spike in the demand for energy resources [20]. This is because emerging nations have lower energy efficiency standards. At the moment, a significant proportion of the energy that is used all over the globe comes from sources that are derived from petroleum [21] (see Figure 3.1). It is possible, and perhaps likely, that the reserves of fossil fuels will run out in the not-too-distant future since these supplies are finite and may be depleted. When compared to the reserves of oil and gas, it is anticipated that coal reserves will continue to be economically viable until the year 2112, but it is anticipated that oil and gas reserves will be depleted by the years 2040 and 2042, respectively [22]. In addition, the environmental ramifications associated with the unequal exploitation of nonrenewable energy sources, which may include but are not limited to air pollution, emissions of greenhouse gasses, and global warming, eventually impede the projected utilization of these resources [23].

3.1.1 The Utilization of Sustainable Energy Sources for the Transportation Industry

The transportation sector is a significant consumer of energy, with petroleum accounting for more than 95% of the energy consumed. In addition, it has been documented that the transportation industry accounts for 27%

of global carbon dioxide emissions [25]. By implementing alternative fuels derived from renewable sources, the transportation sector can significantly reduce its petroleum consumption and CO_2 emissions. The aforementioned approach has been recognized as a viable method for addressing the aforementioned issue [26]. Biofuels are a variety of fuels derived from biomass with the potential to reduce vehicle emissions of carbon dioxide and other greenhouse gasses. Furthermore, it is commonly accepted that liquid biofuels have substantial promise as a viable replacement for petroleum oil, provided that they can be employed in conventional engines with minimum changes. This fact contributes to the widespread acceptance of liquid biofuels as a viable option [27].

Although there has been a consistent increase in the use of biofuels in the transportation industry since 2011, their impact in this sector relative to petroleum-derived products is still relatively insignificant [27]. The use of biofuels in vehicular conveyance is minimal at present. In 2009, the European Union issued a directive mandating a renewable energy target for all member states. The accomplishment of the goal described above necessitated the acquisition of 10% points of the total amount of energy used in the transportation sector [28]. It has been noted that the rates of incorporation vary depending on the continent. Latin America continues to dominate the global market, with a share of over 9.6% attributable predominantly to its robust ethanol industry. The regional energy consumption rates in North America and Europe are 6.4 and 4.7%, respectively [29].

Despite Asia's relatively modest contribution of 1.2% to global biofuel production, the region has made significant investments and implemented government initiatives to promote the development and adoption of biofuels. Several nations have witnessed a substantial expansion of the biofuel industry due to the rising demand for transportation fuels [30]. In accordance with the European Union's Renewable Energy Directive (RED) for the year 2020, it was determined in 2017 that only Finland and Sweden had effectively attained the renewable goal of 10% in the transport sector [1]. The rate of achievement in Finland was 18.8%, while it was 38.6% in Sweden. It was found that Austria's percentage was quite near the cutoff point of 9.7%. With a rate of 9.1% in 2017, France remained in fourth place worldwide when it came to the number of companies it incorporated. The average proportion of the European Union's 28 member states increased from 7.2% in 2016 to 7.6% in 2017, marking a significant year-over-year jump. It is noteworthy that Norway's forward-thinking policies have led to renewable energy rates in the transport sector that are higher than 15% of total energy consumption [1].

As prospective alternatives to petrol and diesel, respectively, bioethanol and biodiesel have been the subject of extensive research among the numerous types of biofuels. The topic of discussion in this chapter is biodiesel, an alternative fuel that is less harmful to the environment and is produced almost entirely from vegetable oils or, to a lesser degree, animal lipids [31].

3.1.2 Origin and Properties of Biodiesel

Rudolf Diesel's initial experimentation with peanut oil as a fuel source for engines has paved the way for the potential utilization of vegetable oils as a viable alternative fuel for engines [32]. Despite the fact that vegetable oils were available as a potential fuel source, their utilization was hindered by the prevalent accessibility and affordability of petroleum fuel during that era. Moreover, the elevated viscosity exhibited by vegetable oils constitutes a significant limitation to their application as a fuel source [33]. The process of reducing the viscosity of vegetable oils through a transesterification reaction with methanol and ethanol was patented by George Chavanne in 1937. Biodiesel gained significant attention following the recognition of fossil fuel depletion [34].

At the moment, the process of making biodiesel involves transesterification, in which vegetable oils or animal fats are subjected to an alcohol (often methanol) in the presence of a catalyst (typically a base) [35]. This operation is known as the transesterification procedure. The aforementioned procedure leads to the creation of Fatty Acid Methyl Esters (FAMEs). The adoption of this sustainable energy resource offers a significant alternative to conventional diesel fuel owing to its analogous chemical properties and harmoniousness with the current commercial diesel engine and refueling technology [36]. Biodiesel exhibits promising potential for effective utilization in two distinct modes: as a blend with conventional diesel fuel and in its unadulterated state [37]. In addition, biodiesel presents significant ecological benefits owing to its elevated biodegradability, which surpasses that of standard diesel by a factor of four. Additionally, it is nontoxic and presents no hazards during storage and handling [38]. Furthermore, the combustion of biodiesel leads to a noteworthy decrease in exhaust emissions, encompassing CO_2, CO, hydrocarbons, and particulate matter, with the exception of NO. Additionally, it is worth noting that biodiesel has the capacity to effectively eradicate the discharge of SO and sulfates, given that it lacks sulfur, thereby rendering it a more environmentally friendly substitute to conventional petroleum diesel [39].

3.1.3 Fabrication of Biodiesel

In recent years, there has been a notable increase in the production of biodiesel. According to a recent report, the worldwide production of biodiesel amounted to approximately 41.3 billion liters in 2018. The production of biodiesel exhibits a greater degree of geographic diversity in comparison to ethanol production, primarily attributable to policy priorities, and is distributed across numerous nations [40]. In 2018, Europe emerged as the largest biodiesel producer, contributing to 53% of global production. The other four countries that made it to the top five list were the United States, Brazil, Indonesia, and Germany, accounting for 17, 13, 10, and 8% of global production, respectively. Argentina is also included in the list, contributing 5% to the overall global production. In 2018, the United States generated 6.9 billion liters of biodiesel, while Brazil produced 5.4 billion liters of the same product. It is anticipated that by the year 2025, the United States will have achieved a production capacity of more than one billion gallons of biodiesel [31].

3.1.4 Municipal Sewage Sludge (MSS)

The stuff that is left behind after municipal wastewater has been treated is called sewage sludge. Sewage sludge is disposed of in landfills. Because sewage sludge biodiesel manufacturing has the potential to be a lucrative business enterprise, there has been a rise in interest in municipal sewage sludge (MSS) across the world [41]. MSS is distinguished by its high organic waste content and its large lipid concentration. The wastewater treatment plant (WWTP) produces two basic types of sludge, which are referred to as primary sludge and secondary sludge respectively [42]. The primary sludge is made up of floating grease and solid materials, while the secondary sludge, which is also known as activated sludge, is predominantly made up of microbial cells and suspended particles [43]. Both the main clarifier and the secondary clarifiers have the ability to collect two different sludge. The waste product known as sewage sludge is obtained from the WWTP. It is a complicated mixture of a wide variety of non-digested organic components, bacteria, oils, lipids, and inorganic minerals [44].

3.1.5 Generation and Control of Sewage Sludge

Figure 3.2 displays a basic municipal sewage treatment plant (STP). Basic and secondary sludge are the two basic types of sludge that make up the WWTP.

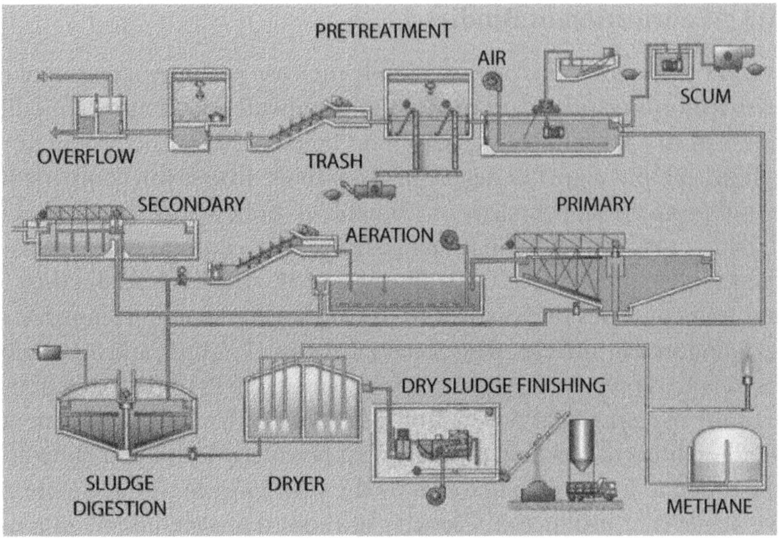

Figure 3.2 Sewage treatment plant. Source:https://chem.libretexts.org/Bookshelves/Introductory_Chemistry/Chemistry_for_Changing_Times_%28Hill_and_McCreary%29/14%3A_Water/14.08%3A_Wastewater_Treatment

There are significant differences between the two types of sludge in terms of their relative compositions [45]. The screening and grit removal process produces the primary sludge, which may be found at the very bottom of the main settler. This sludge is collected when the procedure has been completed [46]. This material is made up of solids, buoyant grease, and basic organic components such as cellulose, lipids, and proteins. Activated sludge, also known as secondary sludge, is a complex biomass that is mostly made up of microbial cells and suspended particles [47]. This kind of sludge goes through a process known as activation. It is manufactured using a procedure known as aerobic biological treatment, which is performed on wastewater that has already been primarily treated. After that, the secondary settling tank is used to collect any remaining silt that may have been present [48].

After the thickening process, the leftover secondary sludge is combined with the thickened primary sludge, which results in a byproduct that is typically composed of sixty percent primary sludge and forty percent secondary sludge [49]. This byproduct is generated as a consequence of the wastewater treatment procedure. A significant proportion of local wastewater treatment facilities employ a combination of sludge as a source for anaerobic digestion with the aim of reducing the prevalence of pathogens, unpleasant smells, and solid waste [50]. This results in the production of stabilized sludge, as. Furthermore, it is anticipated that there will be a rise in sludge production in the coming years as a result of urbanization and industrialization [51].

The residue produced during the process of purifying wastewater originating from sewage is an inescapable byproduct that necessitates specialized processing prior to disposal, resulting in significant expenses in STP operations, accounting for approximately 50 to 60% of the total cost [42]. Moreover, the management of sewage sludge poses numerous environmental difficulties. The traditional methods of sludge disposal, such as incineration, land application as a fertilizer, landfill disposal, and ocean disposal, all have the potential to release hazardous compounds and heavy metals into the surrounding environment [52]. The requirements for the disposal of sludge have been made stricter as a result of the government's regulatory initiatives. Consequently, there is a growing interest in investigating the valorization of raw and untreated sludge as a means of achieving a more sustainable strategy for managing sludge, with a focus on renewable energy recovery [53].

3.1.6 Sewage Sludge as a Potentially Useful Lipid Source

Lipids are a family of chemical substances that include fatty acids and their derivatives, such as glycerides, free fatty acids (FFAs), phospholipids, and sphingolipids. Lipids are not confined to these four categories, though. The lipid fractions may be found in the sewage sludge, and there are a significant number of them. The presence of lipids in sewage sludge may be due to the adsorption of lipids from industrial and residential waste, or it may be the result of the process of lysis, which results in the creation of lipids via the breakdown of cell membranes. Both of these explanations are possible [54]. Figure 3.3 provides a depiction of the biodiesel production process from sewage sludge in the form of a schematic diagram.

It has been indicated, based on the examination of scholarly literature that was carried out by researchers all over the globe, that the lipids that are found in sewage sludge hold the potential to serve as a suitable source material for the manufacturing of biodiesel [55]. This is the conclusion that was reached as a result of the research that was carried out. This approach not only holds promise as an alternative fuel source but also has significant economic implications for waste management solutions, as evidenced by previous studies [56].

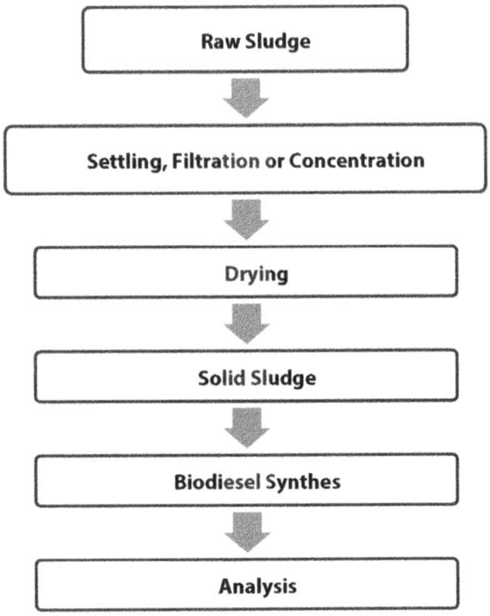

Figure 3.3 Schematic diagram of biodiesel production.

3.1.7 Extraction of Lipids from Sewage Sludge

The removal of lipids from the sludge is the first step in the process of making biodiesel from sewage sludge, which must be completed before commercial production can begin [57]. Therefore, optimizing the process of lipid extraction is a significant problem that has the potential to influence the procedure's ability to be carried out at a profit [58]. Because of their polarity, immiscibility with water, high boiling point, cost-effectiveness, and environmental consequences, organic solvents including methanol, ethanol, methane, hexane, and chloroform are often used for the extraction of lipids from sewage sludge. Other organic solvents include chloroform. Despite the fact that chloroform and toluene exhibit superior performance, their lack of environmental sustainability renders them infrequently employed [59]. According to the research that has been done, nonpolar solvents have the ability to extract nonpolar and saponifiable lipids such as glycerides and FFAs from their respective environments. The conversion of these lipids into biodiesel is considered more suitable for biodiesel production in comparison to polar solvents, as per reference [60]. In addition, it has been reported that the employment of nonpolar solvents such as hexane in the production of biodiesel incurs lower costs compared to the use of polar solvents like methanol and ethanol. Hexane is comparatively more eco-friendly than toluene [61].

3.1.8 Synthesis of Biodiesel

Transesterification: There are many different ways to convert vegetable oil and alcohol into fatty acid alkyl esters and glycerol, but one of the most common ways is by transesterification processes. The reactions that have been described up to this point provide the essential basis for the generation of biodiesel [62]. The catalysis of the reaction is facilitated by potent bases such as sodium and potassium hydroxide, or sodium methylate. The efficacy of acid catalysts in the conversion of triglycerides to biodiesel is limited, but they have demonstrated effectiveness in the conversion of FFAs to biodiesel. As per the literature, it has been suggested that the process of converting triglycerides into biodiesel can be initiated by pre-treating esters with an acid, followed by an alkali treatment [63]. Transesterification reactions have the ability to decrease the viscosity of oil.

Two processes are utilized to convert sewage sludge into biodiesel [64].

- → The process of obtaining biolipids.
- → The process of transforming lipids that have been extracted into a fuel source known as biodiesel.

The initial phase involves the generation of primary sludge specimens via a mechanical process of treating wastewater. The chemical composition of primary sludge is dependent on the particular site of origin from which the sludge is obtained. The substance under consideration exhibits a notable presence of organic compounds. The primary sludge is a substantial reservoir of oil, and the most efficient method of extracting biodiesel from this sludge is highly desirable [65]. The sludge samples undergo a preliminary drying process, followed by the extraction of oil utilizing solvents such as hexane. Subsequently, the lipids undergo transesterification to yield biodiesel (Figure 3.4).

The process of separating biodiesel from the glycerol phase involves the settling of the latter, which is characterized by its weight and rapid settling rate, as illustrated in Figure 3.5. The production of biodiesel at its optimal level presents significant challenges. One of the challenges faced in the oil industry is the optimization of oil extraction to achieve maximum yield. Numerous factors impact the production of biodiesel [66].

Temperature: Variations in temperature have an impact on the yield of biodiesel. The rate of reaction is positively correlated with temperature, while the reaction time is negatively correlated with temperature due to the decrease in the viscosity of oils [66]. However, it should be noted that elevating the reaction temperature beyond the optimal range may result

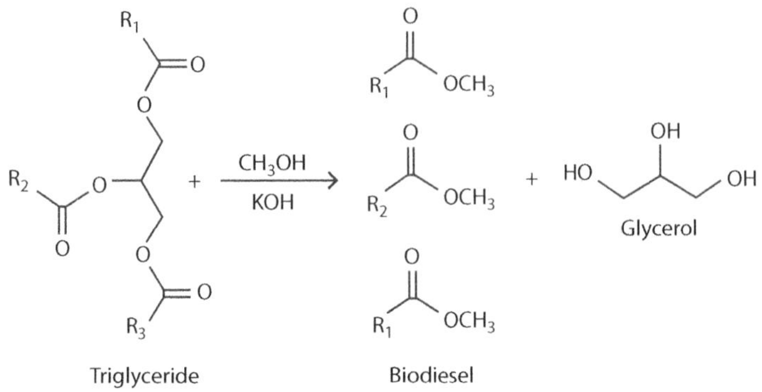

Figure 3.4 Biodiesel production through transesterification using methanol as a base catalyst [24].

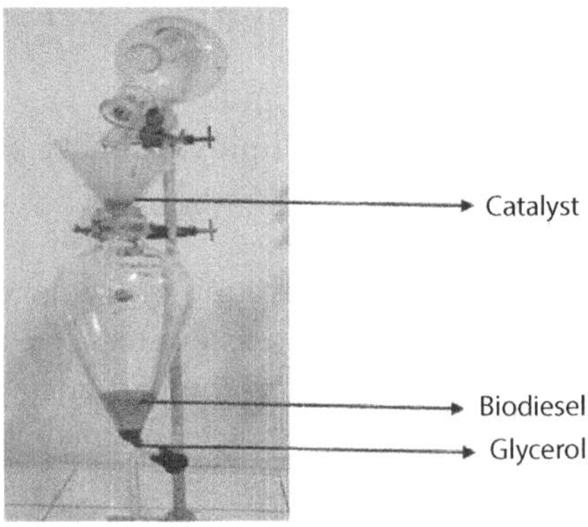

Figure 3.5 The process of isolating biodiesel from glycerol [24].

in a reduction in the production of biodiesel. This is due to the fact that at higher temperatures, the triglycerides undergo saponification, leading to a decrease in yield. Furthermore, elevated temperatures led to the evaporation of methanol, consequently reducing the production of biodiesel [67]. Consequently, the transesterification process takes place at a temperature lower than the boiling point of methanol. It has been determined that an optimal temperature range falls between 50°C and 60°C. It has been determined that ambient temperature is also suitable.

Reaction Time: The conversion of fatty acid esters is significantly impacted by reaction time. The optimal duration for achieving maximum conversion during the reaction is between 60 and 90 min. The reason for the sluggish reaction rate during the early stages is attributed to the mixing and dispersion of alcohol and oil, as reported in reference.

Subsequently, the reaction undergoes rapid progression [68]. An escalation in reaction time may result in a decline in biodiesel production due to the occurrence of reversible transesterification, which can lead to the loss of esters and the formation of soap, as stated in the reference.

Methanol to Oil Molar Ratio: The parameter that exerts the most significant influence on biodiesel yield is the molar ratio between alcohol and triglyceride. The reaction rate reaches its maximum level when methanol is used in its purest form, i.e., 100%. Elevating the methanol concentration results in a corresponding increase in the biodiesel yield [69].

Catalyst: The effectiveness of the process of oil transesterification is notably impacted by the properties and classification of the catalysts employed. At present, catalysts derived from both chemical and biological sources are utilized for catalytic applications. The chemical catalysts that are most frequently utilized include NaOH, CH_3ONa, and KOH [70]. Sodium hydroxide (NaOH) is the preferred option in comparison to potassium hydroxide (KOH) due to its rapid dissolution in methanol. Furthermore, NaOH is obtainable in a state of high purity and at a relatively low expense. Furthermore, in comparison to KOH, a comparatively small amount is required. As hydroxides, it has been discovered that alkali metal oxides are more effective than others. A predetermined volume of water is produced as a byproduct of the chemical reaction that takes place when NaOH and methanol are combined. This water, in conjunction with FFAs, is an essential component in the hydrolysis of triglycerides, which is the first step in the production of soap. The recovery of the product and the resulting yield of biodiesel may be impeded by saponification [71]. In order to inhibit the creation of soap, a sequential process of transesterification is executed, commencing with an acid-catalyzed reaction followed by an alkali-catalyzed reaction. The process of acid esterification results in a lower concentration of FFAs, and the utilization of alkali catalysis, which occurs afterward, improves the quality of the final product. Esterification and transesterification reactions have utilized sulfuric, phosphoric, hydrochloric, and sulphonic acids [72].

In contrast, heterogeneous catalysts have the potential to enhance the transesterification procedure by reducing the additional expenses and pollutants associated with homogeneous catalysis. It has been discovered that a number of different heterogeneous catalysts, including modified zeolites, anionic clays (hydrotalcite), calcium carbonate rock, EST-4 (Eni Slurry Technology), Li/CaO, MgO/KOH, and Na/NaOH/-Al_2O_3, are efficient in promoting the transesterification process. The utilization of nanocatalysts that possess immobilized lipase on magnetic nanoparticles has demonstrated its versatility as a biocatalyst in the production of biodiesel.

According to a study, the employment of lipase nanocatalysts results in a significantly greater rate of biodiesel conversion compared to that of a chemical catalyst [73].

3.1.9 Quality of Sewage Biodiesel

Researchers face a huge obstacle when trying to ensure that the biodiesel they make from lipids extracted from wastewater sludge is of a high enough grade. Sludge is made up of a wide variety of chemical compounds, some of which include wax esters, steroids, terpenoids, polyhydroxy alkanoates, hydrocarbons, linear alkyl benzene, polycyclic aromatic hydrocarbons, and medicinal drugs [74]. This assortment of substances makes up the sludge's makeup. In addition, the sludge has a variety of lipid sources, including triglycerides, fatty acids, phospholipids, and lipids from bacteria. There is a substantial opportunity to extract chemical compounds, such as lipids, which may have an effect on the gravimetric yield of biodiesel. Among these chemicals is the possibility to extract biodiesel. The use of the cracking approach has the ability to convert those contaminants, which would then lead to a significant increase in the amount of FAME that could be extracted from the sludge that is produced by wastewater [74].

The biodiesel made from scum has a substantially higher amount of saturated fatty acids than biodiesel made from vegetable oils that have been treated. This was discovered by comparing the two fuels' levels of these acids [75]. It is the increased concentration of saturated fatty acid that is present in the fuel that is responsible for the enhancement of its cetane number, which in turn results in an improvement in its combustion capabilities. However, this phenomenon also results in the escalation of the cloud number, which has a detrimental impact on the engine's performance during colder temperatures.

3.2 Conclusion

Biodiesel and bio-oil are now regarded as key sources of renewable energy that give various advantages in contrast to traditional fossil fuels. This is due to the fact that biodiesel and bio-oil are produced through biological processes rather than chemical processes. The elevated cost of production can be attributed to the costly feedstock, rendering it less financially viable in comparison to conventional energy sources. Furthermore, the utilization of primary materials for biofuel production, including animal fats and vegetable oils, may potentially compete with human food sources, resulting in an increase in the cost of biofuel. In order to enhance the caliber of bio-oil and chemicals, it is imperative to optimize diverse production parameters. To improve the qualities of bio-oil as a fuel source, it is imperative that cutting-edge processes, such as emulsification and micro-emulsification, be put into practice as quickly as possible. Alternatively, combining it with biodiesel is another alternative that may be taken into consideration as a workable strategy to improve the combustion quality and emission characteristics. In addition, it is essential to perform in-depth research into the kinetic processes underpinning the catalytic activity that is involved in the synthesis of biodiesel and bio-oil from microalgae-based feedstocks such as MSS, with the end objective of improving the effectiveness of the process.

References

1. Bora, A.P., Gupta, D.P., Durbha, K.S., Sewage sludge to bio-fuel: A review on the sustainable approach of transforming sewage waste to alternative fuel. *Fuel*, 259, 116262, 2020.
2. Wang, F., Harindintwali, J.D., Yuan, Z., Wang, M., Wang, F., Li, S., Yin, Z., Huang, L., Fu, Y., Li, L., Chang, S.X., Zhang, L., Rinklebe, J., Yuan, Z., Zhu, Q., Xiang, L., Tsang, D.C.W., Xu, L., Jiang, X., Liu, J., Wei, N., Kästner, M., Zou, Y., Ok, Y.S., Shen, J., Peng, D., Zhang, W., Barceló, D., Zhou, Y., Bai, Z., Li, B., Zhang, B., Wei, K., Cao, H., Tan, Z., Zhao, L., He, X., Zheng, J., Bolan, N., Liu, X., Huang, C., Dietmann, S., Luo, M., Sun, N., Gong, J., Gong, Y., Brahushi, F., Zhang, T., Xiao, C., Li, X., Chen, W., Jiao, N., Lehmann, J., Zhu, Y.G., Jin, H., Schäffer, A., Tiedje, J.M., Chen, J.M., Technologies and perspectives for achieving carbon neutrality. *Innovation*, 2, 100180, 2021.
3. Alper, K., Tekin, K., Karagöz, S., Ragauskas, A.J., Sustainable energy and fuels from biomass: A review focusing on hydrothermal biomass processing. *Sustain. Energy Fuels*, 4, 4390, 2020.
4. Abomohra, A.E.F., Wang, Q., Huang, J., *Waste-to-energy*, Springerplus, 2022.
5. Pikula, K., Zakharenko, A., Stratidakis, A., Razgonova, M., Nosyrev, A., Mezhuev, Y., Tsatsakis, A., Golokhvast, K., The advances and limitations in biodiesel production: feedstocks, oil extraction methods, production, and environmental life cycle assessment. *Green Chem. Lett. Rev.*, 13, 11, 2020.
6. Martins, M.G., da Silva Arouche, T., Neto, A.F.G., da Cruz, J.N., da Costa, F.L.P., Fernandes, L.L., de Carvalho Junior, R.N., da Silva Costa, J.F., de Jesus Chaves Neto, A.M., Density functional theory for the thermodynamic gas-phase investigation of butanol biofuel and its isomers mixed with gasoline and ethanol. *J. Mol. Model.*, 27, 80, 2021.
7. Gautam, R. and Kumar, S., Performance and combustion analysis of diesel and tallow biodiesel in CI engine. *Energy Rep.*, 6, 2785, 2020.

8. Nagtode, V.S., Cardoza, C., Yasin, H.K.A., Mali, S.N., Tambe, S.M., Roy, P., Singh, K., Goel, A., Amin, P.D., Thorat, B.R., Cruz, J.N., Pratap, A.P., Green Surfactants (Biosurfactants): A Petroleum-Free Substitute for Sustainability-Comparison, Applications, Market, and Future Prospects. *ACS Omega*, 8, 11674, 2023.
9. Hill, J., Nelson, E., Tilman, D., Polasky, S., Tiffany, D., Environmental, economic, and energetic costs and benefits of biodiesel and ethanol biofuels. *Proc. Natl. Acad. Sci. U. S. A.*, 103, 11206, 2006.
10. Etim, A.O., Jisieike, C.F., Ibrahim, T.H., Betiku, E., Biodiesel and its properties, in: *Production of Biodiesel from Non-Edible Sources: Technological Updates*, A. Arumugam, (Ed.), pp. 39–79, Elsevier, 2022.
11. Canakci, M. and Sanli, H., Biodiesel production from various feedstocks and their effects on the fuel properties. *J. Ind. Microbiol. Biotechnol.*, 35, 431, 2008.
12. Issariyakul, T. and Dalai, A.K., Biodiesel from vegetable oils. *Renew. Sustain. Energy Rev.*, 31, 446, 2014.
13. Mandolesi De Araújo, C.D., De Andrade, C.C., De Souza E Silva, E., Dupas, F.A., Biodiesel production from used cooking oil: A review. *Renew. Sustain. Energy Rev.*, 27, 445, 2013.
14. Degfie, T.A., Mamo, T.T., Mekonnen, Y.S., Optimized Biodiesel Production from Waste Cooking Oil (WCO) using Calcium Oxide (CaO) Nano-catalyst. *Sci. Rep.*, 9, 2019.
15. Achinas, S., Horjus, J., Achinas, V., Euverink, G.J.W., A PESTLE analysis of biofuels energy industry in Europe. *Sustain.*, 11, 5981, 2019.
16. Santamouris, M. and Vasilakopoulou, K., Present and future energy consumption of buildings: Challenges and opportunities towards decarbonisation. *e-Prime - Adv. Electr. Eng. Electron. Energy*, 1, 100002, 2021.
17. Usman, M., Cheng, S., Cross, J.S., Biodiesel production from wet sewage sludge and reduced CO2 emissions compared to incineration in Tokyo, Japan. *Fuel*, 341, 127614, 2023.
18. Usman, M. and Balsalobre-Lorente, D., Environmental concern in the era of industrialization: Can financial development, renewable energy and natural resources alleviate some load? *Energy Policy*, 162, 112780, 2022.
19. Sewage Sludge as an Alternative to Ordinary Soil in Manufacturing of Bricks. *Int. J. Recent Trends Eng. Res.*, 3, 194, 2017.
20. Li, R., Li, L., Wang, Q., The impact of energy efficiency on carbon emissions: Evidence from the transportation sector in Chinese 30 provinces. *Sustain. Cities Soc*, 82, 103880, 2022.
21. Salvi, B.L., Subramanian, K.A., Panwar, N.L., Alternative fuels for transportation vehicles: A technical review. *Renew. Sustain. Energy Rev.*, 25, 404, 2013.
22. Malode, S.J., Prabhu, K.K., Mascarenhas, R.J., Shetti, N.P., Aminabhavi, T.M., Recent advances and viability in biofuel production. *Energy Convers. Manage. X*, 10, 100070, 2021.

23. Keasling, J., Garcia Martin, H., Lee, T.S., Mukhopadhyay, A., Singer, S.W., Sundstrom, E., Microbial production of advanced biofuels. *Nat. Rev. Microbiol.*, 19, 701, 2021.
24. Bharathiraja, B., Jayamuthunagai, J., Chakravarthy, M., Kumar, R.R., Yogendran, D., Praveenkumar, R., Algae: Promising Future Feedstock for Biofuels, in: *Algae and Environmental Sustainability*, B. Singh, K. Bauddh, F. Bux, (Eds.), pp. 1–8, Springer, India, New Delhi, 2015.
25. Das, P. and Gundimeda, H., Is biofuel expansion in developing countries reasonable? A review of empirical evidence of food and land use impacts. *J. Clean. Prod.*, 372, 133501, 2022.
26. Huang, D., Zhou, H., Lin, L., Biodiesel: An alternative to conventional fuel. *Energy Proc.*, 16, 1874, 2012.
27. Moraes, E.S., Reis, G.F., Cruz, J., Cozzolino, K., Neto, A.F.G., Andrade-Filho, T., Neto, A.M.J.C., Thermodynamics study of biokerosene from coconut and palm kernel oils and JP-8 aircraft fuels in the gas phase by the DFT method. *J. Mol. Model.*, 26, 79, 2020.
28. Wardoyo, Widodo, A.S., Wijayanti, W., Wardana, I.N.G., The Role of Areca catechu Extract on Decreasing Viscosity of Vegetable Oils. *Sci. World J.*, 2021, 1, 2021.
29. Pasha, M.K., Dai, L., Liu, D., Guo, M., Du, W., An overview to process design, simulation and sustainability evaluation of biodiesel production. *Biotechnol. Biofuels*, 14, 2021.
30. Fabiano, B., Reverberi, A.P., Del Borghi, A., Dovi, V.G., Biodiesel production via transesterification: Process safety insights from kinetic modeling. *Theor. Found. Chem. Eng.*, 46, 673, 2012.
31. Ogunkunle, O. and Ahmed, N.A., A review of global current scenario of biodiesel adoption and combustion in vehicular diesel engines. *Energy Rep.*, 5, 1560, 2019.
32. Mitter, E.K., Germida, J.J., de Freitas, J.R., Impact of diesel and biodiesel contamination on soil microbial community activity and structure. *Sci. Rep.*, 11, 2021.
33. Abed, K.A., Gad, M.S., El Morsi, A.K., Sayed, M.M., Elyazeed, S.A., Effect of biodiesel fuels on diesel engine emissions. *Egypt. J. Pet.*, 28, 183, 2019.
34. Mathew, G.M., Raina, D., Narisetty, V., Kumar, V., Saran, S., Pugazhendi, A., Sindhu, R., Pandey, A., Binod, P., Recent advances in biodiesel production: Challenges and solutions. *Sci. Total Environ.*, 794, 148751, 2021.
35. Demirbas, A., Edris, G., Alalayah, W.M., Sludge production from municipal wastewater treatment in sewage treatment plant. *Energy Sources, Part A*, 39, 999, 2017.
36. Ahmad, T., Ahmad, K., Alam, M., Characterization of Water Treatment Plant's Sludge and its Safe Disposal Options. *Proc. Environ. Sci.*, 35, 950, 2016.

37. Markis, F., Baudez, J.C., Parthasarathy, R., Slatter, P., Eshtiaghi, N., Rheological characterisation of primary and secondary sludge: Impact of solids concentration. *Chem. Eng. J.*, 253, 526, 2014.
38. Voutchkov, N., Introduction to Saline Water Pretreatment, in: *Pretreatment for Reverse Osmosis Desalination*, pp. 1–10, Elsevier, 2017.
39. Wang, C., Chen, M., Zhou, L., Hou, Y., Zhao, P., Che, D., Correlations of chemical properties of sludge: A comparison study between municipal sludge and industrial sludge. *J. Energy Inst.*, 108, 101202, 2023.
40. El Moussaoui, T., Belloulid, M.O., Elharbili, R., El Ass, K., Ouazzani, N., Simultaneous assessment of purification performances and wastewater byproducts management plans towards a circular economy: Case of Marrakesh WWTP. *Case Stud. Chem. Environ. Eng.*, 6, 100228, 2022.
41. Tran, H.T., Lesage, G., Lin, C., Nguyen, T.B., Bui, X.T., Nguyen, M.K., Nguyen, D.H., Hoang, H.G., Nguyen, D.D., Activated sludge processes and recent advances, in: *Current Developments in Biotechnology and Bioengineering: Advances in Biological Wastewater Treatment Systems*, X.-T. Bui, D.D. Nguyen, P.-D. Nguyen, H.H. Ngo, A. Pandey, (Eds.), pp. 49–79, Elsevier, 2022.
42. Patziger, M., Kainz, H., Hunze, M., Józsa, J., Influence of secondary settling tank performance on suspended solids mass balance in activated sludge systems. *Water Res.*, 46, 2415, 2012.
43. El Gohary, E.H., El Monayeri, D.S., Emeara, M.S., D.El Monayeri, O., Application of different techniques to enhance sludge drying paved bed drying process. *Alexandria Eng. J.*, 61, 5687, 2022.
44. Anjum, M., Al-Makishah, N.H., Barakat, M.A., Wastewater sludge stabilization using pre-treatment methods. *Process Saf. Environ. Prot.*, 102, 615, 2016.
45. Valderrama, C., Granados, R., Cortina, J.L., Stabilisation of dewatered domestic sewage sludge by lime addition as raw material for the cement industry: Understanding process and reactor performance. *Chem. Eng. J.*, 232, 458, 2013.
46. Nemerow, N.L. (Ed.), Treatment and disposal of sludge solids, in: *Industrial Waste Treatment*, pp. 149–174, Butterworth-Heinemann, Burlington, 2007.
47. Yaashikaa, P.R., Senthil Kumar, P., Varjani, S., Valorization of agro-industrial wastes for biorefinery process and circular bioeconomy: A critical review. *Bioresour. Technol.*, 343, 126126, 2022.
48. Villalobos-Delgado, F.J., di Bitonto, L., Reynel-Ávila, H.E., Mendoza-Castillo, D.I., Bonilla-Petriciolet, A., Pastore, C., Efficient and sustainable recovery of lipids from sewage sludge using ethyl esters of volatile fatty acids as sustainable extracting solvent. *Fuel*, 295, 120630, 2021.
49. Liu, S., Luo, T., Liu, G., Xu, X., Shao, Y., Qi, L., Wang, H., Characterization and reutilization potential of lipids in sludges from wastewater treatment processes. *Sci. Rep.*, 10, 2020.
50. Stančin, H., Mikulčić, H., Wang, X., Duić, N., A review on alternative fuels in future energy system. *Renew. Sustain. Energy Rev.*, 128, 109927, 2020.

51. D'Ambrosio, V., di Bitonto, L., Angelini, A., Gallipoli, A., Braguglia, C.M., Pastore, C., Lipid extraction from sewage sludge using green biosolvent for sustainable biodiesel production. *J. Clean. Prod.*, 329, 129643, 2021.
52. Ren, X., Wei, C., Yan, Q., Shan, X., Wu, M., Zhao, X., Song, Y., Optimization of a novel lipid extraction process from microalgae. *Sci. Rep.*, 11, 2021.
53. Zhang, J., Rao, C., Peng, H., Peng, C., Zhang, L., Xu, X., Liu, W., Wang, Z., Zhang, N., Wang, X., Enhanced toluene combustion performance over Pt loaded hierarchical porous MOR zeolite. *Chem. Eng. J.*, 334, 10, 2018.
54. Vasaki, M., Sithan, M., Ravindran, G., Paramasivan, B., Ekambaram, G., Karri, R.R., Biodiesel production from lignocellulosic biomass using Yarrowia lipolytica. *Energy Convers. Manage. X.*, 13, 100167, 2022.
55. Zarrinmehr, M.J., Daneshvar, E., Nigam, S., Gopinath, K.P., Biswas, J.K., Kwon, E.E., Wang, H., Farhadian, O., Bhatnagar, A., The effect of solvents polarity and extraction conditions on the microalgal lipids yield, fatty acids profile, and biodiesel properties. *Bioresour. Technol.*, 344, 126303, 2022.
56. Robinson, P.K., Enzymes: principles and biotechnological applications. *Essays Biochem.*, 59, 1, 2015.
57. Athar, M., Zaidi, S., Hassan, S.Z., Intensification and optimization of biodiesel production using microwave-assisted acid-organo catalyzed transesterification process. *Sci. Rep.*, 10, 2020.
58. Melero, J.A., Sánchez-Vázquez, R., Vasiliadou, I.A., Martínez Castillejo, F., Bautista, L.F., Iglesias, J., Morales, G., Molina, R., Municipal sewage sludge to biodiesel by simultaneous extraction and conversion of lipids. *Energy Convers. Manage.*, 103, 111, 2015.
59. Hui, K., Tang, J., Lu, H., Xi, B., Qu, C., Li, J., Status and prospect of oil recovery from oily sludge:A review. *Arab. J. Chem.*, 13, 6523, 2020.
60. Furman, K.C., El-Bakry, A.S., Song, J.H., Optimization in the oil and gas industry. *Optim. Eng.*, 18, 1, 2017.
61. Parigi, D., Giglio, E., Soto, A., Santarelli, M., Power-to-fuels through carbon dioxide Re-Utilization and high-temperature electrolysis: A technical and economical comparison between synthetic methanol and methane. *J. Clean. Prod.*, 226, 679, 2019.
62. Cai, X., Zhang, Z., Ye, Y., Wang, D., Li, S., Wang, D., Zheng, Z., Conversion of higher fatty acids or higher fatty acid esters to long-chain alkanes by acid added metal catalyst under mild hydrothermal conditions. *Biomass Bioenergy*, 156, 106328, 2022.
63. Verma, P. and Sharma, M.P., Review of process parameters for biodiesel production from different feedstocks. *Renew. Sustain. Energy Rev.*, 62, 1063, 2016.
64. Ulakpa, W.C., Ulakpa, R.O.E., Egwunyenga, M.C., Egbosiuba, T.C., Transesterification of non-edible oil and effects of process parameters on biodiesel yield. *Clean. Waste Syst.*, 3, 100047, 2022.
65. Atadashi, I.M., Aroua, M.K., Aziz, A.A., Biodiesel separation and purification: A review. *Renew. Energy*, 36, 437, 2011.

66. Veillette, M., Giroir-Fendler, A., Faucheux, N., Heitz, M., Esterification of free fatty acids with methanol to biodiesel using heterogeneous catalysts: From model acid oil to microalgae lipids. *Chem. Eng. J.*, 308, 101, 2017.
67. Parandi, E., Safaripour, M., Abdellattif, M.H., Saidi, M., Bozorgian, A., Rashidi Nodeh, H., Rezania, S., Biodiesel production from waste cooking oil using a novel biocatalyst of lipase enzyme immobilized magnetic nanocomposite. *Fuel*, 313, 123057, 2022.
68. Kargbo, D.M., Biodiesel production from municipal sewage sludges. *Energy Fuels*, 24, 2791, 2010.
69. Shareef, S.M. and Mohanty, D.K., Experimental investigations of dairy scum biodiesel in a diesel engine with variable injection timing for performance, emission and combustion. *Fuel*, 280, 118647, 2020.
70. Biswas, J.K., Mitra, A., Shah, M.P., Rodriguez-Couto, S., Nanoadsorbents for scavenging emerging contaminants from wastewater, in: *Development in Wastewater Treatment Research and Processes: Removal of Emerging Contaminants from Wastewater through Bio-nanotechnology*, M. Shah, S. Rodriguez- Couto, J. Biswas, (Eds.), pp. 1–22, Elsevier, 2021.
71. Charles Rajesh Kumar, J. and Majid, M.A., Renewable energy for sustainable development in India: Current status, future prospects, challenges, employment, and investment opportunities. *Energy. Sustain. Soc.*, 10, 2020.
72. Gardner, R., Fuels for the future. *Aerosp. Eng. Manuf.*, 29, 5, 2009.
73. Bhan, C., Verma, L., Singh, J., Alternative Fuels for Sustainable Development, in: *Environmental Concerns and Sustainable Development*, V. Shukla, and N. Kumar, (Eds.), pp. 317–331, Springer, Singapore, 2020.
74. Maheshwari, P., Haider, M.B., Yusuf, M., Klemeš, J.J., Bokhari, A., Beg, M., Al-Othman, A., Kumar, R., Jaiswal, A.K., A review on latest trends in cleaner biodiesel production: Role of feedstock, production methods, and catalysts. *J. Clean. Prod.*, 355, 131588, 2022.
75. Atadashi, I.M., Aroua, M.K., Aziz, A.A., High quality biodiesel and its diesel engine application: A review. *Renew. Sustain. Energy Rev.*, 14, 1999, 2010.

4

Carbon Emissions, Energy Reduction, and Energy Recovery from Wastewater Treatment Plants

Abas Siraj Hamda[1], Dinsefa Mensur[2], Belay Berhane[2], Sunaina[3] and Tatek Temesgen[2,3]*

[1]*Department of Chemical Engineering, Haramaya Institute of Technology, Haramaya University, Dire Dawa, Ethiopia.*
[2]*Department of Chemical Engineering, Adama Science and Technology University, Adama, Ethiopia*
[3]*Department of Chemistry, University of Calgary, Calgary, AB, Canada*

Abstract
Wastewater treatment units are crucial components of most of the process industry or urban infrastructure since they guarantee influent quality to the aquatic environment. Greenhouse gasses (GHGs) are produced and released directly or indirectly from operations in wastewater treatment plants. Conventional treatment plants consume huge amounts of energy. The consumption of high energy resulted in high emissions of GHGs. To reduce these impacts, facilities should consider resources from the waste stream in order to harness valuable resources such as energy, biofertilizers, and water for various uses. The goal of wastewater treatment systems has just started to change recently to include the concept of resource recovery by adopting creative techniques. These techniques improved plant efficiency, decreased resource wastage, and brought positive effects on both the environment and the economy. This chapter gives a brief summary of energy consumption and alternatives for energy reduction and recovery in the wastewater treatment plant in addition to covering the concepts on carbon emission sources, carbon accounting, and carbon emissions reduction options in conventional wastewater treatment plants.

Keywords: Carbon emissions, energy consumption, energy efficiency, energy reduction, energy recovery, energy use, wastewater, wastewater treatment

**Corresponding author*: tate95et@gmail.com; tatek.temesgen@astu.edu.et

Inamuddin, Tariq Altalhi, Mohammad Luqman and Joseph Kapuku (eds.) Sewage and Biomass from Wastewater to Energy, (93–112) © 2024 Scrivener Publishing LLC

4.1 An Overview

Wastewater treatment plants (WWTPs), from the facility's building to the treatment units use energy extensively. The focus of conventional WWTPs was on reducing or removing pollutants from the influent wastewater to meet the expected water quality limits for public well-being and preservation of the environment. Energy is required for the collection, treatment, and release of wastewater in accordance with legal permission requirements. In most conventional WWTPs, this energy is mostly in the form of electrical power except for some exceptional operations where natural gas or other fuels are used. Wastewater treatment is a resource-intensive process that uses many inputs, primarily electricity. Operating expenses for typical wastewater treatment systems range from 15 to 40% of total costs [1] of which most of the cost goes to energy consumption. Due to the expanding population, stricter discharge regulations, poor and old infrastructure, and the propensity toward quality life, the amount of energy consumed for wastewater treatment is likely to increase in the future. In the event that there are no adjustments made to the procedures, or potential future regulations for removing currently uncontrolled harmful substances, such as pharmaceutical and personal care products, there will be a considerable rise in the consumption of energy at WWTP [2]. Even though the classical thought of treating polluted water resulted in clean water and semi-solid waste (sludge), WWTPs have a significant opportunity for the recovery of resources. Recently, a number of WWTPs now go by the label "water resource recovery facilities". In order to advance sustainability in wastewater treatment a concept of "sustainable wastewater treatment" must be introduced. The classical concept behind wastewater treatment was used to focus only on treating the polluted water and producing a clean one. The solid waste produced was primarily considered to end up in landfills. In sustainable wastewater treatment systems, the overall process must consider recourse recycling, recovery, and the treatment of the solid byproducts before ending up in the landfill. In the current WWTPs, new goals have been formulated and implemented into action. Economic, environmental, and social elements of WWTPs are all addressed under the multifaceted concept of sustainability. Efficient utilization of natural resources through effective models, technology, and improved management is essential for WWTPs to consider environmental, economic, and social elements when making sensible selections [3]. Energy use in WWTPs is regarded as one of the greatest man-made greenhouse gas (GHG) emission sources in addition to its economic burden. By releasing

CO_x and CH_4 into the atmosphere, WWTPs contribute to the emission footprint. Throughout the building, construction, and operation stages of the WWTPs, these main GHGs are produced [4]. The energy utilized in WWTPs is contributing to around 57% of global warming [1]. On-site and off-site sources contribute to the overall quantity of GHGs released into the atmosphere. The on-site sources are from the process of treating wastewater and sludge and burning operations for the generation of energy, whereas the off-site sources are from the manufacturing and transportation of fuel, power, chemicals, and building materials. Off-site GHG emissions also include those from discharges of water and the decomposition of residue [4]. It is always necessary to focus on the mitigation of GHGs and wastewater recovery technologies in order to lower these emissions over the entire life span of the WWTPs. From wastewater treatment facilities, significant resources including solid biomass, chemical and thermal energies may be recovered to create goods with added values like renewable energy, biofertilizers, and water for various uses. In WWTPs with resource recovery, recycling treated effluent water and reusing sludge for agricultural purposes were the most widely used recovery methods. The amount of resources that may have been recovered by these technologies, however, was not fully realized. Particularly in developing countries, there is a lack of precise, current, and comprehensive data on wastewater flow rates, treatment efficacy, and recovery activities [4, 5]. Other recovery technologies focusing on energy recovery, such as anaerobic digestion (AD), microbial fuel cells (MFCs), algae biofuels, and heat pumps, have been tried at WWTPs as part of sustainability policy [3]. Overall, this chapter provides insight into energy consumption, energy reduction, and recovery options in WWTPs and it also highlights information on carbon emission sources, carbon accounting, and carbon reduction and treatment alternatives in WWTP.

4.2 Major Units of Energy Consumption in WWTPs

In general, in WWTPs, the flow, effluent quality, operations used, and flow rate all affect how much energy is required. Approximately 0.3 and 0.6 kWh/m^3 are needed for conventional wastewater treatment. In WWTPs, only the cost of labor comes close to matching the energy cost, which makes up 15 to 40% of the overall costs of sustaining the system [1]. Studies have shown that less than 1% of the total energy spent in WWTPs is used for preliminary treatment, which is relatively minimal [6]. The quantity of energy utilized for secondary and tertiary treatment, nevertheless, varies depending on the type of treatment chosen and how much essential treatment is performed on the effluents [7]. In spite of technological changes, the average energy consumption in WWTPs around the world ranged between 0.38 and 1.122 kWh/m^3. In conventional WWT systems, the cost of energy consumption is higher than the cost of substances utilized for treatment. For instance, the energy consumption in one of the WWTPs in Delhi using the activated sludge method of treatment, 65.5% of the cost goes to electricity and 27.8% of the operation cost goes to the substances employed throughout the treatment procedure. The utilization of energy in WWTPs is affected by different operational factors: type and use of chemicals, type of unit operations, seasonal fluctuations, content of the waste effluent, maintenance scheduling, and others. For example, the application of chemicals utilized for the dewatering of sludge and decontamination accounted for 28% of the cost of energy of operation [1]. Compared to one of the traditional activated sludge systems, the plants with an extra-biological treatment for nutrient removal displayed a 30 to 50% greater energy consumption in the aeration, pumping, and solids processing. The steps of sludge drainage, flow adjustment, and air cleaning all made significant contributions, while aeration used 53% of the energy utilization. According to scale and geography, the pumping energy use, which was around 12%, may often make up a large fraction of the entire amount of energy used [7]. The process for disposing of sludge, which must be stabilized before disposal, accounts for 30% of the average WWTP's costs. Because of the various kinds of management, the sludge that undergoes treatment can use between 0.074 and 0.15 kWh/m^3 before its last disposal [8]. The immense energy-generating capacity of sludge, however, can be a means that can produce more energy than is needed for management.

4.3 Alternatives Management Techniques for Energy Reduction from WWTP

The energy requirements of WWTPs differ from plant to plant. While plants lacking removal of nutrient facilities have energy demands that are less than 0.5 kWh/m^3, WWTPs with nutrient recovery infrastructure require in the range from 0.5 to 2.0 kWh/m^3. The following part covers different tools used to evaluate and reduce energy use, including energy auditing, energy benchmarking, and energy management programs.

4.3.1 Energy Auditing

Energy audit is a crucial step in determining how much energy in WWTPs is required and to how much efficiency is achieved in each unit operation. Energy audits aid in locating the major energy uses (processes and infrastructure) and help the WWTP facility pinpoint its least effective operational areas. Simple energy audits normally involve a thorough examination of the installations (handheld measuring equipment may be employed) and a quick desk review of the usage and cost information for the energy supply. These moderate cost audits are crucial for gaining a fundamental understanding of a WWTP's energy utilization [9]. When conducted properly, an energy audit gives crucial information on how much, where, and how the energy is consumed in the treatment plant. Figure 4.1 illustrates the benefits of energy audits in WWTPs. An energy type evaluation on WWTP should be conducted in order to perform an effective energy auditing of the plant. Creating action plans, goals, targets, and indicators for gauging success as well as evaluating management methods must be also a part of the audit process [10].

4.3.2 Energy Benchmarking

Energy benchmarking is a potent method for WWTP improvement. It can assist the cost-saving endeavor and reduction in emissions of GHGs. Determining the most effective systems and best practices must be a continual and deliberate process of comparing energy usage against the standard

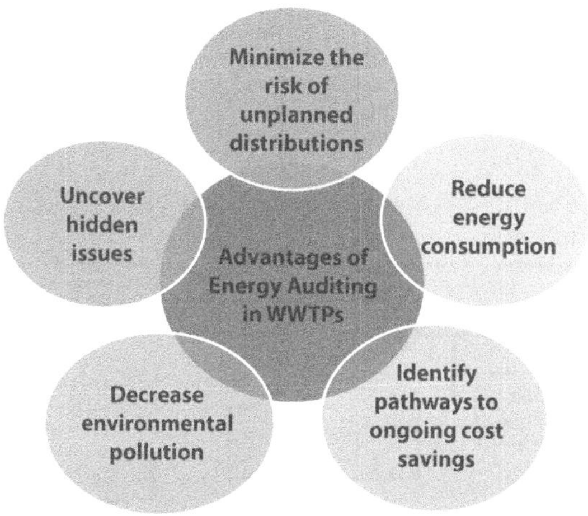

Figure 4.1 Advantages of energy auditing in WWTPs.

of performance. The lower-performing units can then be compared to the accepted and acceptable standards for any specific indicator [11]. The benchmarking research data can be used to assess how other plants in the WWT industry are doing on energy efficiency and organize best practices. More recent developments in this area have helped to include more studies on different types of energy usage that can be expressed in kWh/m^3. Prior to now, methods for energy benchmarking have mainly focused on reporting electricity consumption. To be effective, a WWTP energy benchmarking system must be flexible enough to accommodate various wastewater treatment setups and model types. This option enables the professionals to assess the performance of each infrastructure or set of infrastructures and to determine the most and least effective energy management techniques based on a wide range of criteria throughout the analysis and effectiveness control process [10]. The three primary methods for benchmarking are normalization, statistical techniques like ordinary least squares, and stochastic frontier analysis as well as programming including data envelopment analysis (DEA) and stochastic DEA as presented in Figure 4.2 [11]. Since the benchmarking methodologies are essentially different from one another, it is likely that the results will vary. Each strategy can shed light on various aspects of the energy efficiency of WWTPs. The benchmarking objectives, data accessibility, and the user's willingness to accept the technological requirements have an impact on the model's demand and the technique to choose [11].

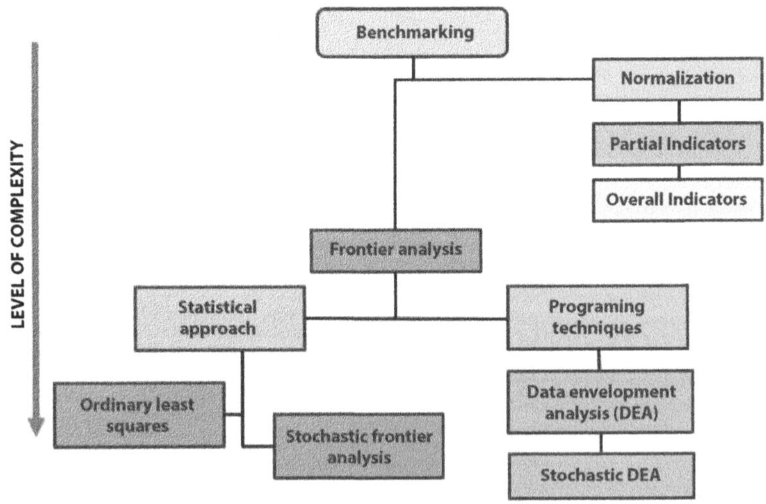

Figure 4.2 Strategies for benchmarking (Longo *et al.*, 2016).

4.3.3 Energy Management Programs

In WWTPs, energy efficiency is a key tool for reducing GHG emissions and achieving other energy management program goals. Different programs can be implemented for WWTPs. The application of renewable energy in WWTPs can, for example, be implemented in WWTPs to reduce fossil fuel energy consumption [12]. Energy-saving as a management program can be considered to lower the cost of generating, transferring, and transporting energy. Moreover, energy management programs can enhance the sustainability of energy efficiency and autonomy, lower GHG emissions and pollution in the environment, and generate employment. The most effective energy policies and management schemes for WWTPs that are suitably framed to meet the legal sector requirements are designed based on an analysis and a framework of input and output data. Despite being relatively new, the incorporation of energy-saving techniques into wastewater treatment systems and facility management schemes has attracted the interest of numerous professionals from a variety of scientific fields in an effort to cut down the energy use of WWTP operations [10]. Implementing different water and wastewater treatment facility management schemes helps to attain better energy efficiency, reduce pollutants emissions, decrease the cost of energy, and expand the market. Showing leadership to improve energy reduction and reduce GHG emissions can be an indication of commitment to protecting public health and the environment. A structural strategy for managing energy must be implemented in WWTPs to address the gaps related to energy efficiency. The world organization for accreditation standard particularly ISO 50001 can serve as a global energy control baseline for any industrial facility to manage energy efficiency. In accordance with Table 4.1, the ISO 50001 framework relies on the Plan-Do-Check-Act ongoing enhancement technique. This can be adopted in energy management programs for the WWTPs.

For WWTPs that have not yet begun an organized strategy for controlling energy use, the first step to take is to start with the collection of internal data on energy, evaluate the data, and enact small/low-cost energy-saving measures. It is also crucial to gain knowledge from other WWTPs that have built viable energy management techniques. To make the initial data collection and analysis, utilizing the machinery's details, using simple data loggers, and modeling are just a few methodologies that can be used for a preliminary estimation and controlling energy consumption in WWTPs. It is possible to improve the energy efficiency of WWTPs by using an energy measurement and directing system. It might have several submeters, a connection to the primary utility meter, controls for particular systems,

Table 4.1 Lists of the ISO 50001 benchmark energy efficiency measures.

ISO 50001	
Plan	• Establish an energy policy and an energy management team • Prepare an action plan on how to address risks and opportunities • Conduct an energy review • Identify high-energy consumption units • Establish key energy performance indicators • Prepare energy baseline • Set objectives and energy targets • Prepare strategy and detail action plans necessary to attain the objectives and goals
Do	• Implement the action plans • Follow each operation and maintenance • Document detailed reports and communications • Ensure competence and consider energy performance in design, procurement, and operation
Check	• Follow up, measure, analyze, audit, and conduct management reviews of energy performance and the energy management system
Act	• Take measures to address nonconformities and work on continual improvement

and an application to display energy usage and modify various settings. Where the price of energy is a significant issue for management and there is already an organizational effort in progress to maximize energy use, it is likely to become accepted and used by WWTPs [11]. Similar to this, several modeling tools, including genetic algorithms, were utilized in WWTPs to enhance the utilization of activated sludge for the effective elimination of pollutants, particularly nitrogen, minimize energy consumption, and reduce sludge volume [13]. Such kind of modeling tool aims to optimize the control process parameters of energy-consuming processes. Machine learning methods are yet another energy management solution to reduce the energy requirements of WWTPs. It is an extremely creative way to lower the energy requirements of WWTPs. An enhanced neural network model in WWTPs which is called the example online sequential extreme learning machine has three anaerobic zones made of activated sludge and two anoxic zones built of activated sludge. Its objective is to improve the flow of dissolved oxygen into the treatment zones while accounting for a number of variables, including the quality of the influent and effluent wastewater and the weather.

4.4 Wastewater Nexus Energy Recovery Alternatives

In recent years, numerous stakeholders have come around to the water-energy nexus approach for unified industrial sectors to successfully accomplish sustainable development goals. WWTPs could be one such industry group to study and evaluate. In WWTPs, energy is needed at practically every level of the treatment step, but the types of energy needed differ between unit operations [14]. The interconnectedness of wastewater and energy is amply illustrated by the recovery of nutrients and energy from the polluted influent.

As a result of the recent optimization efforts of energy in WWTPs, effluent quality is now of utmost importance in order to prevent any harmful effects on our health and the environment. Energy and nutrients can be effectively transported through wastewater. Biomass, charcoal, and recyclable plastics formed from polyhydroxyalkanoates gathered in biomass, cellulose, bioplastics, phosphate, and exopolymers that mimic alginate are the resources that can be harvested from WWTPs. Each unit of garbage worth $0.47 in resources such as water, nutrients (nitrogen, phosphorus, and potassium), energy (biogas), and biofertilizer (processed nutrient-rich sludge) recovered from the waste [5]. An effective and doable method appears to be the reuse of energy from unusual resources related to water, such as wastewater resources. Facilities for wastewater treatment present a major economic opportunity to reduce their energy consumption, operating expenses, and carbon footprint through the utilization of energy recovery. Energy can be recovered in the form of biogas, biodiesel, hydrogen, electric power, and thermal energy using heat pumps, mechanical and thermal pretreatments, and high-temperature streams using heat exchangers. The next sections cover a variety of energy recovery and savings alternatives in WWTPs, including biogas generation, energy conservation, heat recovery, the application of algae biofuels, and combined heat and power (CHP) systems. Table 4.2 below summarizes the benefits and drawbacks of major approaches for energy recovery and cost reduction in WWTPs.

These listed alternative techniques mainly biogas production, thermal energy recovery, electric power recovery, bio-hydrogen production, application of MFC, and algal-based biofuel production are discussed in the following subsections as an energy recovery system from WWTPs.

Table 4.2 A summary of the various energy recovery technologies from wastewater, along with their pros and cons.

Energy recovery options	Advantages	Disadvantage
Biogas recovery	• Recovers operational costs • It has widespread application • Contributes toward achieving energy independence	• Expensive investment
Biodiesel recovery by transesterification	• Affordable raw materials for high levels of lipids	• Requires drying as a pre-treatment • Possibility of producing soap prior to being used as fuel, • It needs extensive refining
Hydrogen	• High levels of energy • Eco-friendly	• Pretreatments and catalysts needed to raise the H_2/CO ratio • Expensive investment
Thermal energy	• Can be used onsite for temperature regulation of different units especially in cold seasons	• Surface fouling of pipelines in the system • The location of the final heat source had to be close to the WWTPs
Renewable energy	• Reduce GHG emissions • The curve for WWTP demand can be used to regulate energy production.	• limited ability to adapt to large places • More costly processes and equipment at low temperatures, Efficiency suffers noticeably

4.4.1 Biogas Production

The process of AD is very effective in producing methane for energy use from wastewater. This method has been developed to decompose biodegradable materials, stabilize sludge, and generate methane as a major biogas from wastewater resources. It is commonly considered to be both cost-effective and environmentally safe. Nevertheless, due to the low effluent quality (4100 mg COD/L), dissolved methane (a potent GHG), and negligible nutrient removal, its use has been restricted in countries with strict effluent discharge limits [15]. The production of biogas, which results from the AD of biosolids to release chemical energy from organic elements in wastewater for the generation of heat and electricity, is one of the most intriguing applications of on-site energy recovery. The most common waste-to-energy method in WWTPs is the conversion of produced sludge into biogas in digesters, which is then converted into electricity and heat in CHP systems. During the production of biogas, methane (50 to 70%), carbon dioxide (30 to 50%), and traces of other gasses, like nitrogen and hydrogen, are all produced. Although CO_2 is a significant GHG, CHP systems are preferable to emitting CH_4 when sludge decomposes anaerobically. It is also possible to upgrade biogas in order to remove CO_2 and produce a high-purity methane stream that is known as "biomethane" and can take the place of conventional natural gas. This is a different process from onsite biogas to electricity and heat conversion (Figure 4.3) [16]. The biomethane produced by the WWTPs is adaptable and may be used to power gas engines and produce both electrical and thermal energy [17]. It can also be used as a fuel for transportation. The financial viability and capacity of plants have an impact on the biogas recovery potential from WWTPs.

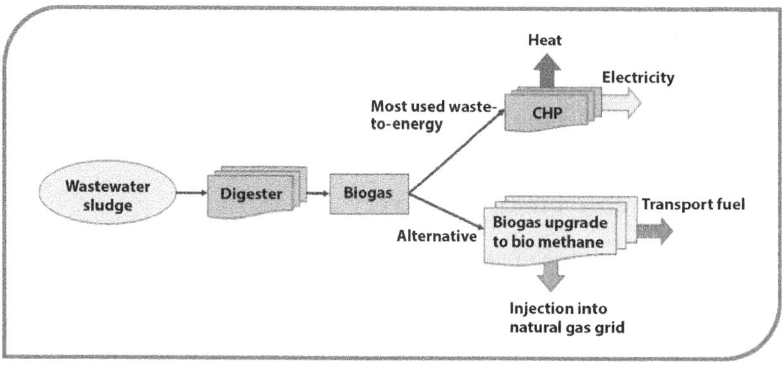

Figure 4.3 Process of Biogas recovery option in WWTP.

Small facilities with a capacity of less than 5 MGD are unlikely to be able to deploy the AD system because the amount of energy that could be recovered through biogas generation does not support the financial sustainability of the process. Large systems with capacities larger than 10 MGD, however, demonstrate that 33% of the energy used can be recovered through the production of biogas, while the 50 MGD plant can recover up to 41% of the energy used [15]. It shows how the economics and energy that can be recovered by the AD system depend on the size of the plant. In general, the AD process might be a good option because it produces a lot of energy while lowering the WWTP's overall running costs.

4.4.2 Thermal Energy Recovery

A variety of sources, water and air, are used by heat pumps to extract energy. Since wastewater from specific sources has sometimes relatively higher temperatures, it can be an offering and cost-effective source of heat for use in heat pumps. Utilizing heat pumps, the thermal energy from the wastewater can be transferred to a hot-water delivery system. Heat pumps use a refrigerant evaporation-compression cycle to transfer heat from a low-quality source (wastewater) to a high-quality source (hot water) [18]. Similar to this, untreated urban wastewater gets heated at municipal pumping stations. The wastewater that enters the pump station is sent to a traveling-screen filter to remove solids before going to a heat pump. The heat recovered from wastewater can be used to heat homes, businesses, public spaces, industrial facilities, water, sludge, infrastructure, and heating and cooling systems. In order to effectively carry out a project, important factors including the quantity of energy that can be extracted from wastewater, the anticipated best use for the heat that can be extracted, the kinds of systems and infrastructure that will be needed, the cost to install, and the funding package must be taken into consideration. Furthermore, due to the temperature sensitivity of the microbes used in downstream wastewater treatment systems, it is important to carefully examine how heat extraction affects the biological process for treating wastewater [15].

4.4.3 Renewable Energy Recovery

Because using fossil fuel–based energy sources has both financial and environmental drawbacks, on-site renewable electricity generation facilities are becoming a source for WWTPs as a source of environmentally friendly energy. In terms of electricity production, biogas from wastewater sludge treatment and AD of wastewater accounted for 67% (187 GWh/y), followed by hydropower (30%, 84 GWh/y), waste-to-energy from AD of organic feedstock (2%, 5.5 GWh/y), and solar photovoltaic (1%, 2.2 GWh/y) [19]. The production of energy from renewable sources (like solar photovoltaics) in WWTPs lowers operating costs and GHG emissions. However, its investment and maintenance cost must be taken into consideration before decision.

4.4.4 Bio-Hydrogen Production from WWTPs

The high specific energy content of hydrogen gas compared to biogas and biodiesel makes it a more preferable energy carrier. Anaerobic wastewater treatment results in the production of higher-value resources like hydrogen gas. To recover these high-value resources from wastewater treatment, both the desired phenotypes and the undesirable ones must be preserved in the anaerobic reactor. The biological digestion of waste materials, such as agricultural and industrial effluent, can be used to produce biohydrogen [20]. Hydrogen can be produced from wastewater using electrohydrolysis and biological processes like microbial fermentation in the dark or photo-fermentation. Other procedures include the use of ultraviolet light and an ultrasonic pre-treatment that increases the hydrogen concentration by 80.6 and 120%, respectively [21]. Even with their distinct qualities, some procedures can only extract hydrogen from wastewater. Because sludge typically contains a high amount of water, drying pre-treatment processes occasionally significantly increase the price of this procedure and the quantity of energy consumption.

4.4.5 Microbial Fuel Cells

In typical WWTPs, the bulk of the energy contained in the organic components is converted to carbon dioxide using activated sludge in WWTPs, which generates less energy but yields more sludge. Energy recovery capacity can be further increased, if the source for carbon-based energy found in the wastewater can be directed toward cutting-edge anaerobic technologies like MFCs. Considering the current WWTP designs, MFCs may be used in conjunction with a membrane bioreactor (MBR) process, submerged in aerobic or anaerobic tanks. These may be even built to function independently following AD in order to make use of its effluent [22]. An energy balance done on such a system suggests that an MFC in conjunction with wastewater treatment and an AD system might have the potential to produce an energy of 17% in the form of electricity and 37% in the form of biogas. In addition, this method reduces energy consumption by 104 kW/MGD of wastewater processed, bringing total energy recovery up to 58% [15].

4.4.6 Alga-Based Biofuel Production

In recent days, there has been a lot of interest in the potential of using microalgae as a renewable energy source. However, studies indicate that further mass culture optimization is necessary if microalgae biofuel manufacturing is to be effective on a commercial level and sustainable over the long term. Potentially affordable and sustainable methods of algae growth for biofuels can be achieved by using wastewater produced from municipal, agricultural, and industrial operations. The concentration of critical nutrients, such as N, P, and organic carbon (and the ratios of these constituents), as well as the accessibility of light, O_2, and CO_2, plays a role in the effective growth of microalgae in wastewater. Additionally, there is an opportunity to combine biofuel production with wastewater treatment by algae to reduce nutrients [23]. Wastewater resources may offer a feasible solution for enhancing the sustainability of the production of algal biofuel. This is because they offer a dual-use process as an efficient growth medium for algae cultivation and a source of freely available nutrients (especially N and P).

4.5 Emission Sources in WWTP

Novel problems are currently being taken into consideration in order to guarantee the sustainability of WWTPs in terms of their financial sustainability and environmental impact. Energy utilization and GHG emissions are the two criteria that have grown to be crucial in determining how well WWTPs operate overall [24]. In terms of GHG emissions, WWTPs are second only to landfills, while it is seventh among stationary sources of biogenic CO_2 emissions [25]. The main processes in WWTP that produce GHG emissions include wastewater extraction, wastewater treatment, and final disposal of the treated wastewater and the resulting sludge. Anthropogenic GHGs including CO_2, CH_4, and N_2O may be produced during the biological treatment of wastewater, and CO_2 is generated during the process of producing the energy required to run the facility. The main source of carbon emissions from the WWTPs is frequently connected to power use, except for a minor amount of fugitive emissions from the treatment and disposal of sludge [25]. As an indirect GHG emission source, WWTPs use additional external carbon sources for improving denitrification, chemical consumption for sludge conditioning, and electrical power used for pumping and aeration. Alternately, both on-site and off-site GHG emissions are calculated using emissions of carbon from WWTPs. The activated sludge process, anaerobic sludge treatment, biogas production, and fuel combustion for digester heating account for the majority of the on-site GHG emissions from the current plant. Off-site generation of energy for supplying WWTPs chemical use, sludge transportation and disposal, and degradation of the residual elements in the effluent are all considered indirect sources of GHG emission from WWTPs [26].

4.6 Carbon Accounting in WWTPs

The approach to carbon accounting is to estimate how much ton of GHGs is produced in WWTPs, both directly and indirectly. In general, there are two methods for calculating carbon emissions. The first one utilizes estimation and calculation of direct and indirect emissions, while the other employs carbon footprint [27]. Despite the fact that the analytical content obtained by different accounting methods is diverse, the essential information is the necessity of integrating the boundaries of sludge carbon emissions and concentrating on the information of carbon sources. During the estimation of emissions, only the total gas emissions from the various unit processes, principally from the use of fuel and energy consumption in the treatment process, can be called direct emissions. Only the additional emissions caused by the equipment used in the emission process are referred to as indirect emissions, such as the carbon source emissions produced when certain energy equipment (like blowers, pumps, and aeration equipment) and other equipment require a significant amount of electricity to operate [28]. The ecological footprint serves as a model for the carbon footprint. It involves calculating the carbon emissions generated during the actual treatment process using operational data on power consumption, sludge formation, chemical use, and biogas production. It consists of carbon dioxide emissions from several steps in the wastewater treatment process.

4.7 Carbon Emissions Reduction Options in WWTP

The key emission reduction goals are to reduce the direct emissions of GHG (CO_2, CH_4, and N_2O) and the indirect emissions from the use of energy in the wastewater treatment process, which accounted for 48 and

40.6%, respectively, of the total emissions for the entire facility. The most economical method to lower GHG emissions is to optimize the operational parameters of WWTP units. In order to lessen direct emissions, it may be possible to reduce GHG emissions by controlling the production and emission of N_2O, CH_4, and CO_2. In order to reduce the amount of N_2O produced, certain operational conditions must be avoided, including utilizing low dissolved oxygen concentrations in the nitrification and denitrification stages, high nitrite concentrations in both stages, a low COD/N ratio in the denitrification stage, abrupt changes in pH, dissolved oxygen, ammonia, and nitrite concentrations, and use low COD [29]. Through the thickening of sludge tanks and covering of sludge disposal tanks to stop gas leaks, the emission of CH4 in WWTPs can be lowered. Hoods, which may be burned with additional biogas in a torch, can capture the emissions. Utilizing the fact that the overall methane load, which has been estimated to be 1% of the input COD load (80%), is primarily oxidized in the activated sludge tanks may help reduce methane emissions from wastewater treatment even more [30]. The oxidation of organic matter in biological reactors and the burning of CH_4 are the sources of the direct emissions of CO_2. The biological reactor's solid retention times (SRTs) have a considerable operational impact on these emissions [24]. When the activated sludge system is run at high levels of SRT, endogenous biomass respiration is encouraged. As a result, more COD is converted to CO_2, and less sludge is produced overall. This decrease in sludge formation suggests a decrease in the production of methane, which, in turn, predicts a decrease in the CO_2 emissions brought on by the combustion of that methane [31]. Thoughtfully designing and effectively employing the machinery reduce the indirect emissions of GHG from the wastewater treatment process's use of power. As an example, working treatment and disposal of sludge (by 35.2%), the recovery of residual thermal energy from secondary effluent (16.6%), and the recuperation of heat from ventilation systems (3.2%) are the main methods used to achieve a decrease in carbon emissions of 252,994 t/y [32].

4.8 Carbon Emissions Capture and Treatment Options in WWTP

Although many technologies exist to remove or capture N_2O, CH_4, and CO_2 from industrial gaseous streams, effective and affordable abatement methods still need to be developed for treating gaseous streams from WWTPs. Alternative bioprocesses using nitrifying and denitrifying bacteria or microalgae have been developed in a number of ways to minimize

NO_x gas emissions. Denitrification-based technologies have been used to successfully remove N_2O with efficiencies ranging from 75 to 99%. Another choice is to burn the methane produced by the anaerobic sludge digester using the exit gaseous stream from the top of the nitrifying unit, which contains N_2O. Multiple biological processes can change methane to CO_2 (1 mol to 1 mol) because CO_2 has a lower climate change factor than methane. This enables a reduction in the total amount of GHG emissions measured in CO_2 equivalents. CO_2 emissions from industrial activity are captured using membrane separation techniques, chemical or physical sorption [33]. However, using these technologies is typically associated with high startup and operating costs as well as waste stream production [34]. These elements have made, for example, the development of microalgae seem like a desirable choice for the storage of CO_2 gas. In order to increase the production of microalgae for use as biofertilizer or as a substrate to increase biogas production, microalgae can be used in WWTPs to collect CO_2 before combustion, as a cheap way to purify biogas, or after combustion [35].

4.9 Challenges and Future Prospective

In response to the rising worldwide energy demand for energy recovery, GHG emission reduction, and wastewater treatment, WWTPs are receiving greater focus with respect to their losses during conversion and energy utilization. It may be impracticable to reduce emissions and recover all the energy from wastewater resources to their full potential. However, the biggest obstacle to energy recovery is the lack of a coordinated framework for operating current wastewater treatment facilities and thinking of wastewater as a resource that needs to be managed at various levels. Research investments and development are also necessary to properly introduce recovered resources into markets for social consumption. It is crucial to draw in the interests of all stakeholders, including business partners, end users, and policymakers who are involved in the planning process, and to develop apps with distinctive selling points. The growth of economies and value chains, environmental pollution, health dangers, social acceptance, and policy concerns are also major obstacles to the successful application of energy recovery and reduction approaches. A roadmap of political decisions and social, economic, and technological factors will be required in the future to overcome the obstacles and achieve all the advantages of wastewater management from WWTPs for energy recovery, reducing GHG emissions and minimizing unnecessary costs.

4.10 Conclusion

This chapter begins with a brief introduction to WWTP's energy demand, consumption, and carbon emissions sources. The major units of energy consumption in WWTPs, and alternatives for reducing energy consumption, including energy auditing, energy benchmarking, and energy management programs, were discussed. Various options for energy recovery and savings in WWTPs, including biogas generation, energy conservation, heat recovery, CHP systems, and the development of algal biofuels, were also highlighted. In addition, carbon emission sources, carbon accounting and reduction, and capturing alternatives are being discussed. The challenges and future prospects in the area are highlighted to show the opportunities for strategic management and handling of WWTPs.

References

1. Soares, R., Comparative analysis of the energy consumption of different wastewater treatment plants. *Int. J. Arch. Arts Appl.*, 3, 79, 2017.
2. Stillwell, A.S., Hoppock, D.C., Webber, M.E., Energy Recovery from wastewater treatment plants in the united states: A case study of the energy-water nexus. *Sustainability*, 2, 945–962, 2010.
3. Maktabifard, M., Zaborowska, E., Makinia, J., Energy neutrality versus carbon footprint minimization in municipal wastewater treatment plants. *Bioresour. Technol.*, 300, 122647, 2020.
4. Nguyen, T.K.L., Environmental impacts and greenhouse gas emissions assessment for energy recovery and material recycle of the wastewater treatment plant. *Sci. Total Environ.*, 784, 147135, 2021.
5. Bora, R.R., Richardson, R.E., You, F., Resource recovery and waste-to-energy from wastewater sludge via thermochemical conversion technologies in support of circular economy: A comprehensive review. *BMC Chem. Eng.*, 2, 8, 2020.
6. Corominas, L., The application of life cycle assessment (LCA) to wastewater treatment: A best practice guide and critical review. *Res. Technol.*, 184, 116058, 2020.
7. Matos, C., Pereira, S., Amorim, E.V., Bentes, I., Wastewater and greywater reuse on irrigation in centralized and decentralized systems — An integrated approach on water quality, energy consumption and CO2 emissions. *Sci. Total Environ.*, 493, 463–471, 2014.
8. Nakkasunchi, S., A review of energy optimization modelling tools for the decarbonisation of wastewater treatment plants. *J. Clean. Prod.*, 279, 123811, 2021.

9. Deshmukh, G.R., Deshmukh, Y.G., Nagdwate, A.B., Energy audit for a waste water treatment process. *Int. Res. J. Eng. Technol.*, 4, 1487–1491, 2017.
10. Esteves, F., Cardoso, J.C., Leitao, S., Piers, E.J.S., Review of energy audit and benchmarking tools to studyenergy efficiency through reducing consumption in wastewater treatment systems. *Cad. Educ., Tecnol. Soc*, 15, 150–165, 2022.
11. Longo, S., Antoni, B.M., Bongards, M., Chaparro, A., Monitoring and diagnosis of energy consumption in wastewater treatment plants. A state of the art and proposals for improvement. *Appl. Energy*, 179, 1251–1268, 2016.
12. Gandiglio, M., Enhancing the energy efficiency of wastewater treatment plants through co-digestion and fuel cell systems. *Front. Environ. Sci.*, 5, 2017.
13. Holenda, B., Domokos, E., Rédey, A., Fazakas, J., Aeration optimization of a wastewater treatment plant using genetic algorithm. *Optim. Control. Appl. Methods*, 28, 191–208, 2007.
14. Feng, L. and Chen, B., Energy-water nexus of wastewater treatment system: Conceptual model and framework. *Energy Proc.*, 104, 141–145, 2016.
15. Gude, V.G., Energy and water autarky of wastewater treatment and power generation systems. *Renew. Sustain. Energy Rev.*, 45, 52–68, 2015.
16. Moreno, F.M., Malico, I., Marques, I.P., Promoting sustainability: Wastewater treatment plants as a source of biomethane in regions far from a high-pressure grid. A Real portuguese case study. *Sustainability*, 13, 8933, 2021.
17. Zarei, M., Wastewater resources management for energy recovery from circular economy perspective. *Water-Energy Nexus*, 3, 170–185, 2020.
18. Hao, X., Energy recovery from wastewater: Heat over organics. *Water Res.*, 161, 74–77, 2019.
19. Strazzabosco, A., Kenway, S.J., Lant, P.A., Quantification of renewable electricity generation in the Australian water industry. *J. Clean. Prod.*, 254, 120119, 2020.
20. Zhu, K., Achieving high-rate hydrogen recovery from wastewater using customizable alginate polymer gel matrices encapsulating biomass. *Environ. Sci., Water Res. Technol.*, 4, 1867–1876, 2018.
21. Elbeshbishy, E., Hafez, H., Nakhla, G., Enhancement of biohydrogen producing using ultrasonication. *Int. J. Hydrogen Energy*, 35, 6184–6193, 2010.
22. Logan, B.E., Microbial fuel cells: Methodology and technology. *Environ. Sci. Technol.*, 40, 5181–92, 2006.
23. Pittman, J.K., Dean, A.P., Osundeko, O., The potential of sustainable algal biofuel production using wastewater resources. *Bioresour. Technol.*, 102, 17–25, 2011.
24. Campos, J.L., Greenhouse gases emissions from wastewater treatment plants: Minimization, treatment, and prevention. *J. Chem.*, 2016, 3796352, 2016.
25. Presura, E. and Robescu, L.D., Energy use and carbon footprint for potable water and wastewater treatment. *Proceedings of the International Conference on Business Excellence*, vol. 11, pp. 191–198, 2017.

26. Wang, Q., Carbon source recovery from waste sludge reduces greenhouse gas emissions in a pilot-scale industrial wastewater treatment plant. *Environ. Sci. Ecotechnol.*, 14, 100235, 2023.
27. Xiao, J., Accounting of carbon emissions in the process of wastewater treatment from the perspective of environmental protection. *IOP Conf. Ser.: Earth Environ. Sci.*, 781, 032021,2021.
28. Lin, Carbon emission assessment of wastewater treatment plant based on accounting perspective, 2020.
29. Law, Y., Ye, L., Pan, Y., Yuan, Z., Nitrous oxide emissions from wastewater treatment processes. *Philos. Trans. R. Soc London, Ser. B*, 367, 1265–1277, 2012.
30. Daelman, M.R.J., Methane emission during municipal wastewater treatment. *Water Res.*, 46, 3657–3670, 2012.
31. Ge, H., Batstone, D.J., Keller, J., Operating aerobic wastewater treatment at very short sludge ages enables treatment and energy recovery through anaerobic sludge digestion. *Water Res.*, 47, 6546–6557, 2013.
32. Pang, C., Carbon emission accounting and the carbon neutralization model for a typical wastewater treatment plant in China. *Int. J. Environ. Res. Public Health*, 20, 1, 2022.
33. Nair, P.S. and Selvi, P.P., Absorption of carbon dioxide in packed column. *Int. J. Sci. Res. Publ.*, 4, 2250–3153, 2014.
34. Heydarifard, M., Reactive absorption of CO2 into Piperazine aqueous solution in a stirrer bubble column: Modeling and experimental. *Int. J. Greenhouse Gas Control*, 79, 91–116, 2018.
35. Wang, K., Khoo, K.S., Chew, K.W., Selvarajoo, A., Chen, W.H., Chang, J.S., Show, P.L., Microalgae: The future supply house of biohydrogen and biogas. *Front. Energy Res.*, 9, 2021.

5

Integrated Use of Biomass to Produce Energy and Construction Material

Samavia Fiaz[1†], Mehwish Khalid[1†], Sumia Akram[2] and Muhammad Mushtaq[1*]

[1]Department of Chemistry, Government College University, Lahore, Pakistan
[2]Division of Science and Technology, University of Education, Lahore, Pakistan

Abstract
Biomass and sewage sludge have been regularly explored and used as fuel and energy sources but, in this chapter, we have tried to overlook this sustainable resource for the production of construction materials. The available literature regarding the composition of agricultural, domestic, algal, and livestock biomass suggests that all of these resources contain bulk materials like polymers, inorganic salts, and cementable ashes. Likewise, the sewage sludge may contain very compact and high-strength polymeric materials which are in particular ready to cement into building blocks. However, the conventional biomass and waste management approaches are no longer appropriate for the eco-friendly, sustainable, and integrated conversion of biomass to energy and construction materials. The thermochemical and biochemical methods need key changes to isolate the polymerizable plastics and inorganic materials. In the overlooks of simultaneous waste removal, pathogen killing, and recovery of energy, thermochemical and biochemical methods have promising future.

Keywords: Biomass, energy, biofuels, cement, ceramics

5.1 Introduction

The planet "Earth" is a living place for all beings not just human beings who make up 0.01% of all the inhabitants; however, their activities have rendered all the inhabitants in danger. For the same reason, environmental

**Corresponding author*: Muhammad.mushtaq@gcu.edu.pk
†*Equal contribution as first author*

Inamuddin, Tariq Altalhi, Mohammad Luqman and Joseph Kapuku (eds.) Sewage and Biomass from Wastewater to Energy, (113–150) © 2024 Scrivener Publishing LLC

safety and protection have become universal challenges. The burning of fossil fuels and woody resources not only produces greenhouse gasses and warms the environment but also causes a wide range of known and concealed health and environmental issues. Another growing concern regarding the use of fossil fuel as an energy source is it comes from consumable/nonrenewable sources (the reservoirs of fossil fuels are depleting). To overcome these problems, scientists have been working to produce biofuels from renewable resources as an alternative source of energy. Meanwhile, wastewater treatment has become an indispensable procedure that generates a huge amount of sewage sludge which can cause serious dumping and environmental issues. These issues include an ongoing rise in sludge output, high sludge treatment expenses, and potential hazards associated with sewage sludge for both the environment and the health of people. The extraction and recycling of valuable products from sludge is an integrated approach to the sludge problem, particularly about the hazardous substances, and reasonable costs are three aspects that may be expected to be the focus for sludge management as well as inquiries into novel treatment technologies [1].

Biomass as a renewable energy could be beneficial to reducing greenhouse gasses and an alternative source of fossil fuel for the long term [2]. Biomass includes a vast variety of materials that are complex in nature having different origins and compositions. Natural sources include woody biomass, herbaceous biomass, aquatic biomass, animal biomass, and human biomass. Woody biomass includes stems, branches, chips, and other parts. Grasses, straws, seeds, fruits, pulps, shells, flowers, bagasse, and others fall in the herbaceous category. All types of algae such as freshwater algae, marine algae, microalgae, and microalgae are categorized as aquatic biomass. Animal biomass and human biomass consist of bones, meat, animal and human dung, and manure. Likewise, the wastewater sludgy may contain many of the abovementioned materials as well as food, paint, plastic, and soil particles. All types of biomass can be used to produce energy, ethanol, fuel, plastics, ceramics, and other construction materials [3].

Biomass and sewage sludge contain complex biogenic materials that are produced due to human activities and natural processes. It contains both organic and inorganic solid materials such as human and animal wastes, crops, and aquatic biomass such as algae. Biomass and sewage sludge can be converted into biofuel and other valuable products like cement, ceramics, packing materials, paper, and boards [4]. The estimated biomass energy harvested from wood and agricultural wastes throughout the world is about 30 EJ/year and the total energy demand is more than 400 EJ [5]. Biomass has emerged as a primary substitute for fossil fuel resources that can produce both energy and a variety of products. Biomass as a renewable energy source offers significant future development potential for the production of biofuels and electricity [6]. The projects that can convert biomass and sewage sludge into energy and other products help to reduce waste that comes from agriculture, human, and animal sources [7].

Figure 5.1 illustrates some eco-friendly and cost-effective ways that produce energy and other useful products from biomass and sewage sludge [9]. For example, the ash generated as waste during the incineration/combustion/pyrolysis of biomass for energy production contains organic and inorganic materials in variable amounts, depending upon the quality and origin of biomass. Biomass especially sewage sludge becomes an environmental issue if not managed in sustainable ways. On the basis of physical and chemical properties, there are several uses for biomass instead of dumping. The depletion of natural resources encourages the construction sector to search for new building materials with more sustainability. Biomass can be used as filler, additive, and raw material for buildings with heat, sound, and fire-insulating properties [10].

The woody biomass has been historically applied as construction material and bringing into practice the latest technologies and solvents can offer more sustainable alternative conventional construction materials. Likewise, the ash and agricultural residues have been incorporated in cement mortars, adhesive mortars, and concrete blocks making it more

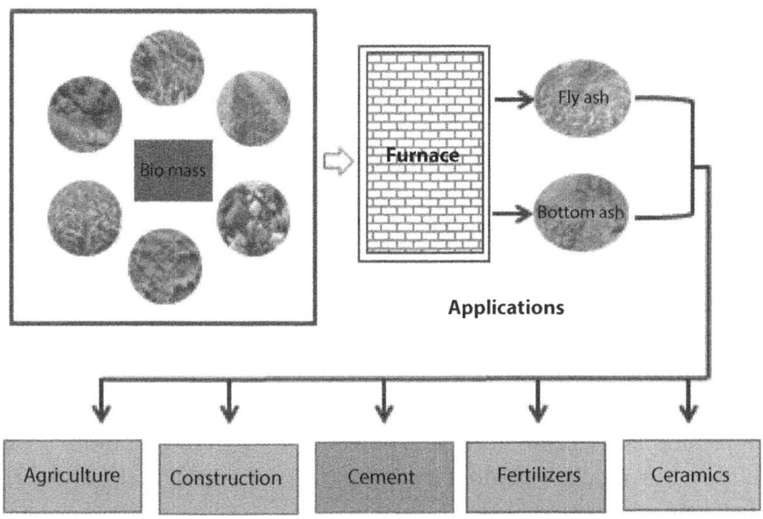

Figure 5.1 Production of useful products including construction material from biomass ash [8].

valuable [11]. Biomass ash can be used as carbonated biomass-based building material instead of hydraulic cement. Another report shows that the incorporation of biomass ash can reduce the consumption of raw materials by up to 10% and minimize the emission of CO_2 [12]. The use of biomass as raw material for the construction of building blocks is more affordable and an emerging area in construction and civil engineering. Despite of many benefits of using biomass-based concrete and cement materials, including low CO_2 emission, good thermal insulation, and low densities, some questions arise about the durability of biomass in an alkali cement matrix, high water absorption, and compatibility between cement and biomass, due to these unanswered question mechanical properties of cement reduce [9]. Overall, it can be speculated that construction materials obtained from biomass and sewage sludge offer better thermal and acoustic insulation, rapid and economical supply of raw materials, and, above all, low cost of construction material. In this chapter, the authors will explore the potential uses of biomass and sewage sludge as construction material in addition to what we have already done *viz* conversion of biomass into fuels and energy.

5.2 Biomass Classification and Utilization

Biomass has been categorized into several types depending upon composition, source, and occurrence but, in general, residues of woody plants, herbaceous plants, aquatic plants, food, sewage sludge comprising organic and inorganics, and livestock manure are often referred to as biomass. These categories can be further classified depending on various parameters such as moisture content, occurrence, and origin. For example, herbaceous plants fall into low (such as wood dust) and high moisture (like sugarcane) containing must be treated by combustion, pyrolysis, gasification, and wet conversion processes like fermentation, respectively [5]. The majority of industrialized and developed countries depend on energy production from staples including corn, wheat, sugar beets, soybeans, and rapeseed [13]. Bamboo, alfalfa hay, tall fescue, reed canary grass, miscanthus, coastal Bermuda grass, sweet sorghum, switchgrass, kochia, and other grasses are included in energy crops. This type of biomass is suitable for the production of boards, structures, fabric, and fertilizers. A German biorefinery company, Biowert, generates green power from agricultural biomass along with useful products including insulation material, polymers, and fertilizers from meadow grass [14].

The leftovers of agricultural and forestry processes and other industrial waste also fall in the category of waste biomass. These wastes are being managed by biorefineries along with the production of useful energy and byproducts [15]. Agricultural residues are considered byproducts of agriculture like cotton stalks, rice husks, coconut and peanut shells, jute sticks, maize, and many more. In Pakistan, the most typical agricultural waste is rice husk which occupies 25% mass of rice. Likewise, corncob is considered a potential source of cellulosic biomass containing 17 to 21% lignin, 25 to 35% xylan, and 39 to 45% glucan. Typically, corncob has been burned for heating and cooking purposes which has caused detrimental health and environmental impacts. Researchers believe that corncob is a potential source of bioenergy or biofuels but could not be completely exploited for its complex structure. Besides, corncob-containing biomass offers a more economical synthesis of antibiotics and certain other drugs. For instance, furfural, glucose, and xylo-oligosaccharides can be more economically synthesized from biomass comprising corncob [16].

The easiest way to get rid of agricultural waste is still considered burning it, but even doing so unintentionally worsens the air quality and other environmental consequences. Agricultural residues mostly consist of hemicellulose, cellulose, ash, and protein. Lignin, cellulose, and hemicellulose combine to make a complex polymer known as lignocellulose [17]. Crop straws are a potential source of lignocellulosic raw material as it contains lignin, cellulose, and hemicellulose. Lignocellulose can be used for the production of biofuels and other chemicals after pre-treatments including physiochemical, biological, and enzymatic procedures. Plant material that remains after harvesting crops is known as crop residue, on the other hand, agro-industrial wastes refer to leftovers of postharvest processes such as cleaning, straining, and grinding. Numerous countries have adopted new policies to control field burning to protect the environment and human health. The "waste to resource" strategy has opted to dispose of rice husk and rice straw and to convert it into energy, building material, fuel, and heavy metal adsorbents. For example, the conversion of peanut shells into bio-carbon is considered a useful way to dispose of it [18].

Livestock wastes fall among untreated waste that contains a large number of microorganisms and has become a worldwide environmental problem. Many of these microorganisms carry infections and may pose a serious risk to the human and animal health. Different physical, chemical, and biological processes can be used to turn animal waste into useful bioenergy and biofertilizers. To deal with animal waste and manure, biological processes are more suitable, economical, and simple to operate [19]. Animal wastes and manure may be utilized as energy sources or raw materials for the production of biogas as these contain high ratio organic matter [20]. One of the best ways to utilize the energy from biowaste biomass is to convert it into biogas, by anaerobic fermentation. Biomass possesses a calorific value of 21 to 24 MJ/m^3 and contains 65 to 70% methane, 30 to 45% CO_2, and traces of other gasses (H_2S, N_2, and H_2) [21]. Europe has expanded its biogas industry in the past decades and generated 16.1 megatons of energy (MTOE). Germany generates half of its energy from biogas; the United Kingdom and Italy produce 2.0 and 2.4 MTOE from livestock wastes, respectively. Biological processes are most widely used to treat animal waste, due to simplicity of operation. Anaerobic digestion can be chosen to convert animal waste into biogas by reducing its volume and toxicity. Interestingly, the livestock biomass left behind after the production of energy can be utilized as construction material and fertilizers [22].

Algae such as microalgae can be found in fresh and seawater. Algal biomass grows faster and has the ability to absorb greenhouse gasses with zero waste production [23]. Algal biomass has gained more attention for biofuel production due to more sustainable, environmentally friendly, and renewable features. It can be an alternative source of energy production and overcome drawbacks related to first- and second-generation biofuel. Meanwhile, the algal biomass has been increasingly focused on the production of insulation material and biopolymers [24].

The sewage sludge obtained during wastewater treatment can work as an inescapable and renewable source of energy, biofuel, biogas, and syngas. Sewage sludge is also a valuable source of biochar, biofertilizer, and construction/building products due to the substantial amount of important organic and inorganic matter [25, 26]. It has been frequently explored for the production of electricity. Sewage sludge with a small proportion of sulfur and phosphorus contains 59 to 88% organic matter, which is the combination of hydrogen (6 to 10%), nitrogen (10 to 15%), carbon (50 to 55%), and oxygen (25 to 30%). Besides, it can have some heavy metals like copper, zinc, chromium, lead, and minerals such as calcium, iron, magnesium, and potassium, and inorganic salts [26, 27]. The primary oxides of aluminum, silicon, and calcium give clay-like characteristics to sewage sludge due to which sludge can be utilized in the ceramic industry [28].

5.3 Integrated Conversion of Biomass into Energy and Other Products

The production of biofuels including bioethanol and biodiesel from plant residues and vegetable oils, known as first-generation biofuels, is associated with high production prices and limited availability of feedstock. Another valid concern arises due to competition between the use of feedstock in the food industry and the fuel production sector. To overcome the issue of production cost, the introduction of an integrated energy and production system is essential which, in the simple sense, recommends the isolation and utilization of each and every component of biomass as a useful product. This approach is equally useful for the production second generation fuels which utilize the lignocellulosic feedstock; an abundant component of agricultural leftovers and byproducts, organic wastes, and material like energy crops [29]. Moreover, the idea of the integrated production system applies to thermochemical, biochemical, and photochemical processes. Thermochemical processes include direct combustion, gasification, pyrolysis, and hydrothermal liquefaction. The use of heat in combination with chemicals and catalysts facilitates the conversion of biomass into fuels, construction materials, and other valuable products. The energy stored in biomass can be converted into energy by directly heating, in the form of solid, liquid, and gaseous fuels such as charcoal, bio-oils, and synthetic gas [30].

Lignocellulosic biomass can be converted into biofuel in less time by hydrothermal liquefaction, and slow and fast pyrolysis which take 90 min, 30 to 180 min, and 3 to 5 s, respectively. In addition to this, hydrothermal liquefaction and pyrolysis are considered more eco-friendly due to no emission of gasses. To improve the efficiency of processes like pyrolysis and gasification, the moisture content and particle size of biomass feedstock must be reduced. On the flip side, pre-treatment of feedstock and post-treatment of biofuel often render the operation more costly.

The biochemical conversion process, however, uses bacteria to convert biomass into biofuel and additional products via fermentation and an anaerobic digestion process. In the biological conversion process, microorganisms, enzymes, and bacteria react on biomass to break cellulosic material and produce biofuel, gaseous fuel, and chemicals. The main drawbacks of the biochemical process are the incapability of the digester to process biomass with higher hydraulic retention time (HRT) and the production of greenhouse gasses. Moreover, the enzymatic hydrolysis of biomass requires the conversion of polysaccharides into monosaccharides. However, few studies revealed that the conversion of biomass through biochemical processes is technically and economically more feasible [31].

Gasification and liquefaction are considered two major categories of thermochemical processes. Gasification involves the conversion of biomass into gaseous and its subsequent hydrocarbon; on the other hand, liquefaction includes the conversion of biomass into liquid form by different methods including high-temperature pyrolysis and high-pressure liquefaction. The selection of conversion technique depends upon the type of biomass feedstock, economic conditions, environmental aspects, and usage of the end product. Major types of thermochemical processes include combustion, gasification, pyrolysis, and hydrothermal liquefaction. Pyrolysis is considered the most abundant technique due to its unique nature of producing more than one kind of product, including solid, liquid, and gas [32]. Pyrolysis takes place under the temperature range from 300°C to 700°C and produces biochar, bio-oil, and syngas. There is another kind of thermochemical process that is known as torrefaction a mild type of pyrolysis with a temperature range of 200°C to 300°C. Hydrothermal liquefaction runs between 250°C and 374°C to produce crude oil. In the gasification process, biomass is converted into synthetic gas at the temperature of 700°C [33]. Thermochemical processes have been used for the complete conversion of various biomasses into biofuel. Algae are one of the most promising feedstock and have advantages over other biomass including renewability, productivity, and sustainability. Because of these benefits, algae biomass is known as third-generation biofuel. It produces bio-oil or char through pyrolysis [34].

5.3.1 Combustion/Incineration

Combustion involves the direct burning of biomass (lignocellulosic biomass) in the air to produce energy in the form of heat and electricity. All types of biomass with a moisture content of less than 50% can be combusted via various homogeneous and heterogeneous reactions. The combustion of biomass depends on the chemical characteristics of feedstock, the nature, and size of biomass particles, temperature, and the environmental conditions [35]. Combustion falls among the most frequently applied techniques to treat biomass; more than 97% of world bioenergy production depends on it [36]. In developing countries, biomass combustion plays a significant role as the primary source of energy for cooking. The process of combustion has been commercialized and well-understood due to its low cost and higher feasibility [37].

The process of combustion/incineration generates heat via thermal breakdown/oxidation of carbon-containing substances [26, 38]. In densely populated communities, incinerating waste sludge has become an effective method for energy production and trash reduction. It aims to completely oxidize the biological waste molecules at elevated temperatures [39]. Incineration converts carbon-containing matter into carbon dioxide and water besides emitting a significant amount of heat and other gasses at high temperatures and in the presence of excess oxygen. Incineration of agricultural biomass primarily produces carbon dioxide, water, ash-like material, and heat energy. The ash produced at the end of the process and the volatile components obtained can be transformed into various byproducts or used as precursors for the synthesis of polymers or other highly valuable/durable substances such as stoneware, bricks, and lightweight concrete [1, 40].

Significant environmental contamination concerns are raised due to combustion byproducts such as nitrogen and sulfur-based compounds, halogen gasses, and contaminants like heavy metals. In the extreme temperature method of incinerating waste, some heat is kept in flue gas, and the remainder departs via the walls of the burner [26]. The complete combustion process occurs in two steps given as follows:

$$2H_2 + O_2 \rightarrow H_2O + 39.5\ \text{kWh/kg} \tag{5.1}$$

$$C + O_2 \rightarrow CO_2 + 9.11\ \text{kWh/kg} \tag{5.2}$$

Total combustion is typically not possible because the carbon may only partially burn, producing carbon monoxide instead of carbon dioxide. Excess air with more oxygen than its stoichiometric amount must be supplied to ensure that there is enough oxygen during the burning process [26]. The typical incineration plant (Figure 5.3) takes six to seven steps to carry out incineration and combustion of biomass or sewage sludgy like drying, devolatilization and auto-gasification, combustion of mass, combustion of char, melting of ash, and finally agglomeration of ash into various products.

Devolatilization takes place along with the drying process of wastewater sludge burning and can begin at extremely low temperatures such as 150°C, based on the setup of a combustor, the conditions of the chemical reaction, and the chemical composition of the sludge being burned [40]. Sewage sludge particles simultaneously dehydrate and devolatilize from the surface to the core, when heat is applied. The emission of steam vapors and volatile substances happens after heating the sludge, which comes in contact with the heated layers of char, by further proceeding to the material created in the process of devolatilization. Under aerobic conditions, volatile substances and light-weight gasses are flamed on the surface of raw material, and the temperature of the layer of material for production is elevated. A thick layer of heated ash particles would build on the outer layer when the ash melted under the highest possible temperature. In the end, the droplet particles of ash are going to combine more easily after the burning of extra char [26, 40].

A logical continuation of drying is incineration, which turns the sludge into inanimate ash through burning. When both gasoline fuel and oxygen are combined, a lot of heat and light energy is emitted during an

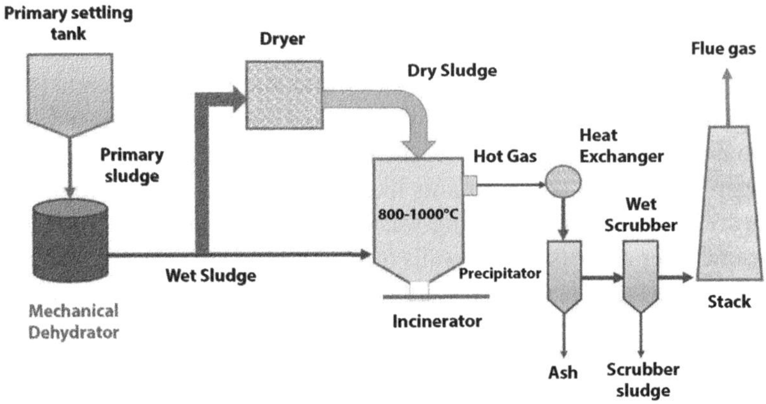

Figure 5.2 Schematic representation of the Incineration process [41].

instantaneous chemical process of combustion. The system (Figure 5.2) involved in the incineration of sewage sludge is made up of the following units; vacuum filter for filtration, centrifuge machine or belt press, incinerator, sludge feeding system, ash-management unit, and pollution control devices [39].

Before being subjected to incineration, the manually dewatered sludge layer can be dried using the energy generated during the combustion procedure, or it can be utilized to generate power. The volume of water in the sludge and the efficiency of its burning have a significant impact on the quantity of energy released. There are two ways to incinerate sewage sludge: mono-combustion and co-combustion, with the former being more common. The most common methods contain fluidized bed and multiple hearth combustion chambers with the latter having a better efficiency due to less fuel use and pollutants. Environmental consequences and high expenditures on operation are the primary concerns with sludge incineration [1, 42].

Heavy metal buildup in burned remains (ash) and atmospheric emissions from combustion engines are factors in environmental issues. By using incinerator ash material as an initial component to make cement, a technique that renders heavy metals in the cement, the primary issue can be resolved. The pollutant gasses released from the combustor are dioxins, furans, and some other major polluting substances like carbon dioxide, oxides of nitrogen, and sulfur. If the exhaust gasses are swiftly chilled beneath 400°C and temperatures beyond 600°C remain intact, dioxin and furan pollutants are typically not an issue. The removal of N_2O can be accomplished by raising the temperature at which it burns to 880°C [39, 42].

Another approach to managing sewage sludge is to co-combust it with other biomass like woody material, coal, lignite, or municipal solid waste [42]. The possibility of incinerating dried waste in a power plant that burns coal has been looked at to avoid the substantial expenses of an isolated sludge incineration facility and to enhance energy harvesting efficacy [1]. Conventional fossil fuel burning and combustion-gas treatment facilities can be used in a positive way. The impact of sludge combustion on the characteristics of the atmosphere and ash can be disregarded because it is insignificant in relation to the quantity of coal burned. Different co-incineration techniques are used in practice, such as when combined with municipal trash. Additionally, sludge from sewage also co-incinerates with fossil fuels such as coal in power plants [1].

The combustion consists of three key stages, namely, drying, pyrolysis, and reduction along with the burning of volatile gasses and solid char. Burning of volatile gasses generates 70% of total heat, and blue flame shows the presence of char [30]. Heat and light are produced after a complex reaction between fuel and oxidizing agent in the combustion chamber. Three types of reactors are being used in the industrial sector for heat and electricity production including the fixed bed reactor, fluidized bed reactor, and entrained flow reactor [43].

5.3.2 Biomass Pyrolysis

When there is no air present or not enough oxygen in the atmosphere, organic materials thermally decompose by the process of pyrolysis. Carbon-based biochar wastes, non-condensable biogas, and condensable bio-oil are all pyrolysis byproducts that have a variety of applications and by adjusting the parameters of the process, these applications can be enhanced [42]. In the absence of oxygen, pyrolysis is a form of thermal degradation that at extreme temperatures, splits the biological waste, and transforms it into gasses, liquids, and solids. No apparent ignited material is required for pyrolysis to distinguish it from the incineration process [1]. Handling sludge and energy pyrolysis of sewage sludge is a novel technique, where the waste material is heated to temperatures between 350°C and 500°C by using pressure and in a system deprived of oxygen. This technique produces oils, ash, vapors, flammable gasses, and biochar (Figure 5.3) as byproducts of sewage sludge [44, 45].

In the process of pyrolysis, the energy from radiation or conduction is used to initially transmit heat to the wastewater/sewage sludge components, after which it progressively moves into the matter. The temperature rises rapidly during the intermittent burning of the sludge material, which

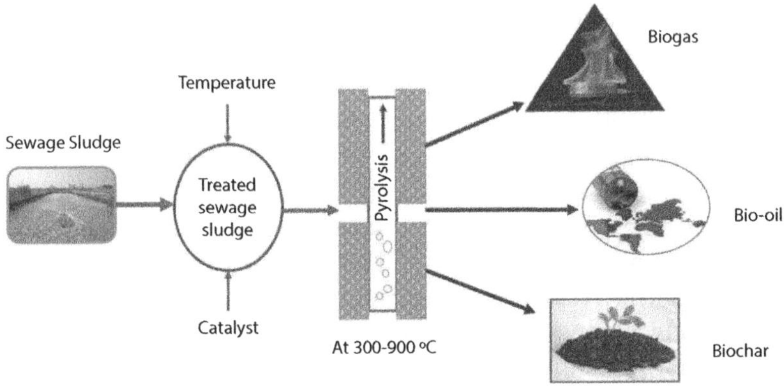

Figure 5.3 The Conventional Pyrolysis scheme and products.

initially causes moisture to evaporate in the drying stage and then gradually emits pyrolytic volatile substances in the initial pyrolysis step. The main source of the volatile substances in sewage sludge is the fragmentation of the chemical interactions of the organic molecules such as carbohydrate molecules, proteins, and fats with the help of heat. Water, carbon dioxide, carbon monoxide, the gas hydrogen, methane gas, and various condensable and non-condensable chemical compounds are constituents of the majority of pyrolytic volatile substances [26, 46].

Pyrolysis has been considered an industrial method to treat biomass. As compared to the conventional combustion of organic matter in the presence of air to produce carbon dioxide and water, pyrolysis is the breakdown of organic matter into components in the absence of oxygen. The conversion of lignocellulosic biomass into bio-oil, char, and biogas through thermal decomposition in the absence of oxygen is known as pyrolysis. Biochar produced as a byproduct of pyrolysis at 550°C is often used as fertilizer. Similarly, the liquids produced by pyrolysis are purified to get commercial-grade bio-oil. Pyrolysis also produces solid matter rich in carbon. The cellulosic part of the plant undergoes two chemical changes in pyrolysis. In the first pathway, bonds between cellulosic chains break to form glucose monomer units specifically levoglucosan. During the second phase of the pathway, it leads to the production of aliphatic hydrocarbon chains, along with carbon monoxide and water. Each constituent of lignocellulosic biomass including cellulose, lignin, and hemicellulose undergoes pyrolysis at different rates by different methods and mechanisms to yield different products [47].

Lignin produces phenolic compounds in bio-oils during pyrolysis, and hemicellulose breaks to produce furfurals. Despite the production of bio-oil, low heating value, high oxygen concentration, and high viscosity are the major drawbacks of pyrolysis. In order to reduce moisture content, acidity, and viscosity, and remove extra oxygen, a catalyst should be added to the pyrolysis reaction. Catalysts should be unique, cheap, recyclable, and selective in nature. High temperatures can be used to assist catalysts, made up of the most affordable materials like aluminum, carbon, and silica [48].

Industrially, depending on the rates of process, pyrolysis can be categorized into three types: slow, flash, and fast pyrolysis. In slow pyrolysis, temperature ranges between 300°C and 500°C with a heating rate of 0.1°C to 1°C per second, and fast pyrolysis takes place at 400°C to 700°C with a heating rate of 10°C to 200°C per second. On the other hand, flash pyrolysis is carried out at a high temperature of 800°C to 1000°C with a high rate of more than 1000°C per second [49]. Charcoal has been produced by slow pyrolysis for more than a thousand years. Slow pyrolysis took too long time for the process to complete with a temperature of less than 500°C. Slow pyrolysis produces char with less heating rate from 0.1°C to 2°C and vapor residence time varying from 5 to 30 min [47]. The vapors in the reaction chamber react with each other to produce solid and liquid biofuel. In this type of pyrolysis, char is considered the most desired product [17]. Industries are focusing on the production of char by utilizing bio-oil and biogas as energy sources. Furthermore, acetone, methanol, and acetic acid have been generated by slow pyrolysis as byproducts [49].

Fast pyrolysis converts biomass into liquids and gaseous fuel more quickly at high temperatures in the absence of oxygen. Biomass decomposes into vapors and aerosols, along with some charcoal and gas. Liquid biofuel may be produced at low residence time under 400°C temperature. After the condensation of vapors, a dark brown liquid is produced with a half heating value than the traditional fuel oil production [50]. It produces 60 to 75% bio-oil, 15 to 25% charcoal, and 10 to 15% gaseous biofuel, based on biomass nature. It involves a high heating value of 10°C to 200°C per second with a short residence time of 0.5 to 10 s [49]. Fast pyrolysis for liquid fuel production has drawn attention as biofuel can be stored, transported, or may be used as an energy carrier [51]. In fast pyrolysis, biomass is converted into liquid fuel in less time, to transfer heat and mass. It produces 75% bio-oil of total dry biomass weight with no waste other than flue gas and ash. Bio-oil yield can be determined by biomass type, vapor residence time, temperature, ash content, and char separation. In fast pyrolysis, the 10% water content of the feedstock is reduced so that the grinding of biomass into small particles takes place which speeds up the reaction. It also produces char 40 to 50% of weight, a thermally stable aromatic polymer with three-dimensional structures [50].

The term flash pyrolysis refers to extremely quick pyrolysis, which involves the quick passage of biomass through the heated tube under gravity, at a higher temperature and the shortest residence time than fast pyrolysis [47]. Flash pyrolysis is considered a more vital and environmentally acceptable method for converting biomass into energy. In flash pyrolysis, biomass is heated from 450°C to 600°C with a heating rate of 103°C to 104°C per second, higher than of other types of pyrolysis, to produce 60 to 75% bio-oil. In this type of pyrolysis, 15% water content from biomass is removed. Flash pyrolysis, also known as ultrafast pyrolysis, is categorized by quick removal of char, high heating rates, and temperature. It consists of the lowest vapor residence time and fast cooling of vapors. Biomass flash pyrolysis has been more common in the past 10 years due to its ease of use, cost-effectiveness, and flexibility [52]. The main constituent of flash pyrolysis products is liquid fuel such as bio-oil. By the pyrolysis of lignin, mainly phenols are produced which include methoxyphenols, dimethoxyphenols, and hydroxyl phenols [52].

5.3.3 Gasification

Gasification is a type of thermochemical process that takes place in the presence of the minimum amount of air at high temperatures. Biomass undergoes each step which involves physical and chemical changes to produce carbon monoxide, carbon dioxide, hydrogen, nitrogen methane, and other gasses. These gasses are further used to produce biofuel including methanol, methane, and pure hydrogen generation. In gasification, carbon dioxide is generated from solid carbon in the airtight and closed chambers [53]. Gasification of biomass is widely researched and applied economically, due to its easy accessibility and less post-recovery treatment than any other technique. The main product of gasification is syngas, which may be utilized as feedstock for reaction or burned as fuel for electricity production. The main objective of biomass gasification is the generation of low-heating gas that can be used as fuel in internal combustion engines to produce electricity. It produces hydrogen in a cheaper way with 4 to 6 MJ/m^3 and 8 to 14% H_2 of total biomass content. Major steps of gasification include drying, pyrolysis, oxidation, and reduction mainly. The gasification procedure uses various gasifiers, from traditional updraft or downdraft gasifiers to advanced gasifiers like plasma. Heating rate, temperature, and residence time also control the reaction like other techniques. Additionally, the environment of the reaction chamber also influences the quality of produced syngas [54]. Gasification can be categorized into conventional and hydrothermal gasification. Both procedures produce the same end products but differ in syngas composition, gasifying agent, and working conditions. Syngas produced by conventional and hydrothermal gasification depends upon various factors including reactor type, temperature, pressure, heating rate, and residence time. Furthermore, hydrothermal gasification involves a number of intermediate reactions with less understood mechanisms [55].

The gasification technique (Figure 5.4) has been developed further by the addition of plasma. Plasma gasification operates at high temperatures likely above 5000°C with a high heating rate, fuel flexibility, and low emission of gasses, inertness, and inert ash generation. Plasma gasification of sawdust and wood produces syngas with 51 vol % CO and 42 vol % H_2 at 1200°C to 1400°C.

When gasification occurs with air, the end result is a blend of tar, carbon monoxide, carbon dioxide, hydrogen gas, and methane. The gas has no existence of nitrogen when oxygen is utilized as a gasifying ingredient. The proportion of the different elements in the gaseous state is likely altered when steam is used as a gasifying substance, which increases the amount of methane and hydrocarbon compounds in the resulting mixture of gas. The initial phase in the gasification process is the evaporation of water from the raw material, which then proceeds by pyrolysis in order to generate gas, vapors of tars or oils, and solid char byproducts. Then, gasification or partial oxidation of the resulting burned gasses, char, and tar components occurs [39].

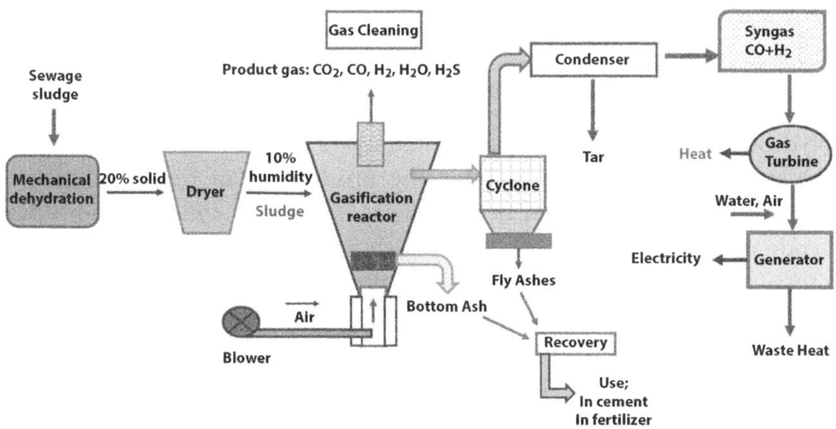

Figure 5.4 Gasification of biomass to produce biogas [56–58, 113].

The process of drying, the initial phase in removing the moisture, happens between 70°C and 200°C. The proportion of humidity in the drying gas, the relative humidity of the substance being processed, and the degree of heat differential involving the particulate matter and the burning gasses all affect how quickly the feedstock particle dries. The decomposition of the organic materials occurs in the second or pyrolysis phase at around 350°C to 500°C when roughly 60 to 70 wt% of the raw material can be transformed into a dense char, gasses, and liquid portion. The solid form of char is transformed into gasses, such as hydrogen gas, carbon monoxide, carbon dioxide, and methane gas, in the third phase by the interaction of gas and solid. The material that is used for the process of drying, burning, and other endothermic reactions involving solids and gasses requires a large amount of energy, while the complete or partial oxidation of carbonaceous substances is highly exothermic in nature. Furthermore, the energy needs of the gasifier are enhanced by the reactions involving hydrogenation. The fourth and last stage of gasification involves the reactions of gas and gas which are homogeneous in nature and have the ability to alter the chemistry of resultant gas phase molecules [26, 59].

In addition to other products, gasification produces syngas mixture of carbon monoxide and hydrogen gas. Syngas works as a precursor and is a crucial source of chemical substances and fuels that are safe for the environment. It is also an appropriate fuel for the generation of electricity. Methanol or dimethyl ether can also be produced from syngas, with the help of zeolites, and these can be converted into fuel. Syngas is converted into hydrogen using a process known as the chemical reaction known as the water gas shift reaction, and it has a wide range of uses, which might involve fuel cells [59].

5.3.4 Hydrothermal Liquefaction

Hydrothermal liquefaction has gained more attention and is considered the most prominent method for the production of high-quality biofuels [60]. In this method, wet biomass is directly converted into biocrude more quickly and effectively with a high heating value [61]. High pressure and low temperature are applied to break biomass into small fragments in the presence of solvents like water [62]. Hydrothermal liquefaction is a kind of thermochemical conversion in which biomass is converted into liquid biofuel by applying high temperature and pressurized water in order to break solid biopolymer into liquid fuel at 250°C to 400°C temperature and 4- to 22-MPa pressure in the presence of suitable catalyst [63]. Hydrothermal liquefaction (Figure 5.5) has been divided

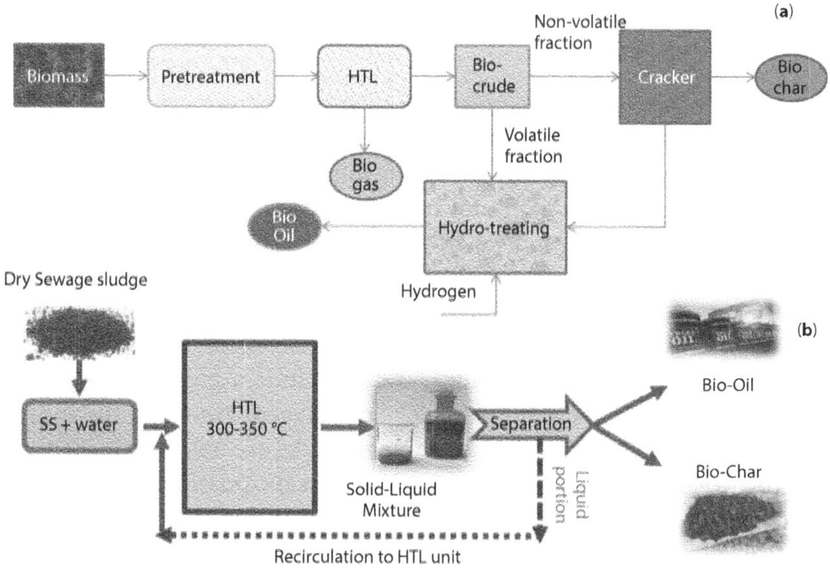

Figure 5.5 Hydrothermal liquefaction (HTL) of biomass [68] and sewage sludge [69, 70].

into three categories depending on working conditions. For instance, at a temperature less than 245°C, hydrothermal liquefaction is known as hydrothermal carbonization and produces hydrochar. Microalgae produce hydrochar from carbohydrates and proteins by hydrothermal carbonization [64]. At the temperature between 245°C and 400°C bio-crude synthesizes by hydrothermal liquefaction, which is further distilled to generate high-quality liquid fuel. When the temperature reaches more than 400°C, hydrothermal gasification starts to produce synthetic gas [65]. Hydrothermal liquefaction has advantages over any other processes in terms of short reaction time, high energy density, and wide raw material applicability [66]. Mainly four types of products are obtained from hydrothermal liquefaction such as bio-oil, biochar, biogas, and aqueous extract. Bio-oil is the most abundant product of hydrothermal liquefaction with 24 to 64% yield. The yield of bio-oil can be further enhanced up to 60% by using a suitable catalyst. Catalysts can boost up the yield by reducing the temperature required for bio-oil synthesis. For instance, in the absence of a catalyst, sorghum bagasse produces 29% bio-oil at high temperatures up to 340°C; on the other hand, by using a catalyst, yield can be enhanced to 39% at temperatures less than 300°C. Catalyst provides an opportunity to enhance the yield and upgrade the quality of bio-oil as compared to non-catalytic liquefaction [67].

5.3.5 Supercritical Wet Oxidation/Supercritical Water Oxidation

A viable approach for sludge disintegration is supercritical water (Figure 5.6) oxidation denoted by SCWO, which occurs at high temperatures (usually 600°C) and pressures (25 MPa). In this process, in organic and biological materials, the carbon and hydrogen atoms undergo oxidation to generate carbon dioxide and water, respectively. Additionally, the nitrogen is transformed into nitrogen gas; sulfur and phosphorus are transformed into their respective oxides; conversion of organic chlorides into chloride and heavy metals into the appropriate oxides also occurs [45]. Special features that characterize supercritical water include a greater capacity for the dissolution of chemical molecules and oxygen. However, the mineral-based molecules, comprising sodium chloride, are not completely soluble. The organic molecules undergo supercritical oxidation, which results in their oxidation in a complete manner.

Nitrogen gas is created when nitrogen, which is present in nitrogen-containing substances like ammonia and amino acids, is broken down. Compared to subcritical preferences, the rate of oxidation is significantly higher under supercritical conditions. It follows that a relatively small reactor size is needed for the combustion of sewage sludge in a supercritical reactor, which requires a retention time of only a few seconds to one minute. Through direct heat exchange within the reactor or from the reactor's exhaust circulation, energy can be recovered from this process of oxidation. The advantage of supercritical oxidation over the incineration method for sludge management is that exhaust gas treatment is fairly simple to comprehend. Ash can be produced by simply removing the inorganic substances from the aqueous phase of processed sludge. Supercritical wet oxidation

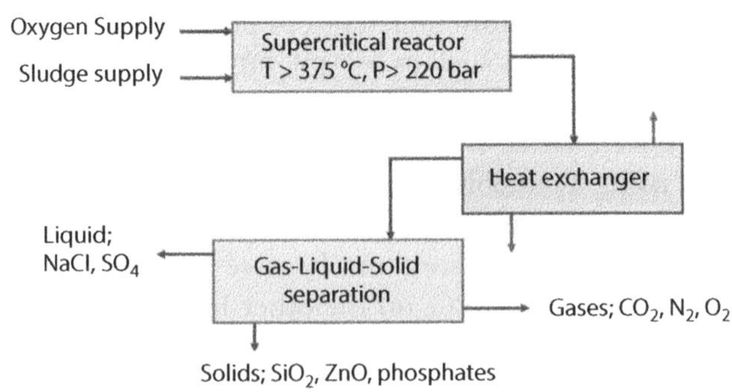

Figure 5.6 Flow sheet of the supercritical/subcritical wet oxidation process.

looks to be economically compatible with conventional sludge processing methods under high temperatures and offers a significant perspective to address the sludge issue in a sustainable manner [1].

5.3.6 Biochemical Conversion Process

The biochemical conversion processes can be defined as the enzymatic hydrolysis of biomass in the presence of microorganisms through aerobic or anaerobic fermentation and digestion, to produce biofuels such as bioethanol and other useful products [71]. In biochemical processes, lignocellulosic biomass is converted into sugar which is further fermented and purified to synthesize cellulosic ethanol. Biochemical processes are highly selective and require low maintenance temperatures [72]. The majority of industrial processes work on biochemical principles, such as the production of pharmaceutical products, enzymes, biofuel, and bulk chemical production [73]. By biochemical conversion, biomass can be converted into useful products in three ways such as biogas production through anaerobic digestion, bioethanol production by anaerobic fermentation, and bio-hydrogen production [74]. Biogas is mainly produced from animal waste by anaerobic digestion in an airtight digester. This produced biogas can be utilized for cooking and heating purposes [73]. Bioethanol and bio-hydrogen are produced through alcoholic and dark fermentation, respectively [75]. Lignocellulosic biomass first hydrolyzes polysaccharides into disaccharides and monosaccharides [76].

Conversion of polysaccharides into monosaccharides from lignocellulosic to produce ethanol, in the presence of enzymes, requires costly operating conditions. The on-spot production of enzymes is the biggest challenge for biorefineries in terms of capital costs [77]. Starch- and sugar-based biomass can be converted into bioethanol and considered as world's largest process to produce 85 million tons of ethanol globally since 2018. Bioethanol as fuel cannot compete with fossil fuel or first-generation biofuel as further development is required. Bioethanol production from the biochemical conversion process is costly due to expensive enzymes and the pre-treatment of biomass [78]. It is suitable when bioethanol is used as raw material for the production of high-value products such as the production of ethylene from ethanol to synthesize polyethylene [79]. The overall procedure for the formation of biogas consists of three major steps including hydrolysis, fermentation, and methanogenesis. First of all, biomass is converted into soluble sugar and then bacteria attack them to produce methane and carbon dioxide [74].

5.3.7 Anaerobic Digestion of Biomass for Biogas Production

Anaerobic digestion is one of the prominent methods for the conversion of biomass into biogas like methane [80]. In this method, complex biomass is digested to produce methane [81] in the presence of microorganisms and the absence of oxygen. It is an extensively used method that can be used to process any type of organic waste for biogas production. This produced biogas can be used to generate heat, electricity, CNG, and LPG [82]. Production of biogas by anaerobic digestion could be beneficial for the environment, as it takes renewable feedstock like energy crops and bio-waste as substrate. It can be helpful to reduce global warming, eutrophication, and acid rain. Methane can be used for the synthesis of other useful substances like ethanol [83]. Anaerobic digestion can take place in four steps including hydrolysis, acidogenic, acetogenesis, and methanogenesis. In the first step, complex biomolecules in biomass such as polysaccharides, lipids, proteins, and carbohydrates are hydrolyzed into their monomers like sugar, fatty acids, and amino acids by bacteria. In the next step, fermentative bacteria convert them to intermediate propionic acid and butyric acid. In the third step, a product of acidogenesis is converted into acetate and the mechanism is known as acetogenesis. In the last step, methanogenesis occurs to produce methane and carbon dioxide from acetate [84]. Biogas production by anaerobic digestion is one of the best alternatives to the traditional method for energy production and helps to manage various kinds of biomass waste including agriculture, food, and animal manure. Biogas production through anaerobic digestion of biomass requires less cost as compared to wind, solar, and hydro [85]. Biogas can be obtained from cellulosic biomass including agricultural wastes, aquatic weeds, plant residues, and other kinds of biomass [86]. Animal manure can be converted into biogas for electricity production and digested substrate can be used as fertilizer as well. The digested substrate after further refining produces fertilizers and fiber products [87]. Microalgae could be a good choice for biomass to produce biofuel due to its eco-friendly and environmental feasibility [88].

5.3.8 Anaerobic Fermentation

Fermentation is a biochemical process for the conversion of biomass waste into useful bio-products such as bioethanol by using microorganisms as shown in Figure 5.7. First of all, complex biomolecules are simplified into simple sugars by microorganisms and then fermented into ethanol and carbon dioxide. Two types of microorganisms take part during the entire

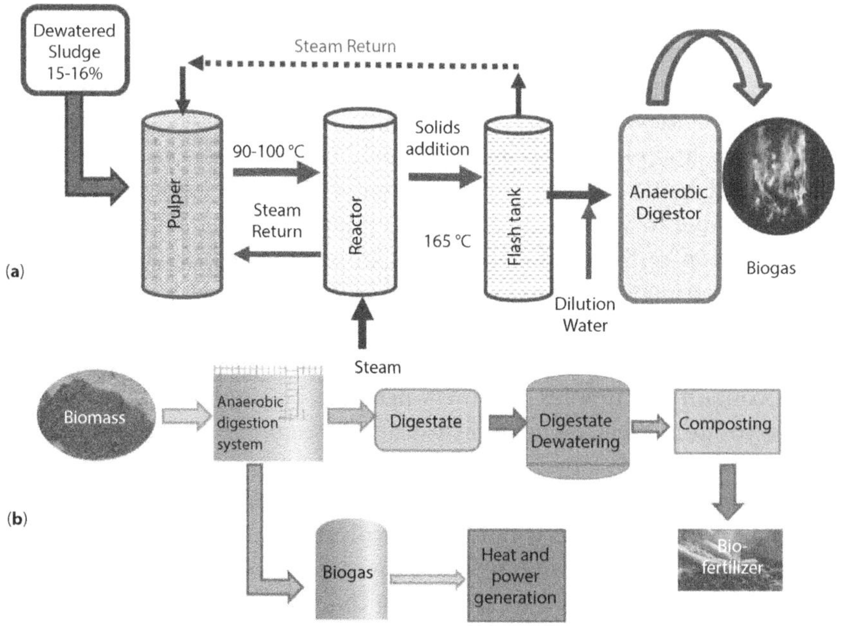

Figure 5.7 Enhanced anaerobic digestion of sewage sludge (a) and biomass (b) [89].

procedure: the first one converts biomass into ethanol and the other one generates enzymes involved in the hydrolysis of complex compounds into simpler ones [90]. Bioethanol production can be produced through first generation or second generation depending on biomass. First-generation bioethanol is mainly produced by the fermentation of carbohydrates such as sugar or starch from biomass like sugarcane, sugar beets, corn, and others. Sugarcane juice containing simple sugar can be extracted through fermentation by using yeast such as *Saccharomyces cerevisiae*. Ethanol produced can be recovered by distillation using downstream processing.

$$(C_6H_{10}O_5)n + nH_2O \rightarrow C_6H_{12}O_6 \rightarrow 2C_2H_5OH + 2CO_2$$

Agricultural biomass having starch can be used as a substrate for ethanol production such as wheat, oats, corn, and potato. The fermentation of starch is complex as compared to the fermentation of sugar. Fermentation of starch takes place at high temperatures (140°C to 180°C) in the presence of α-amylase to hydrolyze the substrate, which is further hydrolyzed into glucose by gluco-amylase. After that, microorganisms are further fermented into ethanol and carbon dioxide. Industrially, ethanol has been produced from starchy materials such as wheat, potatoes, corn, starch, and corn stover [91]. In the case of second-generation bioethanol, lignocellulosic biomass is considered the main substrate, which is renewable and highly abundant. The lignocellulosic biomass consists of cellulose, lignin, and hemicellulose. Cellulosic materials are considered the most abundant due to worldwide availability than any other type of biomass. Fermentation of lignocellulosic occurs in several steps such as crushing and pre-treatment of biomass, hydrolysis, fermentation, and recovery of biofuel. Second-generation bioethanol production is considered more complex than first-generation. In second-generation bioethanol, feedstock containing complex polymers needs to be simplified by extracting carbohydrates. Then, complex hydrocarbons undergo hydrolysis to yield glucose in multiple steps. For instance, hemicellulose consists of branched polymers such as xylose and traces of mannose, galactose, and arabinose and liner cellulose covers with it. To hydrolyze cellulose, it is mandatory to remove this barrier and lignin [92]. However, lignocellulosic biomass can be effectively converted into bioethanol due to seasonal availability, and high transport and storage costs. The enzymatic hydrolysis of cellulosic biomass is the most favorable technology but still has technological issues industrially [93].

5.4 Conclusion and Future Prospects

Tables 5.1 and 5.2 provide a summary of the construction material that can be obtained as a direct byproduct during the physical, chemical, or biological conversion of biomass and sewage sludge into energy and fuels. The biomass for the production of biofuel through a thermochemical process is considered a useful method for the reduction of greenhouse gasses. Besides the production of renewable energy, biomass thermochemical processes such as combustion produce a large amount of ashes. Ash can be defined as an incombustible portion of biomass that is leftover after the burning of biomass for energy production. Every year 7 billion tons of biomass undergoes burning, which generates 475 million tons of ash [8]. Ash produced by thermochemical processes may cause various problems in handling and utilization of ash as direct disposal to soil can alternate the soil nutrients ratio. The dumbing of biomass ash has become an economic and environmental problem [94], so should be managed in an environmentally friendly way to minimize its environmental impact and pollution; as it contains fine particles that can act as filler in the cement mixture.

Table 5.1 Summary of integratable for the conversion of sewage sludge into energy and construction materials.

Raw material	Methods of conversion to energy	Conditions/ requirements	Byproducts/ energy products	Processing of products and byproducts	Use in building material	References
Sewage sludge	Pyrolysis	T = 100°C to 1300°C 0.05-s to 4-hour residence time	Pyrolysis gasses: 2 to 29% Pyrolysis oil: 5 to 20% Biochar: 10 to 80% Ashes	Sludge ash Biochar	Concrete blocks, modifier in cement and asphalt, sediment-based material	[44, 103–111]
	Fast pyrolysis	300°C to 1200°C <20-s residence time No oxygen (anaerobic)	Pyrolysis gas: 15 to 25% Bio-oil: 50 to 70% Biochar: 10 to 30% Ash	Ash and biochar	Bricks, cement, aggregates, fillers in concrete, ceramics, mortar	[103–106, 110]
	Gasification	T = 800°C to 1200°C 10- to 20-s residence time Gasifying agent (air) Partial oxidation	Syngas: >85% Bio-oil: <5% Char: < 10% Tar Fly ash Bottom ash	Char after carbon sequestering Fly ash	In concrete and mortar composite, cement Concrete, bricks, cement mortar	[103, 105–107, 109–115]

(*Continued*)

Table 5.1 Summary of integratable for the conversion of sewage sludge into energy and construction materials. (*Continued*)

Raw material	Methods of conversion to energy	Conditions/ requirements	Byproducts/ energy products	Processing of products and byproducts	Use in building material	References
	Combustion/ incineration	Excess oxygen, T= >1000°C	Flue gasses, slag, fly ashes	Fly ash, Sludge ash residues, slag	Cement, bricks, concrete, clinker, mortar, road pavements	[105, 111–115]
	Supercritical water oxidation	T = 374.2°C P = 22.1 MPa	Gasses, solid residues No ash	N/A	N/A	[45]
	Hydrothermal carbonization	T = 175°C to 295°C 5-min to 12-hour residence time, Saturated pressures	Gasses: 2 to 5% Liquid products: 5 to 25% Hydrochar: 45 to 70%	Hydrochar/ biochar	Cement, concrete, soil amendments,	[108, 110, 116, 117]
	Hydrothermal liquefaction	T = 250°C to 375°C P = 5 to 20 MPa	Gasses Bio-oil/crude oil Biochar	Biochar	Concrete, modifier in cement and asphalt	[106, 109, 113, 118]

Table 5.2 Integrated use of various biomasses for energy production and construction material production.

Type of biomass	Conversion process	Reaction condition	Product obtained	Use of product and byproduct	Construction material	Reference
Rice husk	Pyrolysis	900°C, 10 bar, 20 min	Bio-oil: 30% Biochar: 38% Syngas: 32%	Rice husk ash Rice husk biochar	Zeolites concrete	[119–121]
Wheat straw waste	Combustion	500°C to 750°C, 5 hours	Heat for electricity production	Wheat straw ash	Modified concrete	[122, 123]
Sugarcane bagasse	Fast pyrolysis Fermentation	500°C *Saccharomyces cerevisiae*, 39°C	Bio-oil: 42.5% 4.8 g L^{-1} of ethanol	Sugarcane bagasse ash	Cement	[124–127]
Corncob residue	Fast pyrolysis	(400°C to 700°C) 550°C and 1.46 bar	Bio-oil	Corncob ash	Concrete	[128, 129]
Canola residue	Microwave torrefaction	450 W, 15 min, 306°C	Biochar 76.3% Bio-oil 18.2% Biogas 5.6%	Ash	Concrete	[130]

(Continued)

Table 5.2 Integrated use of various biomasses for energy production and construction material production. (*Continued*)

Type of biomass	Conversion process	Reaction condition	Product obtained	Use of product and byproduct	Construction material	Reference
Conifer wood chips	Hydrothermal carbonization	180°C to 250°C, 3 to 8 hours	Hydrochar 73 to 78%	Ash	Cement	[131, 132]
Switchgrass	Pyrolysis	450°C to 650°C	Bio-oil 70.2% Biochar 19.8% Biogas 9.9%	Ash	N\A	[133]
Forestry waste *pinus pinaster*	Slow pyrolysis	749°C to 900°C 32 to 64 min	Biochar 20 to 30% Bio-oil 6 to 22% Biogas >47%	Biochar	Biochar concrete Biochar bricks	[134, 135]
Pinewood (*Pinus sylvestris*	Catalytic pyrolysis	Catalyst: acidic zeolite Reactor: fluidized bed 400°C	Bio-oil 43.5 to 52.7% Biochar 20 to 30%	Biochar	Composites	[136, 137]

Woody biomass ash can be used for the production of construction materials like cement which is technically possible. Biomass ash is considered a raw material for cement production, an alternative to conventional supplementary cementing materials [95]. The utilization of biomass ash to get cheaper construction materials is a developing field in construction material production. It has advantages over traditional concrete and cement due to low density, low carbon dioxide emission, and thermal insulation but less durability than mortar cement [96]. Ashes can be of two types of ash fly and bottom ash: fly ash escapes through the chimney and is released in the air and, on the other hand, bottom ash, is non-combustible and produced as residues in the bottom of the furnace. Instead of dumping, biomass ash can be utilized as fertilizer, soil improvement, concrete, ceramics, and construction products [97]. In developed countries like Germany, Denmark, Japan, and Netherlands, bottom ash is used for concrete, soil improvement, and soundproof wall production [98]. Agricultural waste ash such as rice husk ash and palm oil fuel ash has gained enough attention to produce cement, due to its low cost and excellent reactivity of ash. Besides economic benefits, inducing biomass ash in concrete produces zero waste and sustainable development [99].

During the processing of paddy rice, husk is produced as a byproduct in mills which is combusted to generate heat that is used as fuel at the same time. Paddy rice production throughout the world has reached up to 678 million tons which produces 149 million tons of rice husks [100]. By combustion of these rice husks, 37 million tons of rice husk ash has been produced that should be managed carefully [101]. Rice husk ash can be used as cement in concrete with more strength and durability [102].

Burning of rice straw and husk by different thermochemical conversion processes produces ash. Besides organic biomass, it also contains inorganic materials like silicon, calcium, magnesium, potassium, phosphorus, aluminum, and iron [138]. Likewise, the rice husk ash composition depends on the amount of these elements [139]. Rice straw ash has high silica content with a wide range of applications including ceramics and construction material production [140]. It can be used as an alternative to cement material with 15 to 20% rice husk ash in concrete, providing more strength than traditional cement concrete [141]. Rice husk contains 83 to 90% silicon that can be used as silicon chips, lightweight construction material production, carbon capturing, drug delivery, and catalyst [142]. Conventional Portland cement production involves high carbon dioxide emissions and more energy consumption, as carbon dioxide emission is 7% throughout the world. Rice husk ash is considered an eco-friendly construction material with less energy consumption and low carbon dioxide emissions.

Rice husk ash is used to produce construction material like cement in an environmentally friendly way by replacing Portland cement as it contains aluminum and silica essential for cement production. Rice husk ash produces concrete and bricks have more durability [143]. Concrete bricks are formed by Portland cement and adobe bricks are produced from organic matter light in weight with more durability. Rice straw ash is considered as the main component for adobe brick manufacturing due to its low thermal conductivity and more water resistance [144].

Biomass waste is the most essential raw material for energy and construction material production by minimizing greenhouse gas emissions. It generates renewable energy as the biomass can be regenerated such as agricultural wastes, animal wastes, algal biomass, and woody biomass. Biomass can undergo thermochemical and biochemical conversion for renewable energy production. Utilization of various thermochemical conversion processes including combustion, pyrolysis, hydrothermal liquefaction, and gasification produces biogas such as methane and hydrogen, and bio-oils as a biofuel. The biochemical conversion process involves the production of bioethanol and biogas. Burning of biomass also produces biomass ash that can cause pollution, so it is used for the production of construction materials with more durability and other useful byproducts.

References

1. Rulkens, W., Sewage Sludge as a Biomass Resource for the Production of Energy: Overview and Assessment of the Various Options. *Energy Fuels*, 22, 9, 2008.
2. Okolie, J.A., Epelle, E.I., Nanda, S., Castello, D., Dalai, A.K., Kozinski, J.A., Modeling and process optimization of hydrothermal gasification for hydrogen production: A comprehensive review. *J. Supercrit. Fluids*, 173, 105199, 2021.
3. Kumar, A., Kumar, N., Baredar, P., Shukla, A., A review on biomass energy resources, potential, conversion and policy in India. *Renew. Sustain. Energy Rev.*, 45, 530, 2015.
4. Vassilev, S.V., Baxter, D., Andersen, L.K., Vassileva, C.G., An overview of the chemical composition of biomass. *Fuel*, 89, 913, 2010.
5. McKendry, P., Energy production from biomass (part 1): overview of biomass. *Bioresour. Technol.*, 83, 37, 2002.
6. Zang, G., Zhang, J., Jia, J., Lora, E.S., Ratner, A., Life cycle assessment of power-generation systems based on biomass integrated gasification combined cycles. *Renew. Energy*, 149, 336, 2020.

7. Sajjakulnukit, B., Yingyuad, R., Maneekhao, V., Pongnarintasut, V., Bhattacharya, S.C., Abdul Salam, P., Assessment of sustainable energy potential of non-plantation biomass resources in Thailand. *Biomass Bioenergy*, 29, 214, 2005.
8. Munawar, M.A., Khoja, A.H., Naqvi, S.R., Mehran, M.T., Hassan, M., Liaquat, R., Dawood, U.F., Challenges and opportunities in biomass ash management and its utilization in novel applications. *Renew. Sustain. Energy Rev.*, 150, 111451, 2021.
9. Hinojosa, M.J.R., Galvín, A.P., Agrela, F., Perianes, M., Barbudo, A., Potential use of biomass bottom ash as alternative construction material: Conflictive chemical parameters according to technical regulations. *Fuel*, 128, 248, 2014.
10. Carrasco-Hurtado, B., Corpas-Iglesias, F.A., Cruz-Pérez, N., Terrados-Cepeda, J., Pérez-Villarejo, L., Addition of bottom ash from biomass in calcium silicate masonry units for use as construction material with thermal insulating properties. *Constr. Build. Mater.*, 52, 155, 2014.
11. da Costa, T.P., Quinteiro, P., Tarelho, L.A.C., Arroja, L., Dias, A.C., Environmental assessment of valorisation alternatives for woody biomass ash in construction materials. *Resour. Conserv. Recycl.*, 148, 67, 2019.
12. Tripathi, N., Hills, C.D., Singh, R.S., Atkinson, C.J., Biomass waste utilisation in low-carbon products: harnessing a major potential resource. *NPJ Clim. Atmos. Sci.*, 2, 35, 2019.
13. Koçar, G.Civaş, N.s,, An overview of biofuels from energy crops: Current status and future prospects. *Renew. Sustain. Energy Rev.*, 28, 900, 2013.
14. Maity, S.K.s, Opportunities, recent trends and challenges of integrated biorefinery: Part I. *Renew. Sustain. Energy Rev.*, 43, 14272015.
15. Zhou, C. and Wang, Y., Recent progress in the conversion of biomass wastes into functional materials for value-added applications. *Sci. Technol. Adv. Mater.*, 21, 787, 2020.
16. Balat, M.s, Biomass Energy and Biochemical Conversion Processing for Fuels and Chemicals. *Energy Sources Part A*, 28, 517, 2006.
17. Rahimi, Z., Anand, A., Gautam, S., An overview on thermochemical conversion and potential evaluation of biofuels derived from agricultural wastes. *Energy Nexus*, 7, 100125, 2022.
18. de Almeida Moreira, B.R., Barbosa Júnior, M.R., de Brito Filho, A.L., da Silva, R.P., Production of high-quality biogenic fuels by co-pelletization of sugarcane bagasse with pinewood sawdust and peanut shell. *Biomass Convers. Biorefin.*, 2022.
19. Atelge, M.R., Krisa, D., Kumar, G., Eskicioglu, C., Nguyen, D.D., Chang, S.W., Atabani, A.E., Al-Muhtaseb, A.H., Unalan, S., Biogas Production from Organic Waste: Recent Progress and Perspectives. *Waste Biomass Valori.*, 11, 1019, 2020.

20. Afazeli, H., Jafari, A., Rafiee, S., Nosrati, M., An investigation of biogas production potential from livestock and slaughterhouse wastes. *Renew. Sustain. Energy Rev.*, 34, 380, 2014.
21. Azadbakht, M., Safieddin Ardebili, S.M., Rahmani, M., Potential for the production of biofuels from agricultural waste, livestock, and slaughterhouse waste in Golestan province, Iran. *Biomass Convers. Biorefin.*, 13, 3123, 2023.
22. Cheng, D., Liu, Y., Ngo, H.H., Guo, W., Chang, S.W., Nguyen, D.D., Zhang, S., Luo, G., Liu, Y., A review on application of enzymatic bioprocesses in animal wastewater and manure treatment. *Bioresour. Technol.*, 313, 123683, 2020.
23. Ravichandran, S.R., Venkatachalam, C.D., Sengottian, M., Sekar, S., Kandasamy, S., Ramasamy Subramanian, K.P., Purushothaman, K., Lavanya Chandrasekaran, A., Narayanan, M., A review on hydrothermal liquefaction of algal biomass on process parameters, purification and applications. *Fuel*, 313, 122679, 2022.
24. Vassilev, S.V., Vassileva, C.G., Composition, properties and challenges of algae biomass for biofuel application: An overview. *Fuel*, 181, 1, 2016.
25. Raheem, A., Sikarwar, V.S., He, J., Dastyar, W., Dionysiou, D.D., Wang, W., Zhao, M., Opportunities and challenges in sustainable treatment and resource reuse of sewage sludge: A review. *Chem. Eng. J.*, 337, 616, 2018.
26. Hu, M., Ye, Z., Zhang, H., Chen, B., Pan, Z., Wang, J., Thermochemical conversion of sewage sludge for energy and resource recovery: technical challenges and prospects. *Environ. Pollut. Bioavailab.*, 33, 145, 2021.
27. Tyagi, V.K., Lo, S.-L., Sludge: A waste or renewable source for energy and resources recovery. *Renew. Sustain. Energy Rev.*, 25, 708, 2013.
28. Mao, H., Zhang, Y., Wang, H., Cui, K., Yu, L., Tan, T., Recycling sewage sludge into ceramic materials: A review. *Environ. Chem. Lett.*, 2023.
29. Damartzis, T., Zabaniotou, A., Thermochemical conversion of biomass to second generation biofuels through integrated process design—A review. *Renew. Sustain. Energy Rev.*, 15, 366, 2011.
30. Zhang, L., Xu, C., Champagne, P., Overview of recent advances in thermo-chemical conversion of biomass. *Energy Convers. Manage*, 51, 969, 2010.
31. Awasthi, M.K., Sar, T., Gowd, S.C., Rajendran, K., Kumar, V., Sarsaiya, S., Li, Y., Sindhu, R., Binod, P., Zhang, Z., Pandey, A., Taherzadeh, M.J., A comprehensive review on thermochemical, and biochemical conversion methods of lignocellulosic biomass into valuable end product. *Fuel*, 342, 127790, 2023.
32. Goyal, H.B., Seal, D., Saxena, R.C., Bio-fuels from thermochemical conversion of renewable resources: A review. *Renew. Sustain. Energy Rev.*, 12, 504, 2008.
33. Ong, H.C., Chen, W.-H., Farooq, A., Gan, Y.Y., Lee, K.T., Ashokkumar, V., Catalytic thermochemical conversion of biomass for biofuel production: A comprehensive review. *Renew. Sustain. Energy Rev.*, 113, 109266, 2019.
34. Das, P., V.P, C., Mathimani, T., Pugazhendhi, A., Recent advances in thermochemical methods for the conversion of algal biomass to energy. *Sci. Total Environ.*, 766, 144608, 2021.

35. Patel, M., Zhang, X., Kumar, A., Techno-economic and life cycle assessment on lignocellulosic biomass thermochemical conversion technologies: A review. *Renew. Sustain. Energy Rev.*, 53, 1486, 2016.
36. Liu, Q., Chmely, S.C., Abdoulmoumine, N., Biomass Treatment Strategies for Thermochemical Conversion. *Energy Fuels*, 31, 3525, 2017.
37. Saravanakumar, A., Vijayakumar, P., Hoang, A.T., Kwon, E.E., Chen, W.-H., Thermochemical conversion of large-size woody biomass for carbon neutrality: Principles, applications, and issues. *Bioresour. Technol.*, 370, 128562, 2023.
38. Liang, S., Yang, L., Chen, H., Yu, W., Tao, S., Yuan, S., Xiao, K., Hu, J., Hou, H., Liu, B., Yang, J., Phosphorus recovery from incinerated sewage sludge ash (ISSA) and reutilization of residues for sludge pretreated by different conditioners. *Resour. Conserv. Recycl.*, 169, 105524, 2021.
39. Liu, Y.-Q., Tay, J.-H., Liu, Y., *Biological Sludge Minimization and Biomaterials/Bioenergy Recovery Technologies*, pp. 405–427, 2012.
40. Syed-Hassan, S.S.A., Wang, Y., Hu, S., Su, S., Xiang, J., Thermochemical processing of sewage sludge to energy and fuel: Fundamentals, challenges and considerations. *Renew. Sustain. Energy Rev.*, 80, 888, 2017.
41. Liang, Y., Xu, D., Feng, P., Hao, B., Guo, Y., Wang, S., Municipal sewage sludge incineration and its air pollution control. *J. Cleaner Prod.*, 295, 126456, 2021.
42. Chen, H., Yan, S.H., Ye, Z.L., Meng, H.J., Zhu, Y.G., Utilization of urban sewage sludge: Chinese perspectives. *Environ. Sci. Pollut. Res.*, 19, 1454, 2012.
43. Solarte-Toro, J.C., González-Aguirre, J.A., Poveda Giraldo, J.A., Cardona Alzate, C.A., Thermochemical processing of woody biomass: A review focused on energy-driven applications and catalytic upgrading. *Renew. Sustain. Energy Rev.*, 136, 110376, 2021.
44. Stolarek, P., Ledakowicz, S., Thermal processing of sewage sludge by drying, pyrolysis, gasification and combustion. *Water Sci. Technol.*, 44, 333, 2001.
45. Tyagi, V.K., Lo, S.L., *Environmental Materials and Waste*, pp. 221–244, Academic Press, 2016.
46. Naqvi, S.R., Tariq, R., Hameed, Z., Ali, I., Naqvi, M., Chen, W.-H., Ceylan, S., Rashid, H., Ahmad, J., Taqvi, S.A., Shahbaz, M., Pyrolysis of high ash sewage sludge: Kinetics and thermodynamic analysis using Coats-Redfern method. *Renew. Energy*, 131, 854, 2019.
47. Chhiti, Y. and Kemiha, M., Thermal conversion of biomass, pyrolysis and gasification. *Int. J. Eng. Sci. (IJES)*, 2, 75, 2013.
48. Jha, S., Okolie, J.A., Nanda, S., Dalai, A.K., A Review of Biomass Resources and Thermochemical Conversion Technologies. *Chem. Eng. Technol.*, 45, 791, 2022.
49. Fahmy, T.Y.A., Fahmy, Y., Mobarak, F., El-Sakhawy, M., Abou-Zeid, R.E., Biomass pyrolysis: past, present, and future. *Environ. Dev. Sustain.*, 22, 17, 2020.
50. Diebold, J. and Bridgwater, A., Overview of fast pyrolysis of biomass for the production of liquid fuels, in: *Developments in Thermochemical Biomass Conversion*, pp. 5–23, Springer, Dordrecht, 1997.

51. Bridgwater, A.V., Review of fast pyrolysis of biomass and product upgrading. *Biomass Bioenergy*, 38, 68, 2012.
52. Ighalo, J.O., Iwuchukwu, F.U., Eyankware, O.E., Iwuozor, K.O., Olotu, K., Bright, O.C., Igwegbe, C.A., Flash pyrolysis of biomass: a review of recent advances. *Clean Technol. Environ. Policy*, 24, 2349, 2022.
53. Balat, M., Gasification of Biomass to Produce Gaseous Products. *Energy Sources Part A*, 31, 516, 2009.
54. Akhtar, A., Krepl, V., Ivanova, T., A Combined Overview of Combustion, Pyrolysis, and Gasification of Biomass. *Energy Fuels*, 32, 7294, 2018.
55. Umenweke, G.C., Afolabi, I.C., Epelle, E.I., Okolie, J.A., Machine learning methods for modeling conventional and hydrothermal gasification of waste biomass: A review. *Bioresour. Technol. Rep.*, 17, 100976, 2022.
56. Migliaccio, R., Brachi, P., Montagnaro, F., Papa, S., Tavano, A., Montesarchio, P., Ruoppolo, G., Urciuolo, M., Sewage Sludge Gasification in a Fluidized Bed: Experimental Investigation and Modeling. *Ind. Eng. Chem. Res.*, 60, 5034, 2021.
57. Oladejo, J., Shi, K., Luo, X., Yang, G., Wu, T., A Review of Sludge-to-Energy Recovery Methods. *Energies*, 12, 60, 2018.
58. Giglio, E., Vitale, G., Lanzini, A., Santarelli, M., Integration between biomass gasification and high-temperature electrolysis for synthetic methane production. *Biomass Bioenergy*, 148, 106017, 2021.
59. Sikarwar, V.S., Zhao, M., Clough, P., Yao, J., Zhong, X., Memon, M.Z., Shah, N., Anthony, E.J., Fennell, P.S., An overview of advances in biomass gasification. *Energy Environ. Sci.*, 9, 2939, 2016.
60. Mishra, R.K., Kumar, V., Kumar, P., Mohanty, K., Hydrothermal liquefaction of biomass for bio-crude production: A review on feedstocks, chemical compositions, operating parameters, reaction kinetics, techno-economic study, and life cycle assessment. *Fuel*, 316, 123377, 2022.
61. Akhtar, J., Amin, N.A.S., A review on process conditions for optimum bio-oil yield in hydrothermal liquefaction of biomass. *Renew. Sustain. Energy Rev.*, 15, 1615, 2011.
62. Ni, J., Qian, L., Wang, Y., Zhang, B., Gu, H., Hu, Y., Wang, Q., A review on fast hydrothermal liquefaction of biomass. *Fuel*, 327, 125135, 2022.
63. Elliott, D.C., Biller, P., Ross, A.B., Schmidt, A.J., Jones, S.B., Hydrothermal liquefaction of biomass: Developments from batch to continuous process. *Bioresour. Technol.*, 178, 147, 2015.
64. Heilmann, S.M., Jader, L.R., Harned, L.A., Sadowsky, M.J., Schendel, F.J., Lefebvre, P.A., von Keitz, M.G., Valentas, K.J., Hydrothermal carbonization of microalgae II. Fatty acid, char, and algal nutrient products. *Appl. Energy*, 88, 3286, 2011.
65. Brown, T.M., Duan, P., Savage, P.E., Hydrothermal Liquefaction and Gasification of Nannochloropsis sp. *Energy Fuels*, 24, 3639, 2010.

66. Xue, Y., Chen, H., Zhao, W., Yang, C., Ma, P., Han, S., A review on the operating conditions of producing bio-oil from hydrothermal liquefaction of biomass. *Int. J. Energy Res.*, 40, 865, 2016.
67. Nagappan, S., Bhosale, R.R., Nguyen, D.D., Chi, N.T.L., Ponnusamy, V.K., Woong, C.S., Kumar, G., Catalytic hydrothermal liquefaction of biomass into bio-oils and other value-added products – A review. *Fuel*, 285, 119053, 2021.
68. Pedersen, T.H., Hansen, N.H., Pérez, O.M., Cabezas, D.E.V., Rosendahl, L.A., Renewable hydrocarbon fuels from hydrothermal liquefaction: A techno-economic analysis. *Biofuels Bioprod. Biorefin.*, 12, 213, 2018.
69. Shah, A.A., Toor, S.S., Seehar, T.H., Nielsen, R.S., H. Nielsen, A., Pedersen, T.H., Rosendahl, L.A., Bio-Crude Production through Aqueous Phase Recycling of Hydrothermal Liquefaction of Sewage Sludge. *Energies*, 13, 493, 2020.
70. Chen, G., Hu, M., Du, G., Tian, S., He, Z., Liu, B., Ma, W., Hydrothermal Liquefaction of Sewage Sludge by Microwave Pretreatment. *Energy Fuels*, 34, 1145, 2020.
71. Williams, C.L., Dahiya, A., Porter, P., *Bioenergy*, pp. 5–36, Academic Press, 2015.
72. Singh, R., Prakash, A., Balagurumurthy, B., Bhaskar, T., *Recent Advances in Thermo-Chemical Conversion of Biomass*, pp. 269–291, Elsevier, 2015.
73. Demirbaş, A., Biomass resource facilities and biomass conversion processing for fuels and chemicals. *Energy Convers. Manage.*, 42, 1357, 2001.
74. Hossain, S.M.Z., Biochemical Conversion of Microalgae Biomass into Biofuel. *Chem. Eng. Technol.*, 42, 2594, 2019.
75. Osman, A.I., Mehta, N., Elgarahy, A.M., Al-Hinai, A., Al-Muhtaseb, A.a.H., Rooney, D.W., Conversion of biomass to biofuels and life cycle assessment: a review. *Environ. Chem. Lett.*, 19, 4075, 2021.
76. Bilal, M., Iqbal, H.M.N., Hu, H., Wang, W., Zhang, X., Metabolic engineering and enzyme-mediated processing: A biotechnological venture towards biofuel production – A review. *Renew. Sustain. Energy Rev.*, 82, 436, 2018.
77. Guo, H., Chang, Y., Lee, D.-J., Enzymatic saccharification of lignocellulosic biorefinery: Research focuses. *Bioresour. Technol.*, 252, 198, 2018.
78. Morschbacker, A., Bio-Ethanol Based Ethylene. *Polym. Rev.*, 49, 79, 2009.
79. Yu, I.K.M., Chen, H., Abeln, F., Auta, H., Fan, J., Budarin, V.L., Clark, J.H., Parsons, S., Chuck, C.J., Zhang, S., Luo, G., Tsang, D.C.W., Chemicals from lignocellulosic biomass: A critical comparison between biochemical, microwave and thermochemical conversion methods. *Crit. Rev. Environ. Sci. Technol.*, 51, 1479, 2021.
80. Bedoić, R., Čuček, L., Ćosić, B., Krajnc, D., Smoljanić, G., Kravanja, Z., Ljubas, D., Pukšec, T., Duić, N., Green biomass to biogas – A study on anaerobic digestion of residue grass. *J. Clean. Prod.*, 213, 700, 2019.

81. Rasapoor, M., Young, B., Brar, R., Sarmah, A., Zhuang, W.Q., Baroutian, S., Recognizing the challenges of anaerobic digestion: Critical steps toward improving biogas generation. *Fuel*, 261, 116497, 2020.
82. Ge, X., Matsumoto, T., Keith, L., Li, Y., Biogas energy production from tropical biomass wastes by anaerobic digestion. *Bioresour. Technol.*, 169, 38, 2014.
83. Chynoweth, D.P., Owens, J.M., Legrand, R., Renewable methane from anaerobic digestion of biomass. *Renew. Energy*, 22, 1, 2001.
84. Kainthola, J., Kalamdhad, A.S., Goud, V.V., A review on enhanced biogas production from anaerobic digestion of lignocellulosic biomass by different enhancement techniques. *Process Biochem.*, 84, 81, 2019.
85. Rao, P.V., Baral, S.S., Dey, R., Mutnuri, S., Biogas generation potential by anaerobic digestion for sustainable energy development in India. *Renew. Sustain. Energy Rev.*, 14, 20862010.
86. Mathew, A.K., Bhui, I., Banerjee, S.N., Goswami, R., Chakraborty, A.K., Shome, A., Balachandran, S., Chaudhury, S., Biogas production from locally available aquatic weeds of Santiniketan through anaerobic digestion. *Clean Technol. Environ. Policy*, 17, 1681, 2015.
87. Holm-Nielsen, J.B., Al Seadi, T., Oleskowicz-Popiel, P., The future of anaerobic digestion and biogas utilization. *Bioresour. Technol.*, 100, 5478, 2009.
88. Ward, A.J., Lewis, D.M., Green, F.B., Anaerobic digestion of algae biomass: A review. *Algal Res.*, 5, 204, 2014.
89. Deena, S.R., Vickram, A.S., Manikandan, S., Subbaiya, R., Karmegam, N., Ravindran, B., Chang, S.W., Awasthi, M.K., Enhanced biogas production from food waste and activated sludge using advanced techniques – A review. *Bioresour. Technol.*, 355, 127234, 2022.
90. Lin, Y., Tanaka, S., Ethanol fermentation from biomass resources: current state and prospects. *Appl. Microbiol. Biotechnol.*, 69, 627, 2006.
91. Kadam, K.L., McMillan, J.D., Availability of corn stover as a sustainable feedstock for bioethanol production. *Bioresour. Technol.*, 88, 17, 2003.
92. Kennes, D., Abubackar, H.N., Diaz, M., Veiga, M.C., Kennes, C., Bioethanol production from biomass: carbohydrate vs syngas fermentation. *J. Chem. Technol. Biotechnol.*, 91, 304, 2016.
93. Yu, Z. and Zhang, H., WITHDRAWN: Ethanol fermentation of acid-hydrolyzed cellulosic pyrolysate with Saccharomyces cerevisiae. *Bioresour. Technol.*, 93, 199, 2004.
94. Melotti, R., Santagata, E., Bassani, M., Salvo, M., Rizzo, S., A preliminary investigation into the physical and chemical properties of biomass ashes used as aggregate fillers for bituminous mixtures. *Waste Manage.*, 33, 1906, 2013.
95. Demis, S., Tapali, J.G., Papadakis, V.G., An investigation of the effectiveness of the utilization of biomass ashes as pozzolanic materials. *Constr. Build. Mater.*, 68, 291, 2014.
96. Vo, L.T.T., Navard, P., Treatments of plant biomass for cementitious building materials – A review. *Constr. Build. Mater.*, 121, 161, 2016.

97. Carrasco, B., Cruz, N., Terrados, J., Corpas, F.A., Pérez, L., An evaluation of bottom ash from plant biomass as a replacement for cement in building blocks. *Fuel*, 118, 272, 2014.
98. Cabrera, M., Galvin, A.P., Agrela, F., Carvajal, M.D., Ayuso, J., Characterisation and technical feasibility of using biomass bottom ash for civil infrastructures. *Constr. Build. Mater.*, 58, 234, 2014.
99. Thomas, B.S., Yang, J., Mo, K.H., Abdalla, J.A., Hawileh, R.A., Ariyachandra, E., Biomass ashes from agricultural wastes as supplementary cementitious materials or aggregate replacement in cement/geopolymer concrete: A comprehensive review. *J. Build. Eng.*, 40, 102332, 2021.
100. Khan, R., Jabbar, A., Ahmad, I., Khan, W., Khan, A.N., Mirza, J., Reduction in environmental problems using rice-husk ash in concrete. *Constr. Build. Mater.*, 30, 360, 2012.
101. Chindaprasirt, P., Rukzon, S., Strength and chloride resistance of the blended Portland cement mortar containing rice husk ash and ground river sand. *Mater. Struct.*, 48, 3771, 2015.
102. Tahami, S.A., Arabani, M., Foroutan Mirhosseini, A., Usage of two biomass ashes as filler in hot mix asphalt. *Constr. Build. Mater.*, 170, 547, 2018.
103. Racek, J., Sevcik, J., Chorazy, T., Kucerik, J., Hlavinek, P., Biochar – Recovery Material from Pyrolysis of Sewage Sludge: A Review. *Waste Biomass Valori.*, 11, 3677, 2020.
104. Fonts, I., Gea, G., Azuara, M., Ábrego, J., Arauzo, J., Sewage sludge pyrolysis for liquid production: A review. *Renew. Sustain. Energy Rev.*, 16, 2781, 2012.
105. Gao, N., Kamran, K., Quan, C., Williams, P.T., Thermochemical conversion of sewage sludge: A critical review. *Prog. Energy Combust. Sci.*, 79, 100843, 2020.
106. Zhang, Y., He, M., Wang, L., Yan, J., Ma, B., Zhu, X., Ok, Y.S., Mechtcherine, V., Tsang, D.C.W., Biochar as construction materials for achieving carbon neutrality. *Biochar*, 4, 59, 2022.
107. Singhal, S., Biochar as a cost-effective and eco-friendly substitute for binder in concrete: a review. *Eur. J. Environ. Civ. Eng.*, 27, 984, 2023.
108. Tay, J.-H., Show, K.-Y., Utilization of municipal wastewater sludge as building and construction materials. *Resour. Conserv. Recycl.*, 6, 191, 1992.
109. Akinyemi, B.A., Adesina, A., Recent advancements in the use of biochar for cementitious applications: A review. *J. Build. Eng.*, 32, 101705, 2020.
110. Gupta, S., Kua, H.W., Low, C.Y., Use of biochar as carbon sequestering additive in cement mortar. *Cem. Concr. Compos.*, 87, 110, 2018.
111. Hamood, A., Khatib, J.M., Williams, C., The effectiveness of using Raw Sewage Sludge (RSS) as a water replacement in cement mortar mixes containing Unprocessed Fly Ash (u-FA). *Constr. Build. Mater.*, 147, 27, 2017.
112. Quan, L.M., Kamyab, H., Yuzir, A., Ashokkumar, V., Hosseini, S.E., Balasubramanian, B., Kirpichnikova, I., Review of the application of gasification and combustion technology and waste-to-energy technologies in sewage sludge treatment. *Fuel*, 316, 123199, 2022.

113. Zaharioiu, A.M., Bucura, F., Ionete, R.E., Marin, F., Constantinescu, M., Oancea, S., Opportunities regarding the use of technologies of energy recovery from sewage sludge. *SN Appl. Sci.*, 3, 775, 2021.
114. Rutkowska, G., Wichowski, P., Fronczyk, J., Franus, M., Chalecki, M., Use of fly ashes from municipal sewage sludge combustion in production of ash concretes. *Constr. Build. Mater.*, 188, 874, 2018.
115. Esmeray, E., Atıs, M., Utilization of sewage sludge, oven slag and fly ash in clay brick production. *Constr. Build. Mater.*, 194, 110, 2019.
116. Wang, L., Chang, Y., Li, A., Hydrothermal carbonization for energy-efficient processing of sewage sludge: A review. *Renew. Sustain. Energy Rev.*, 108, 423, 2019.
117. Tasca, A.L., Puccini, M., Gori, R., Corsi, I., Galletti, A.M.R., Vitolo, S., Hydrothermal carbonization of sewage sludge: A critical analysis of process severity, hydrochar properties and environmental implications. *Waste Manage.*, 93, 1, 2019.
118. Fan, Y., Hornung, U., Dahmen, N., Hydrothermal liquefaction of sewage sludge for biofuel application: A review on fundamentals, current challenges and strategies. *Biomass Bioenergy*, 165, 106570, 2022.
119. Unrean, P., Lai Fui, B.C., Rianawati, E., Acda, M., Comparative techno- economic assessment and environmental impacts of rice husk-to-fuel conversion technologies. *Energy*, 151, 581, 2018.
120. Prasetyoko, D., Ramli, Z., Endud, S., Hamdan, H., Sulikowski, B., Conversion of rice husk ash to zeolite beta. *Waste Manage.*, 26, 1173, 2006.
121. Muthukrishnan, S., Gupta, S., Kua, H.W., Application of rice husk biochar and thermally treated low silica rice husk ash to improve physical properties of cement mortar. *Theor. Appl. Fract. Mech.*, 104, 102376, 2019.
122. Bheel, N., Sohu, S., Awoyera, P., Kumar, A., Abbasi, S.A., Olalusi, O.B., Effect of Wheat Straw Ash on Fresh and Hardened Concrete Reinforced with Jute Fiber. *Adv. Civ. Eng.*, 2021, 6659125, 2021.
123. Shar, I., Fareed, A., Memon, M., Bheel, N., Hussain, Z., Dayo, A., Use of Wheat Straw Ash as Cement Replacement Material in the Concrete. *J. Build. Pathol. Rehabil.*, 6, 4, 2020.
124. Schmitt, C.C., Moreira, R., Neves, R.C., Richter, D., Funke, A., Raffelt, K., Grunwaldt, J.-D., Dahmen, N., From agriculture residue to upgraded product: The thermochemical conversion of sugarcane bagasse for fuel and chemical products. *Fuel Process. Technol.*, 197, 106199, 2020.
125. Jain, A., Wei, Y., Tietje, A., Biochemical conversion of sugarcane bagasse into bioproducts. *Biomass Bioenergy*, 93, 227, 2016.
126. Shabbirahmed, A.M., Haldar, D., Dey, P., Patel, A.K., Singhania, R.R., Dong, C.-D., Purkait, M.K., Sugarcane bagasse into value-added products: a review. *Environ. Sci. Pollut. Res.*, 29, 62785, 2022.

127. Quedou, P.G., Wirquin, E., Bokhoree, C., Sustainable concrete: Potency of sugarcane bagasse ash as a cementitious material in the construction industry. *Case Stud. Constr. Mater.*, 14, e00545, 2021.
128. Demiral, İ., Eryazıcı, A., Sensoz, S., Bio-oil production from pyrolysis of corncob (Zea mays L.). *Biomass Bioenergy*, 36, 43, 2012.
129. Ahmad, J., Arbili, M.M., Alabduljabbar, H., Deifalla, A.F., Concrete made with partially substitution corn cob ash: A review. *Case Stud. Constr. Mater.*, 18, e02100, 2023.
130. Sarker, T.R., Azargohar, R., Dalai, A.K., Meda, V., Enhancement of fuel and physicochemical properties of canola residues via microwave torrefaction. *Energy Rep.*, 7, 6338, 2021.
131. Sermyagina, E., Saari, J., Kaikko, J., Vakkilainen, E., Hydrothermal carbonization of coniferous biomass: Effect of process parameters on mass and energy yields. *J. Anal. Appl. Pyrolysis*, 113, 551, 2015.
132. Gharibi, H., Mostofinejad, D., Teymouri, M., Impacts of Conifer Leaves and Pine Ashes on Concrete Thermal Properties. *Constr. Build. Mater.*, 377, 131144, 2023.
133. Ighalo, J.O., Adeniyi, A.G., Modelling of thermochemical energy recovery processes for switchgrass (Panicum virgatum). *Indian Chem. Eng.*, 63, 240, 2021.
134. Solar, J., de Marco, I., Caballero, B.M., Lopez-Urionabarrenechea, A., Rodriguez, N., Agirre, I., Adrados, A., Influence of temperature and residence time in the pyrolysis of woody biomass waste in a continuous screw reactor. *Biomass Bioenergy*, 95, 416, 2016.
135. Gupta, S., Kua Harn, W., Factors Determining the Potential of Biochar As a Carbon Capturing and Sequestering Construction Material: Critical Review. *J. Mater. Civ. Eng.*, 29, 04017086, 2017.
136. Pang, S., Advances in thermochemical conversion of woody biomass to energy, fuels and chemicals. *Biotechnol. Adv.*, 37, 589, 2019.
137. Das, O., Sarmah, A.K., Bhattacharyya, D., A novel approach in organic waste utilization through biochar addition in wood/polypropylene composites. *Waste Manage.*, 38, 132, 2015.
138. Kaknics, J., Defoort, F., Poirier, J., Inorganic Phase Transformation in Miscanthus Ash. *Energy Fuels*, 29, 6433, 2015.
139. Beidaghy Dizaji, H., Zeng, T., Hölzig, H., Bauer, J., Klöß, G., Enke, D., Ash transformation mechanism during combustion of rice husk and rice straw. *Fuel*, 307, 121768, 2022.
140. Hossain, S.S., Mathur, L., Roy, P.K., Rice husk/rice husk ash as an alternative source of silica in ceramics: A review. *J. Asian Ceram. Soc*, 6, 299, 2018.
141. Shukla, S.S., Chava, R., Appari, S., A, B., Kuncharam, B.V.R., Sustainable use of rice husk for the cleaner production of value-added products. *J. Environ. Chem. Eng.*, 10, 106899, 2022.
142. Pode, R., Potential applications of rice husk ash waste from rice husk biomass power plant. *Renew. Sustain. Energy Rev.*, 53, 1468, 2016.

143. Kumar, N., Kupwade-Patil, K., Higuchi, R., Ferrell, D.P., Luttrull, V.A., Lynam, J.G., Use of Biomass Ash for Development of Engineered Cementitious Binders. *ACS Sustain. Chem. Eng.*, 6, 13122, 2018.
144. Morsy, M.I., Alakeel, K.A., Ahmed, A.E., Abbas, A.M., Omara, A.I., Abdelsalam, N.R., Emaish, H.H., Recycling rice straw ash to produce low thermal conductivity and moisture-resistant geopolymer adobe bricks. *Saudi J. Biol. Sci.*, 293759, 2022.

6

Technical and Economic Evaluation of Old and Novel Technologies for Energy Resources Production from Wastewater Treatment Plants

Shubhankar Mishra, Tohira Banoo, Yogendra Kumar and Subbiah Nagarajan*

Department of Chemistry, National Institute of Technology Warangal (Institute of National Importance), Hanumakonda, Telangana, India

Abstract

Wastewater treatment plants (WWTPs) are potential sources of renewable energy due to the organic substance that is present in wastewater. It is quite interesting to see how this organic material may be turned into energy because it not only produces clean, renewable energy but also lessens the environmental effects of wastewater discharge. In this report, we go over technical and financial assessments of both traditional and cutting-edge energy resource production systems from WWTPs. We talk about the many energy-producing technologies, such as anaerobic digestion, gasification, microbial fuel cells, and hydrothermal liquefaction. We also examine the many economic features of these technologies, such as payback time, operational expenses, and capital costs.

Keywords: Wastewater treatment, energy resource production, microbial fuel cell, anaerobic digestion, sewage treatement

6.1 Introduction

The growing population of humans throughout the world causes havoc on the environment, such as contamination of water, soil, and air. Urbanization and industrialization are caused by the ongoing assessment

**Corresponding author*: snagarajan@nitw.ac.in; snrajannpt@yahoo.co.in

Inamuddin, Tariq Altalhi, Mohammad Luqman and Joseph Kapuku (eds.) Sewage and Biomass from Wastewater to Energy, (151–174) © 2024 Scrivener Publishing LLC

of human socialization, which encourages anthropogenic pollution and negatively impacts ecological stability and climatic change. To fulfill the needs of the growing population, the demand for energy continues to grow day by day at an unexpected rate. However, the resources which we consume from Earth are available in sufficient amount and their extraction and processing can have massive negative impacts on the environment. So, the researchers are trying to explore alternative sources of energy and resources that are sustainable and environmentally friendly [1–6]. One such source is wastewater, which is generated by households, industries, and municipalities and contains a significant amount of organic matter and nutrients that can be recovered and converted into energy and resources [7–10].

A system of circular economy is gaining popularity in cities as a way to manage natural resources sustainably. It says, "Keep products, components, and materials at their highest utility and resources available." So, nowadays, the research area is more focused on organic and nutrient recovery in a more economical way. Urban circular development plans aim to reduce the environmental effect on urban infrastructures while targeting a wide range of resource flows and local resource availability. Water treatment facilities are a crucial part of urban infrastructure [11, 12].

6.2 Wastewater Treatment

Wastewater treatment is the process of removing pollutants and contaminants from sewage or wastewater before it is discharged into the environment or reused. Before being released into the environment or used again, wastewater must be treated to remove toxins and other impurities. Since untreated wastewater can contain dangerous pathogens, chemicals, and other pollutants that can harm aquatic life and public health, wastewater treatment serves to protect both human health and the environment.

The process of treating wastewater involves a number of steps (Figure 6.1), including the following:

Preliminary treatment: During this phase, big materials like sticks, rags, and trash that could harm treatment equipment are removed.

Primary treatment: Using physical and chemical techniques including sedimentation and flocculation, this stage involves the removal of suspended solids, organic debris, and some nutrients from the wastewater.

Secondary treatment: Using biological techniques like activated sludge or trickling filter systems, this stage involves the removal of dissolved and suspended organic debris and nutrients from the wastewater.

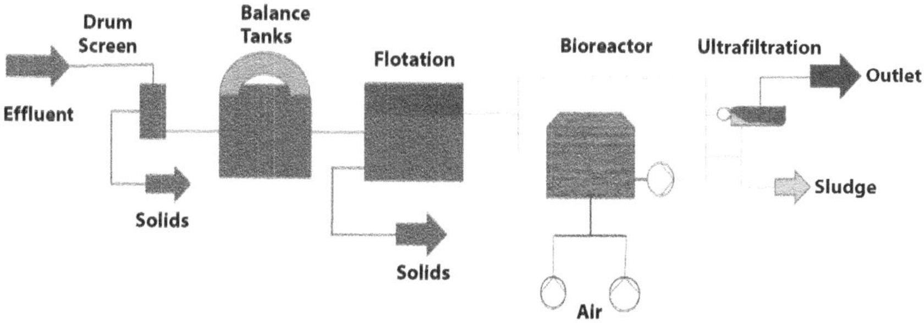

Figure 6.1 A simple overview of the wastewater treatment process in the industry.

Tertiary treatment: In this phase, extra treatment procedures including filtration, disinfection, or chemical precipitation are used to remove any leftover nutrients and other pollutants.

Following treatment, wastewater can be safely released into the environment or recycled for uses other than drinking, like industrial processes or irrigation. The sustainability of water supplies and the preservation of human health and the environment depend on effective wastewater treatment [13–20].

6.3 Technical and Economical Evaluation

There are different technologies that have been used for energy production from WWTPs, including anaerobic digestion (AD), gasification, microbial fuel cells, and hydrothermal liquefaction. These technologies have different technical and economic aspects that need to be evaluated to determine their feasibility and viability [21–30].

6.3.1 Anaerobic Digestion

AD involves the degradation and stabilization of organic materials under anaerobic conditions by microbial organisms and leads to the formation of biogas (a mixture of carbon dioxide and methane, a renewable energy source) and microbial biomass. Anaerobic treatment provides a method of reducing pollution from agricultural and industrial operations while at the same time offsetting the operations' usage of fossil fuels [1]. As one of the most efficient waste and wastewater treatment technologies, AD has been widely used for the treatment of municipal sludge and limited application in the treatment of organic industrial wastes including fruit and vegetable processing wastes, packinghouse wastes, and agricultural wastes.

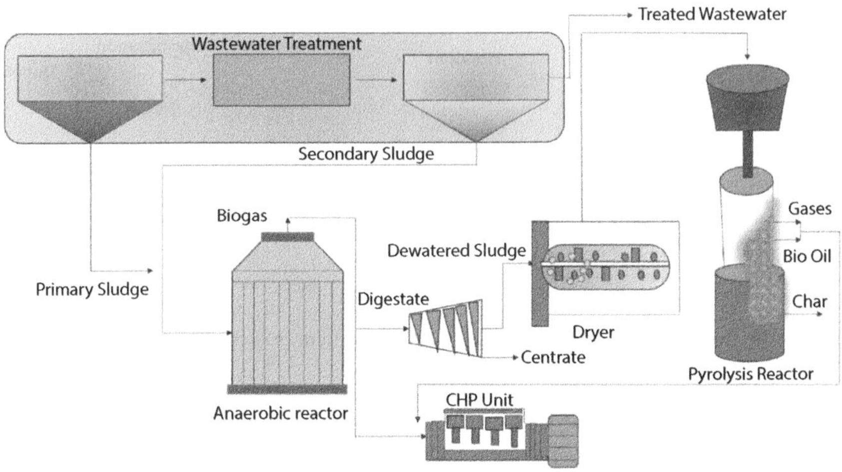

Figure 6.2 Industrial method of bio-oil and gas production from wastewater treatment by anaerobic digestion [2].

The benefits of anaerobic wastewater treatment include lower energy uses compared to aerobic treatment processes, as well as the production of biogas which can be beneficial as a source of renewable energy (Figure 6.2). In addition, the process can lower the amount of sludge produced compared to aerobic treatment processes [2, 31].

However, anaerobic treatment can be more sensitive to changes in influent wastewater quality and requires careful monitoring and management to ensure optimal performance. The process may also be limited by the presence of certain contaminants, such as high levels of heavy metals or toxic organic compounds [32].

6.3.1.1 Anaerobic Digestion Process and Application

For comprehension of the technical constraints of AD chemistry, a comprehensive understanding of the processes underpinning its theory and applications is required.

6.3.1.2 AD Process Chemistry

Hydrolysis, acidogenesis, acetogenesis, and methanogenesis are the four main phases of AD. The feedstock is broken down in the absence of oxygen by a mixture of microorganisms that are present at each stage of the digestion process, producing digestate (decomposed feedstock) and a range of gasses, with CH_4 gas being the main component [1]. The following subsections provide

Table 6.1 The range of the percentage composition of the products of biogas.

Product	(%) of composition
CH_4	51–57
CO_2	26–50
N_2	0–11
H_2	0–2
H_2S	0–4
O_2	0–3

a description of the chemistry of the four essential stages of AD. The main response patterns in these crucial processes are also shown in Table 6.1 [1–3].

6.3.1.3 Hydrolysis Step

From a chemical perspective, hydrolysis describes the disruption of chemical bonds caused by the addition of water. Water molecules with cations and anions interact, changing the pH and leading to the rupturing of H-O bonds. The AD process begins with hydrolysis. The reaction which is associated with this step is given in Equation 6.1:

$$(C_6H_{10}O_5)n + nH_2O \rightarrow nC_6H_{12}O_6 + nH_2 \tag{6.1}$$

The hydrolysis stage produces a variety of products, some of which may be used directly by methanogens (such as H_2 and CH_3COO^-), while others, which are composed of comparatively bigger molecules, are converted into smaller molecules, like acetic acid CH_3COOH (ethanoic acid). At the subsequent stage, where higher chain organic molecules like volatile fatty acids (VFAs) are made, fermentative microbes consume CH_3COO^-, and H_2 generation occurs during the hydrolysis stage [1].

6.3.1.4 Acidogenesis Step

Here is the fermentation step, in which acidogenic bacteria (fermentative microorganisms) break down soluble chemicals produced during the hydrolysis stage to produce CO_2 and H_2.

$$C_6H_{12}O_6 \leftrightarrow C_2H_5OH + 2CO_2 \tag{6.2}$$

$$C_6H_{12}O_6 + 2H_2 \leftrightarrow 2C_2H_5OH + 2H_2O \tag{6.3}$$

$$C_6H_{12}O_6 \rightarrow 3CH_3COOH \tag{6.4}$$

6.3.1.5 *Acetogenesis Step*

Since the H_2 gasses released in the acidogenic stage of the process are the waste product of the acetogenesis step, this stage is also known as the dehydrogenation stage. This is accurate since H_2 gas generated reduces the rate at which acetogenic bacteria metabolize [3]. However, some of the bacteria can be converted to CH_4 by the CH_4-producing bacteria, which behave as hydrogen-scavenging bacteria.

$$CH_3CH_2COO^- + 3H_2O \leftrightarrow CH_3COO^- + H^+HCO_3^- + 3H_2 \tag{6.5}$$

$$C_6H_{12}O_6 + 2H_2O \leftrightarrow 2CH_3COOH + 2CO_2 + 4H_2 \tag{6.6}$$

$$CH_3CH_2OH + 2H_2O \leftrightarrow CH_3COO^- + 3H_2 + H^+ \tag{6.7}$$

6.3.1.6 *Methanogenesis*

The AD process' fourth and last stage is represented by this. Bacteria at this step transform CH_3COOH and H_2 into CO_2 and CH_4. Due to their sluggish growth and sensitivity to environmental changes, methanogens are crucial to AD processes. The simplest substrates are absorbed and digested by them. Methanobrevibacter ruminantium, M. bryantic and M. thermoautotrophicum, Methanogenium cariaci, and M. marinsnigri are only a few of the well-known methanogen species [2].

$$CH_3COOH \rightarrow CH_4 + CO_2 \tag{6.8}$$

$$CO_2 + 4H_2 \rightarrow CH_4 + 2H_2O \tag{6.9}$$

$$2CH_3CH_2OH + CO_2 \rightarrow CH_4 + 2CH_3COOH \tag{6.10}$$

Equation 6.8 demonstrates the conversion of CH_3COOH into CH_4 and CO_2 [1].

6.3.2 Chemical Characterization of Two-Phase Anaerobic Digestion–Produced Municipal Wastewater Sludges for Biogas Generation

In the current study, chemical analysis, stable carbon isotope ratios (^{13}C), HS-SPME-GC-MS, TG-DTA analysis, and DRIFT spectroscopy were used to characterize the chemical properties of municipal wastewater sludges treated in two-phase separate digesters (one for acetogenesis and the other for methanogenesis) [4].

Researchers have really tried to tackle the problem by focusing on a specific bacterial strain or by creating artificial conditions. To give an example, VFA production and utilization in a single reactor used for the entire process must be balanced because an excessive amount of VFA production could cause the reactor to fail. VFA and methane-producing phases have been divided into two stages in two-stage anaerobic processes in order to maximize each stage. In fact, doing so would enable a more thorough examination of the associated chemical process. Each step of sludge decomposition is distinguished by qualitative and quantitative fluctuations in organic carbon [4]. Recent developments have expanded the application of thermogravimetric analysis (TGA) to the examination of sewage sludge. This method makes it possible to track the development of organic matter stabilization throughout composting operations when paired with differential thermal analysis (DTA). The ease of sample preparation is a common benefit of these approaches. Stable carbon isotope analysis is another potent monitoring technique that can be used to identify organic matter transformation. Since organic carbon is transformed, used, and stabilized during AD of wastes, the ^{13}C approach may help us better understand these processes. Controlled laboratory tests are needed to interpret ^{13}C fractionation because a variety of factors affect it [4].

6.3.2.1 Resources and Techniques

6.3.2.1.1 Sludge

The urban wastewater treatment facility in Bologna used a series of stabilization operations, carried out in aerobic and anaerobic conditions, to produce the raw sludge (influent) [4].

6.3.2.1.2 Experimental Approach

Using an elemental analyzer, overall, C and N were determined on lyophilized sludge. The amount of carbon in the sample was estimated using the recognized standard acetanilide, which contains 71.09% carbon.

Continuous flow isotope ratio mass spectrometry was used to measure the C isotopic ratio.

$$\delta^{13}C\% = \frac{R_{sample} - R_{stdPDB}}{R_{stdPDB}} \text{ where R} = \frac{13_C}{12_C}$$

Each analysis was carried out three times. Ninety percent of the time, the samples' 13*C* readings had repeatability better than 0.1. It was 0.1% for the variation coefficient [4].

6.3.2.1.3 Thermal Analysis (TG-DTA)

Strong exothermic reactions above 300°C are often shown to be induced by the thermal degradation of thermolabile organic matter (proteins and carboxyl groups) by thermo-analytical data. While at higher temperatures (over 450°C), exothermic reactions are started by the breakdown of refractory carbon (refractory C), such as aromatic aliphatic chains, rings, and long-chain N-alkyl structures.

In the DTA curve of the acetogenic sludge, two strong exothermic reactions at 293°C (weight loss of 35.56%) and 456°C (weight loss of 17.61%) were actually discernible [4].

6.3.2.1.4 DRIFT Spectroscopy Data

DRIFT spectra provide support for the structural alterations detected by TG-DTA analysis. The OH stretching vibrations in acids and alcohols cause a broad band at around 3300 cm^{-1} to predominate in the spectra. At 3000 and 2800 cm^{-1}, hydrocarbon chains displayed robust bands. The asymmetric C-H stretching of the methyl (2957 cm^{-1} shoulder) and methylene (2923 cm^{-1}) groups, as well as the symmetric C-H stretching of the methylene (2850 cm^{-1}) groups, are credited with causing these bands. The presence of double bonds (CH_2) within the carbon backbone was suggested by the band at 3060 cm^{-1}. Also, the bands between 1450 and 731 cm^{-1} in the fingerprint region were caused by CH_2 scissoring deformation and CH_2 rocking vibrations, respectively. While the peak at around 1551 cm^{-1} was mostly caused by asymmetric stretching of carboxylate groups, the sharp peak at around 1651 cm^{-1} was attributed to the stretching vibrations of the -C O group in ketones and amide I. The bands at 1420 and 1240 cm^{-1} attributed to symmetrical stretching of carboxylate groups and C-O stretching in carboxylic acids, respectively, were also included in the fingerprint region [4, 5]. Moreover, O-H stretching in mineral substances was mostly responsible for the bands between 1170 and 1000 cm^{-1}.

This finding was corroborated by the VFA concentration and DTA analysis. For odd and even carbon numbers, the carbonyl frequency in fatty acids typically rises and falls alternately. In general, the role of acetogenesis in lipid hydrolysis and the ensuing generation of VFA is fundamental. Furthermore, the findings demonstrated that, if kept apart from methanogenesis, VFA synthesis in acetogenesis might be boosted [4, 5]. Methane composition in various feedstock is given in Table 6.2.

6.3.2.2 Economical Evaluation for Anaerobic Digestion

AD has been shown to be economically viable in many cases. The capital costs depend on the size and complexity of the digester, and the operating costs depend on the cost of energy and maintenance. The payback period is typically around 3 to 5 years [6].

Table 6.2 CH_4 percentage of various feedstocks.

Feedstock	***CH_4* composition (%)**
Cattle manure	51–61
Pig manure	61
Poultry waste	69
Sheep dung	64
Horse dung	67
Grass	85
Wheat straw	79.5
Dried leaves	59
Barley straw	78
Beet leaves	85.8
Corn silage	55.5

6.3.3 Microbial Fuel Cells (MFCs)

Microbial fuel cells (MFCs) are a comparatively new technology with the help of bacteria to generate electricity from wastewater. MFCs have the advantage of producing electricity directly from wastewater without the need for intermediate processes. MFCs have been tested at lab scale and their technical feasibility has been demonstrated. However, their efficiency is still low and their scalability and cost-effectiveness need to be evaluated but they can operate continuously without the need for sludge removal.

6.3.3.1 Principles Involving in MFC

A galvanic cell is an MFC. The electrochemical reactions are exergonic, meaning they have a Gibbs free energy that is negative (also known as Gibbs free energy) and spontaneous on their own electron release. As demonstrated in Equation 6.1, the standard Gibbs free energy is simply transformed into a standard EMF [7].

$$\Delta E^0 = \frac{-\left[\sum v_i \Delta G^0{}_{i,products} - \sum v_i \Delta G^0{}_{i,reactans}\right]}{nF} = -\frac{\Delta G}{nF} \tag{6.1}$$

The system's ability to generate electricity is determined by the theoretical cell voltage or electromotive force (emf) of the entire reaction (Equation 6.2) [7].

$$\Delta E^0 = \Delta E^0_{cathode} - \Delta E^0_{anode} \tag{6.2}$$

$$\text{Anode: } CH_3COO^- + 4H_2O \rightarrow 2HCO_3^- + 9H^+ + 8e^-$$

$$(E^O = -0.289V \ vs \ SHE) \tag{6.3}$$

$$\text{Cathode: } 2O_2 + 8H^+ + 8e^- \rightarrow 4H_2O$$

$$(E^O = 0.805 \ vs \ SHE) \tag{6.4}$$

Total: $CH_3COO^- + 2O_2 \rightarrow 2HCO_3^- + H^+$

$$\left(\Delta G = -847.60 \frac{kJ}{mol}\right) \tag{6.5}$$

6.3.3.2 *Oxidation-Reduction Reactions in MFCs*

6.3.3.2.1 Oxidation Reactions at Anode

$$\text{Glucose: } C_6H_{12}O_6 + 12H_2O \rightarrow 6HCO_3^- + 30H^+ + 24e^- \quad E^O = -0.429V \; vs \; SHE \tag{6.6}$$

$$\text{Glycerol: } C_3H_8O_3 + 6H_2O \rightarrow 3HCO_3^- + 17H^+ + 14e^- \quad E^0 = -0.288V \; vs \; SHE \tag{6.7}$$

$$\text{Malate: } C_4H_5O_5^- + 7H_2O \rightarrow 4H_2CO_3 + 11H^+ + 12e^- \quad E^0 = -0.288V \; vs \; SHE \tag{6.8}$$

$$\text{Sulfur}: HS^- \rightarrow S^0 + H^+ + 2e^- \quad E^0 = -0.231V \; vs \; SHE \tag{6.9}$$

6.3.3.2.2 Reduction Reactions (Cathode)

$$O_2 + 4H^+ + 4e^- \rightarrow 2H_2O \quad E^0 = +1.330V \; vs \; SHE \tag{6.10}$$

$$O_2 + 2H^+ + 4e^- \rightarrow 2H_2O_2 \quad E^0 = +0.259V \; vs \; SHE \tag{6.11}$$

$$NO_3^- + 2e^- + 2H^+ \rightarrow NO_2^- + H_2O \quad E^0 = +0.443V \; vs \; SHE \tag{6.12}$$

$$NO_2^- + e^- + 2H^+ \rightarrow NO^- + H_2O \quad E^0 = +0.350V \; vs \; SHE \tag{6.13}$$

$$NO + e^- + H^+ \rightarrow \frac{1}{2}N_2O + \frac{1}{2}H_2O \quad E^0 = +1.175V \text{ vs } SHE \tag{6.14}$$

$$\frac{1}{2}N_2O + e^- + H^+ \rightarrow \frac{1}{2}N_2 + \frac{1}{2}H_2 \quad E^0 = +1.344V \text{ vs } SHE \tag{6.15}$$

$$2NO_3^- + 10e^- + 12H^+ \rightarrow N_2 + 6H_2O \quad E^0 = +0.754V \text{ vs } SHE \tag{6.16}$$

$$Fe^{3+} + e^- + H^+ \rightarrow Fe^{2+} + \frac{1}{2}H_2O \quad E^0 = +0.773V \text{ vs } SHE \tag{6.17}$$

$$Mno_2 + 4H^+ + 3e^- \rightarrow Mn^{+2} + 2H_2O \quad E^0 = +0.602V \text{ vs } SHE \tag{6.18}$$

6.3.3.3 *Removal of Organic Substance in MFCs*

In earlier investigations, the performance of MFCs using synthetic (artificial) wastewater was examined in order to determine the viability of the underlying mechanisms and the enhancement of energy recovery and organic removal efficiency actual wastewater differs significantly from that of wastewater [7].

6.3.3.4 *MFCs with Actual Wastewater Using as Substrates*

Due to the anaerobic conditions in the anode chamber, MFCs are also capable of treating high-strength wastewaters (high energy density) with BOD concentrations reaching 2000 mg/L. These wastewater sources with high strength are produced by the food processing industry, breweries, dairy farms, and animal feeding facilities. Although brewery wastewater concentrations can vary, they are typically between 3000 and 5000 mg/L COD, which is roughly 10 times more concentrated than domestic wastewater. Due to its high carbohydrate content (high energy density) and low ammonium nitrogen concentration, it might possibly be the perfect substrate for MFCs. Researchers explored the use of air cathode MFC for the treatment of wastewater from beer breweries [7, 8].

6.3.3.5 *Biocathode in MFCs*

For several environmental problems including nutrient removal and waste degradation, in particular, biocathodes provide an economical and eco-friendly solution. They do away with the necessity for costly chemical catholytes, as well as the need for their recovery and proper disposal. Algae can act as a simple biocatalyst by providing the electron acceptors needed in the cathode chamber (Figure 6.3).

Algae-based MFCs imitate the syntrophic interaction between bacteria and photosynthetic algae that are frequently seen in natural water bodies, algae release oxygen while converting carbon dioxide into organic matter during the hours of sunshine (Figure 6.4). The organic matter (which includes naturally occurring algae and human waste) is broken down by the benthic heterotrophs into CO_2 and O_2. Algae use the resultant CO_2 to generate organic matter and oxygen [7].

Similar to this, algae are crucial to the cycles of phosphorus and nitrogen in water. Particularly, it has been observed that the electrochemical

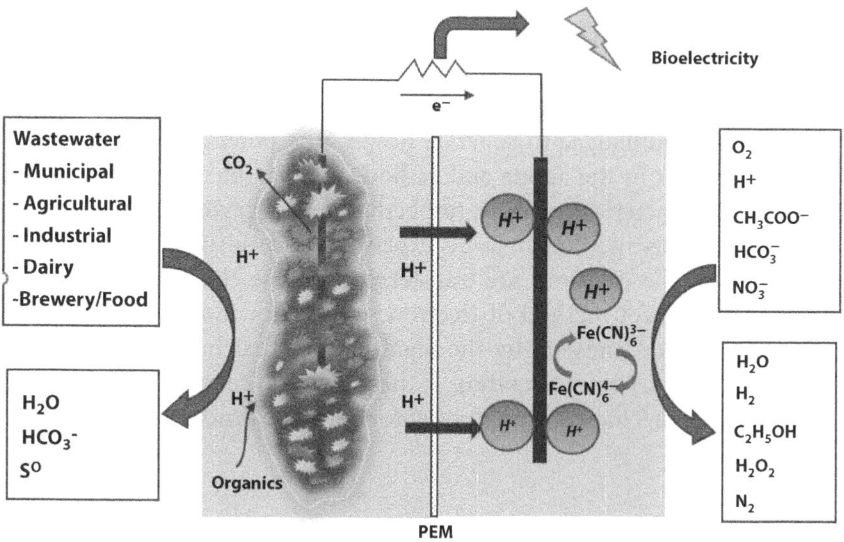

Figure 6.3 The anode chamber of a microbial fuel cell for wastewater treatment with a chemical cathode may be fed with different wastewater sources, and the cathode chamber can be utilized to make valuable chemicals or remove contaminants from the environment [7].

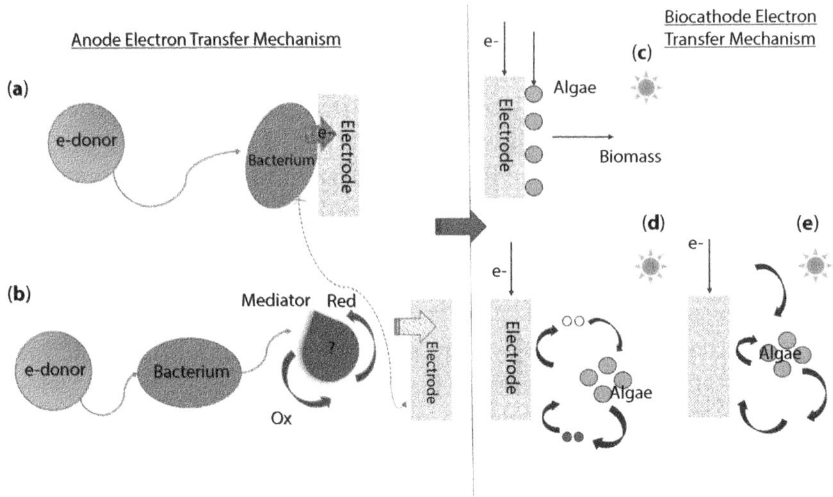

Figure 6.4 (a) Electron flow direction from bacterium to electrode; (b) transfer of electron from metabolic product redox reaction; (c)cathode to algae electron transfer directly; (d) mediator (metabolic products) assisted electron transfer; and (e) Reduction of *in-situ* generated oxygen [8].

processes between microorganisms and algae produce electricity that may be stored in MFCs [7].

It is possible to synchronize the electron transport between exoelectrogenic bacteria and algae to generate power. The potential electron transfer mechanisms in the anode and cathode compartments that have been injected with bacteria and algae, respectively, are depicted in Figure 6.4.

However, this method may be expensive and unsustainable. On the cathode side, the electrons are transferred into the circuit either directly (known as a direct method of electron transfer) or indirectly (known as indirect electron transfer) by the algal biofilm that has adhered to the electrode. A better understanding of these electron release, transport, and acceptance of mechanisms is essential for MFCs to produce more electricity [7, 8].

6.3.3.6 *Nitrogen Removal in MFCs*

The latter allows the bacteria to use the nitrate/nitrite reduction process as an energy source by requiring an electron donor, often organic molecules. When reduced to nitrogen gas, nitrate can receive electrons from organic substances (as in a typical denitrification process), and the amount of electricity needed in an MFC to supply the required reduction power for denitrification can be significantly lowered. More recently, the anodic oxidation of organic carbon utilizing MFCs was combined with cathodic denitrification without intermediate H_2 generation. Nitrogen's impacts on microorganisms, its ability to change pH, and its competition with other elements for electron donors and acceptors can all have an impact on the BES performance, particularly power generation. It was determined that the anode-respiring bacteria's activity had been impeded by a significant amount of free ammonia nitrogen [7].

The denitrification process is hindered by high DO levels, much as high pH circumstances. Processes involving nitrification and denitrification in MFCs work best under neutral pH values. When the cathode is subjected to high pH conditions, aeration, and subsequent acid solution absorption, ammonia recovery is also conceivable [7, 8].

6.3.3.7 *Current Challenges and Potential Opportunities*

The technology of MFCs was explored in the sections above as a viable substitute for environmentally friendly wastewater treatment that also produces electricity. However, in order to achieve this capability and make it actual, practical challenges must be overcome in a number of ways. The primary drawbacks of modern wastewater treatment technologies compared to conventional ones include low power densities, limited pollutant degradation, poor air cathode performance, costs associated with electrode and membrane materials, and environmental (life cycle) effects (Figure 6.5).

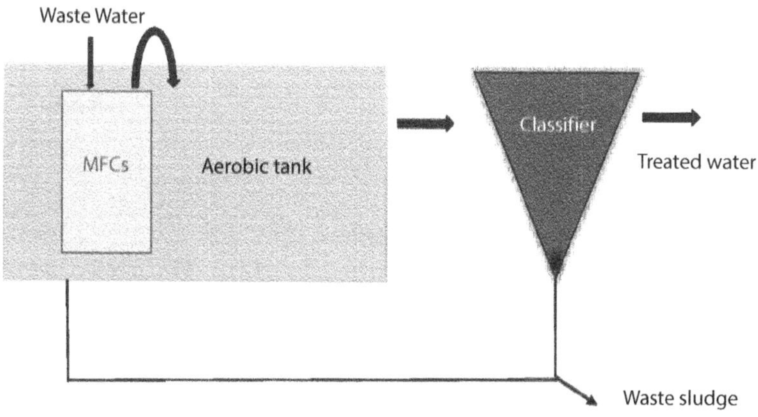

Figure 6.5 MFCs used in large-scale activated sludge systems for centralized wastewater systems.

6.3.3.8 *Technological Advancement in MFCs*

Any old technology should outperform the alternatives now available in comparison. Now, MFCs need to be able to offer improved advantages in addition to energy savings and advanced medical care. Metal removal in MFCs was the subject of many investigations. It is important to assess the removal of additional hazardous and micropollutants originating from radioactive substances, pharmaceuticals, and personal care items.

Depending on the chemistry of the process, the removal mechanisms change from system to system. To accurately examine the true capability and capability of MFCs in eliminating these hazards as well as dependability difficulties, research efforts in this field need to be improved.

6.3.3.9 *Economics and Life Cycle Assessment*

To advertise MFC technology as an alternative to traditional wastewater treatment, cost, and environmental performance goals should be carefully considered. Analysis of MFCs' economic and environmental impacts was covered in a very small number of research studies. According to reports, the economics of such a system for producing electricity are on par with those of other traditional options. It was obvious that the activated sludge system could not generate income. The MFC technique is not cost-effective; however, the AD system proved capable of generating cost-positive revenue. While MECs now appear to be cost-competitive, the MFC alternative is able to cover 50% of the entire cost. The distribution of cost reveals that anode and cathode components make up the majority of the cost, followed by current collector components and other crucial components (Figure 6.6).

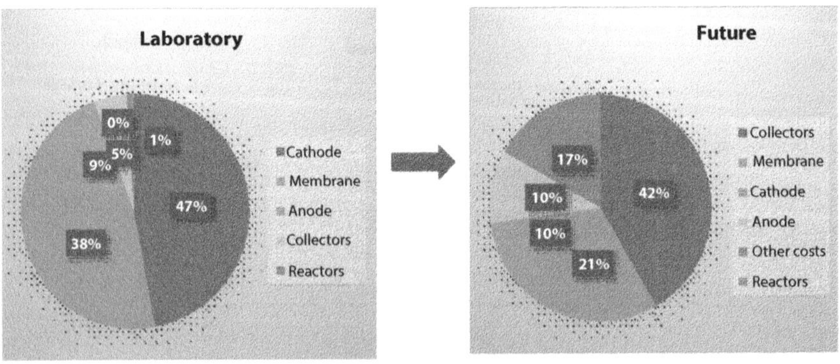

Figure 6.6 MFC cost allocations in present-day laboratory vs. large-scale assessment [8].

That is why methane, biomass rich in lipids, and bioelectricity should all be produced as by-products of energy recovery systems from wastewater. The process' total economics might be greatly enhanced with this cutting-edge plan. Powell and Hill (2010) [33] reported that complementary power production using photobioreactors and yeast fermenters is economically feasible [8].

6.3.4 Hydrothermal Liquefaction

Hydrothermal liquefaction is a promising technology that uses high-pressure and high-temperature conditions to convert organic matter into liquid fuel. HTL has the advantage of producing a liquid fuel that can be easily transported and used as a substitute for fossil fuels. HTL has been tested at a lab scale, and its technical feasibility has been demonstrated (Figure 6.7, Figure 6.8). However, its scalability and cost-effectiveness need to be evaluated. HTP uses significantly less energy because it can directly use water as a medium of reaction and does not require moisture removal before processing [9].

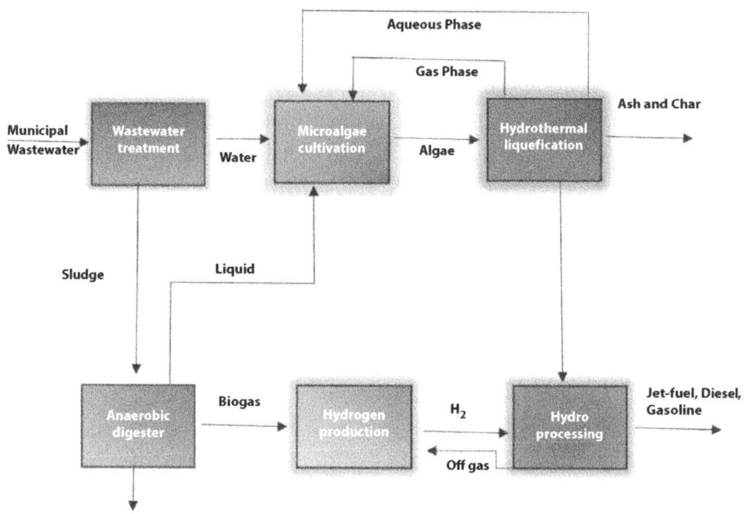

Figure 6.7 Process block diagram of microalgae-based hydrothermal liquefaction [9].

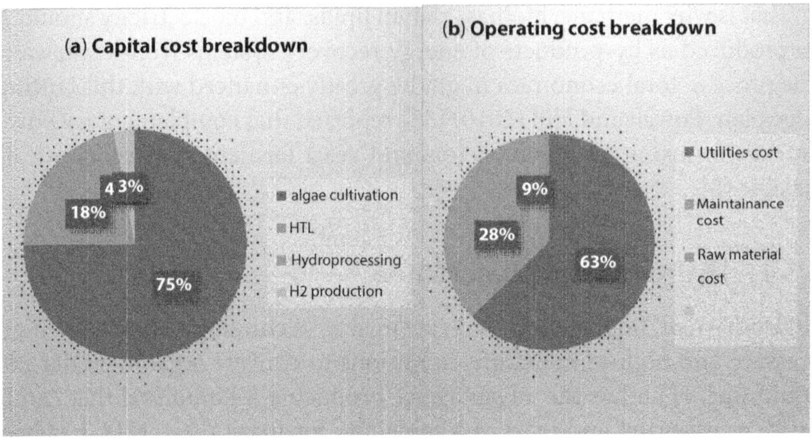

Figure 6.8 Cost breakdown and operating cost analysis of hydrothermal liquefaction process [9].

6.3.4.1 *Requirement and Process*

Very high temperatures (270°C to 480°C) and pressures (10 to 30 MPa), where water is present in the subcritical or supercritical state, are necessary for HTL. The main reactions in HTL are thought to be ionic and free-radical reactions. Due to its low diffusivity, subcritical water interacts with sewage sludge only minimally and primarily serves as a reaction medium. As opposed to this, supercritical water has higher diffusivity and speeds up the rate at which the atmosphere and sewage sludge interact. When the water reaches the supercritical point, the conversion efficiency is greatly improved. In summary, pressure has different effects on subcritical and supercritical water. HTL turns different sewage sludge types into bio-crude oil (Figure 6.7). HTL has additionally been used to remove chemicals of emerging concerns (CECs), dechlorinate, and denitrogenate water. This viewpoint focuses on utilizing HTL to promote environmental health [9, 10].

6.3.4.2 *Evolution of HTL*

Due to the working simplicity, batch-scale tests were the main focus of the publications. There is still little publicly accessible information about the continuous/pilot HTL process. The most recent publications on continuous HTL (mostly from the last two years) are listed in. Temperatures ranging from 300°C to 400°C were used to conduct the continuous reactions listed. Except for some fast HTL processes that were completed in less than 10 min, the time needed for the reaction was typically between 15 and 60 min. The yield of bio-crude oil ranged from 12.1 to 62.6 wt% [9].

Diverse feedstocks have been evaluated under continuous HTL over the past 5 years, although algae accounted for the bulk. The biochemical makeup of the feedstock significantly influenced the oil output. Most oil was produced utilizing a feedstock with a high concentration of lipids, according to a comparison of several feedstocks. For the majority of the feedstocks, including sewage sludge (which resulted in a greater heating value of 28.8 MJ/kg), the greater heating value of the bio-crude oil generated via continuous HTL generally ranged from 34.2 to 40.3 MJ/kg. The two primary methods for managing sewage sludge are biochemical and thermochemical procedures [9, 10].

Sludge from sewage systems was previously disposed of through cremation, the production of biomethane through AD, a free oxygen microbial fermentation of organic materials, could be taken into consideration to deal with the growing amount of sewage sludge. For energy recovery, AD is typically superior to incineration. The most popular technique for stabilizing sewage sludge and reducing volatile solids is AD. However, AD frequently experiences low conversion efficiencies (50%), sluggish response times, and high capital expenditures [9].

6.3.4.3 *Environmental Effects of Treating Sewage Sludge with HTP Technology*

In order to lower the negative impact of HTP on the environment, more research on low-temperature treatment of sludge is needed. For maximizing reaction yields at lower temperatures, a cosolvent mixture for HTP made up of a variety of organic solvents with different solvent-like water ratios can be beneficial. Additionally, a greater comprehension of general chemical pathways and the production of HTP will result from the kinetic research of HTP reactions, enabling the development of more efficient and environment friendly treatment process.

6.3.5 Gasification

Gasification is a process that converts organic matter into a gas by exposing it to high temperatures in the presence of a gasifying agent such as steam or air the resulting gasses can be used to generate electricity or heat. Gasification can be an effective way to produce energy from wastewater but the technology is more complex and expensive than AD or MFCs. These are only means of disposal, and as such, their disposal produces no energy. New possibilities exist for converting sewage sludge into clean fuel with minimal environmental impact through pyrolysis and gasification (Figure 6.9). Many studies have looked at pyrolysis and gasification, and they all imply that if processed in gasification or pyrolysis systems, wastewater sludge has the potential to be used as an energy source [11].

6.3.5.1 *Economical Evaluation in Gasification*

Gasification has higher capital costs than AD, but it can also generate more revenue. The operating costs depend on the cost of energy and maintenance, and the payback period is typically around 5 to 7 years [11].

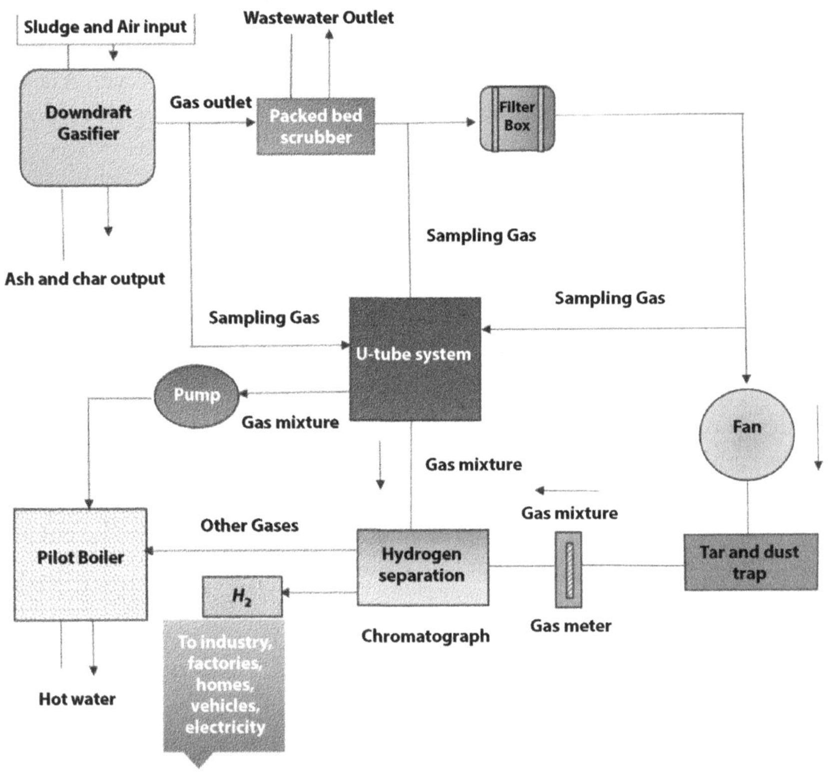

Figure 6.9 Simplified illustration of downdraft gasification and the process flow of hydrogen production [11].

6.4 Conclusion

The technical and financial analysis of traditional and innovative techniques for producing energy resources from wastewater treatment facilities demonstrates their potential as long-term fixes. AD is an outdated technology that has been shown to be successful in recovering energy but may have limits in terms of efficiency and environmental impact. Innovative technologies with the potential to increase energy output and resource extraction, such as MFCs, gasification, and hydrothermal liquefaction present interesting alternatives. These technologies collect rich resources like phosphorus and nitrogen from wastewater by using microbial processes and cutting-edge designs. From an economic perspective, the analysis demonstrates that the upfront investment and ongoing expenses for implementing breakthrough technology may be more than for conventional methods. However, these expenditures may result in long-term cost reductions and enhanced sustainability due to the possibility of higher energy output and resource recovery. It is significant to remember that a technology's appropriateness is dependent on a number of elements, including the size of the wastewater treatment facility, regional laws, and the available resources. Therefore, it is essential to conduct a thorough analysis that takes into account technical viability, environmental impact, and economic viability in order to make informed judgments about the adoption of new or old technologies in the creation of energy resources from wastewater treatment plants.

References

1. Anukam, A., Mohammadi, A., Naqvi, M., Granström, K., A review of the chemistry of anaerobic digestion: Methods of accelerating and optimizing process efficiency. *Processes*, 7, 1, 2019.
2. Pantaleo, A., De Gennaro, B., Shah, N., Assessment of optimal size of anaerobic co-digestion plants: An application to cattle farms in the province of Bari (Italy). *Renew. Sustain. Energy Rev.*, 20, 57, 2013.
3. Montero, A., Tojo, Y., Matsuo, T., Matsuto, T., Yamada, M., Asakura, H., Ono, Y., Gypsum and organic matter distribution in a mixed construction and demolition waste sorting process and their possible removal from outputs. *J. Hazard. Mater.*, 175, 747, 2010.
4. Francioso, O., Rodriguez-Estrada, M.T., Montecchio, D., Salomoni, C., Caputo, A., Palenzona, D., Chemical characterization of municipal wastewater sludges produced by two-phase anaerobic digestion for biogas production. *J. Hazard. Mater.*, 175, 740, 2010.

5. Angenent, L.T., Karim, K., Al-Dahhan, M.H., Wrenn, B.A., Domíguez-Espinosa, R., Production of bioenergy and biochemicals from industrial and agricultural wastewater. *Trends Biotechnol.*, 22, 477, 2004.
6. Dang, Y., Holmes, D.E., Zhao, Z., Woodard, T.L., Zhang, Y., Sun, D., Wang, L.Y., Nevin, K.P., Lovley, D.R., Enhancing anaerobic digestion of complex organic waste with carbon-based conductive materials. *Bioresour. Technol.*, 220, 516, 2016.
7. Gude, V.G., Wastewater treatment in microbial fuel cells - An overview. *J. Clean. Prod.*, 122, 287, 2016.
8. He, L., Du, P., Chen, Y., Lu, H., Cheng, X., Chang, B., Wang, Z., Advances in microbial fuel cells for wastewater treatment. *Renew. Sustain. Energy Rev.*, 71, 388, 2017.
9. Zhou, Y., Schideman, L., Yu, G., Zhang, Y., A synergistic combination of algal wastewater treatment and hydrothermal biofuel production maximized by nutrient and carbon recycling. *Energy Environ. Sci.*, 6, 3765, 2013.
10. Zhuang, L., Zheng, Y., Zhou, S., Yuan, Y., Yuan, H., Chen, Y., Scalable microbial fuel cell (MFC) stack for continuous real wastewater treatment. *Bioresour. Technol.*, 106, 82, 2012.
11. Nipattummakul, N., Ahmed, I., Kerdsuwan, S., Gupta, A.K., High temperature steam gasification of wastewater sludge. *Appl. Energy.*, 87, 3729, 2010.
12. Chen, W.T., Haque, M.A., Lu, T., Aierzhati, A., Reimonn, G., A perspective on hydrothermal processing of sewage sludge. *Curr. Opin. Environ. Sci. Health*, 14, 63, 2020.
13. Abourached, C., Catal, T., Liu, H., Efficacy of single-chamber microbial fuel cells for removal of cadmium and zinc with simultaneous electricity production. *Water Res.*, 51, 228, 2014.
14. Hernández-Fernández, F.J., Pérez De Los Ríos, A., Salar-García, M.J., Ortiz-Martínez, V.M., Lozano-Blanco, L.J., Godínez, C., Tomás-Alonso, F., Quesada-Medina, J., Recent progress and perspectives in microbial fuel cells for bioenergy generation and wastewater treatment. *Fuel Process. Technol.*, 138, 284, 2015.
15. Salar-García, M.J., Ortiz-Martínez, V.M., Baicha, Z., de los Ríos, A.P., Hernández-Fernández, F.J., Scaled-up continuous up-flow microbial fuel cell based on novel embedded ionic liquid-type membrane-cathode assembly. *Energy*, 101, 113, 2016.
16. Hu, M., Gao, L., Chen, Z., Ma, C., Zhou, Y., Chen, J., Ma, S., Laghari, M., Xiao, B., Zhang, B., Guo, D., Syngas production by catalytic in-situ steam co-gasification of wet sewage sludge and pine sawdust. *Energy Convers. Manage.*, 111, 409, 2016.
17. Paolini, V., Petracchini, F., Carnevale, M., Gallucci, F., Perilli, M., Esposito, G., Segreto, M., Occulti, L.G., Scaglione, D., Ianniello, A., Frattoni, M., Characterisation and cleaning of biogas from sewage sludge for biomethane production. *J. Environ. Manage.*, 217, 288, 2018.

18. Akbaş, H., Bilgen, B., Turhan, A.M., An integrated prediction and optimization model of biogas production system at a wastewater treatment facility. *Bioresour. Technol.*, 196, 566, 2015.
19. Chen, Y.C. and Liu, H.M., Evaluation of greenhouse gas emissions and the feed-in tariff system of waste-to-energy facilities using a system dynamics model. *Sci. Total Environ.*, 792, 148445, 2021.
20. Zhao, Z., Zhang, Y., Quan, X., Zhao, H., Evaluation on direct interspecies electron transfer in anaerobic sludge digestion of microbial electrolysis cell. *Bioresour. Technol.*, 200, 235, 2016.
21. Luo, G., Wang, W., Angelidaki, I., A new degassing membrane coupled upflow anaerobic sludge blanket (UASB) reactor to achieve in-situ biogas upgrading and recovery of dissolved CH4 from the anaerobic effluent. *Appl. Energy*, 132, 536, 2014.
22. Jiménez, A., Aller, M., Martín, R., Pérez-Martín, M., Pérez-Fígares, J., Fernández-López, A., Fernández-Llebrez, P., The subcommissural organ of the frog Rana perezi is innervated by nerve fibres containing GABA. *Cell Tissue Res.*, 299, 253, 2000.
23. Ou, S., Kashima, H., Aaron, D.S., Regan, J.M., Mench, M.M., Multi-variable mathematical models for the air-cathode microbial fuel cell system. *J. Power Sources*, 314, 49, 2016.
24. Mei, Z., Chen, D., Zhang, J., Yin, L., Huang, Z., Xin, Q., Sewage sludge pyrolysis coupled with self-supplied steam reforming for high quality syngas production and the influence of initial moisture content. *Waste Manage.*, 106, 77, 2020.
25. Karimi Alavijeh, M., Mardanpour, M.M., Yaghmaei, S., One-dimensional Conduction-based Modeling of Bioenergy Production in a Microbial Fuel Cell Engaged with Multi-population Biocatalysts. *Electrochim. Acta*, 184, 151, 2015.
26. Zhao, Z., Zhang, Y., Wang, L. *et al.* Potential for direct interspecies electron transfer in an electric-anaerobic system to increase methane production from sludge digestion. *Sci Rep* 5, 11094, 2015. https://doi.org/10.1038/srep11094
27. Hauck, M., Maalcke-Luesken, F.A., Jetten, M.S.M., Huijbregts, M.A.J., Removing nitrogen from wastewater with side stream anammox: What are the trade-offs between environmental impacts? *Resour. Conserv. Recycl.*, 107, 212, 2016.
28. Lin, Y., Guo, M., Shah, N., Stuckey, D.C., Economic and environmental evaluation of nitrogen removal and recovery methods from wastewater. *Bioresour. Technol.*, 215, 227, 2016.
29. Smol, M., Adam, C., Preisner, M., Circular economy model framework in the European water and wastewater sector. *J. Mater. Cycles Waste Manage.*, 22, 682, 2020.
30. Roots, P., Sabba, F., Rosenthal, A.F., Wang, Y., Yuan, Q., Rieger, L., Yang, F., Kozak, J.A., Zhang, H., Wells, G.F., Integrated shortcut nitrogen and

biological phosphorus removal from mainstream wastewater: Process operation and modeling. *Environ. Sci. Water Res. Technol.*, 6, 566, 2020.

31. Juan-Díaz, X., Olmo, L., Pérez, J., Carrera, J., Coupling anammox and heterotrophic denitrification activity at mainstream conditions in a single reactor unit. *Chem. Eng. J.*, 431, 134087, 2022.
32. Moško, J., Jeremiáš, M., Skoblia, S., Beňo, Z., Sikarwar, V.S., Hušek, M., Wang, H., Pohořelý, M., Residual moisture in the sewage sludge feed significantly affects the pyrolysis process: Simulation of continuous process in a batch reactor. *J. Anal. Appl. Pyrolysis*, 161, 105387, 2022.
33. Powell, E.E., Hill, G.A., Carbon dioxide neutral, integrated biofuel facility. *Energy*, 35, 12, 2010. Pages 4582-4586, ISSN 0360-5442, https://doi.org/10.1016/j.energy.2010.02.029.

7

Recovery of Cellulose and Extracellular Polymers from Sewage Sludge

Muthumari Perumal[1], Varalakshmi Varatharajan[1], V. Karthik[2], Selvakumar Periyasamy[3,4] and Beula Isabel J.[5]*

[1]Department of Biotechnology, P.S.R Engineering College, Sivakasi, Tamil Nadu, India
[2]Department of Industrial Biotechnology, Government College of Technology, Coimbatore, Tamil Nadu, India
[3]Department of Chemical Engineering, School of Mechanical, Chemical and Materials Engineering, Adama Science and Technology University, Adama, Ethiopia
[4]Department of Biomaterials, Saveetha Dental College and Hospitals, SIMATS, Saveetha University, Chennai, India
[5]Department of Energy and Environment, National Institute of Technology, Tiruchirappalli, Tamil Nadu, India

Abstract

The disposal of sewage sludge poses significant environmental challenges due to its high organic content and potential for releasing harmful substances into the ecosystem. This study focuses on recovering valuable cellulose and extracellular polymers from sewage sludge, aiming to explore sustainable solutions for sludge management and resource utilization. The chapter employs physical, chemical, and biological methods to extract cellulose and extracellular polymers from the sludge matrix. Initially, the sludge is subjected to mechanical pretreatments, such as size reduction and solid-liquid separation, to enhance the accessibility of target polymers. Subsequently, chemical treatments involving alkaline and acidic hydrolysis selectively dissolve cellulose and extract extracellular polymers. The recovered cellulose and extracellular polymers are characterized using various analytical techniques, including Fourier transform infrared spectroscopy (FTIR), scanning electron microscopy (SEM), and X-ray diffraction (XRD). These analyses provide insights into the extracted polymer structure, composition, and purity.

**Corresponding author*: beulaisabelbiotech@gmail.com

Inamuddin, Tariq Altalhi, Mohammad Luqman and Joseph Kapuku (eds.) Sewage and Biomass from Wastewater to Energy, (175–194) © 2024 Scrivener Publishing LLC

The potential applications of the recovered cellulose and extracellular polymers are investigated, considering their biodegradability, biocompatibility, and physicochemical properties. Cellulose can be a renewable feedstock for producing biofuels, bioplastics, and high-value chemicals. At the same time, extracellular polymers hold promise in areas like bioremediation, pharmaceuticals, and materials science. Moreover, the environmental impact of the recovery process is evaluated, encompassing energy consumption, waste generation, and emissions. Life cycle assessments (LCAs) are conducted to determine the overall sustainability of the proposed approach, considering both economic and environmental aspects. These contribute to developing sustainable strategies for sewage sludge management, providing a valuable alternative to traditional disposal methods. The recovery of cellulose and extracellular polymers not only minimizes waste generation but also harnesses the potential of these biomaterials, fostering the transition toward a circular economy and reducing reliance on fossil-based resources.

Keywords: Sewage sludge, cellulose recovery, extracellular polymers, sustainable resource utilization, biodegradability, circular economy

7.1 Introduction

The demand for freshwater is increasing day by day due to the formation of wastewater from various sources such as domestic, municipal, agricultural, medical, and industrial wastes. To meet the water scarcity, the treatment of waste reclamation is needed [1–3]. The primary municipal waste sources comprise toxic elements, microbes, and metal ions. During municipal or industrial wastewater treatment, semi-solid residues were formed as byproducts referred to as sewage sludge [4, 5]. In most countries, various agencies were created to manage sewage sludge from municipal or industrial wastewater. The agencies allocated form permissible limits for sewage disposal. Due to these acceptable sludge limits, it is disposed into landfills, which may minimize the environmental crisis [6]. Recently the disposal of sewage sludge comprises 14% landfills, 42% agri applications, 27% incineration, and the remaining 17% utilized by the industrial sector for energy recovery. It mainly consists of heterogeneous mixtures of toxic compounds, organic and inorganic matter, and macro and microelements [7, 8].

Due to toxic compounds, the rapid release of sewage sludge into the environment is an emerging threat and human concern. The risk factors in wastewater are heavy metals (HMs), organic matter, and nonbiodegradable compounds. Accumulation of these HMs may affect plant growth in agricultural conditions [9–11]. According to the World Health Organization, the guidelines for using wastewater for human health protection were adopted in 1973. The treatment of this sludge depends on efficient disposal and waste valorization. The poor disposal management of sewage sludge causes toxic effects on water quality and environmental issues. Municipal wastewater treatment is vital in wastewater reuse and human health protection [12–14]. The disposal of sewage sludge depends on physical, chemical, and biological methods, which in turn causes severe environmental effects. Biological sludge disposal depends on the microbial sources under aerobic and anaerobic conditions. Among them, anaerobic digestion is the most effective method for sludge disposal and recovering the value-added products from sewage sludge. This method utilizes higher-energy sewage sludge waste to retrieve the energy through methane production [15, 16].

Pyrolysis is a thermochemical process that may be used to use wastewater sludge further. This pyrolysis removes harmful microorganisms and produces goods with added value. This technology has an opportunity to improve the environment by mitigating climate change and minimizing ecological problems related to partially solid disposal [17]. The emerging sewage sludge treatment methods are geared toward enhancing the value of sludge wastes. Waste sewage sludge constitutes many compounds, namely, organic and inorganic matter, carbohydrates, HMs, polymeric substances, and minerals [18, 19]. Among these, cellulose and extracellular polymers are helpful in sewage sludge disposal. The recovered cellulose from the sewage sludge is a renewable source that produces chemical products, feedstuffs, and industrial goods with or without biological processing. However, there are few possibilities to convert the wastewater into value-added products. Still, improvement is needed to retrieve and reuse the compounds present in wastewater. There was a lack of polymer recovery, mainly cellulose and extracellular polymeric substances (EPSs) from the sewage sludge. Primarily phosphorus, magnesium, and ammonium ions were quickly recovered from the sewage sludge. The recovery of polymeric substances from the sewage sludge is essential to avoid the formation of microplastics.

7.2 Stages of Wastewater Treatment

The municipal sewage sludge contains many HMs, organic materials, toxic elements, and pathogenic and nonpathogenic microbes. Releasing this sludge wastewater into the river may cause severe health issues for humans and the environment. Hence, an attempt is made to treat the wastewater before it is released into the atmosphere. The wastewater treatment plant (WWTP) achieved a potential method that collects wastewater from various sources such as industrial, municipal, household, and pharmaceutical wastes, as represented in Figure 7.1. The raw sludge of sewage comprises a

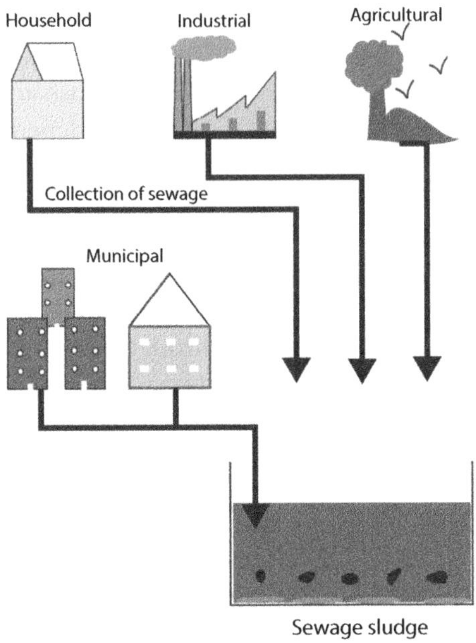

Figure 7.1 Collection of sludge from various sources.

large quantity of organic matter which is usually exposed to COD (chemical oxygen demand), BOD (biological oxygen demand), DOC (dissolved organic carbon), and some traces of contaminants—primary, secondary, and tertiary treatments performed for the municipal sewage sludge.

Because the first step in wastewater treatment is the primary treatment, it is often called a physical process. The sludge contains garbage, grit, sand, colloidal, and solid particles. The screening process is to remove the solid particles in the sludge. Then, the sand and gravel particles were settled using a grit tank. Solid-liquid separation was carried out by mechanical process [20]. After that, the suspended organic solid particles were determined through a clarifier to quickly improve the sludge water flow in pumps. This primary treatment minimizes the wastewater's BOD. Furthermore, the immediate treatment is enhanced through dissolved air flotation (DAF). The DAF acts as a potential clarifier to produce microbubbles that attach the suspended solids, making the particles float on the water surface through air compression. This DAF is a more efficient process than the sedimentation process. But then, air compression requires more power than conventional clarifiers, making this process expensive [21].

The next secondary step is to remove the dissolved organic matter which could not be eliminated by primary treatment. Aerobic and anaerobic technologies carry out this secondary treatment to reduce COD and release nitrogen and phosphorus. The secondary treatment process comprises microbial aerobic digestion (MAD), anaerobic digestion (AD), aerated activated sludge (AAS), and trickling filters. The anaerobic treatment of sewage sludge produces biogas and many value-added products due to the organic matter breakdown via microorganisms [22]. The well-known sewage sludge treatment is MAD which reduces the COD and releases odorless water without microbial hazards. This treatment occurs when dissolved oxygen is injected into the open sludge tank to oxidize the organic matter, thereby minimizing COD [23]. Whereas in AAS, the sludge from the primary level was passed into the aerated bioreactor tank with compressed air, resulting in low COD. Most activated sludge is subjected to several treatment procedures, including AD, MAD, dewatering, burning, and reusing as value-added products [24]. Among them, MAD potential reduces the BOD, and volatile solids, is user-friendly, and is less expensive than other treatments. However, the significant bottlenecks of MAD needed to include value-added products from the sludge [24].

Furthermore, the suspended solids and BOD can be eliminated through biological processes. The sewage treatment of the trickling filter aerobically decomposes the organic matter and reduces the BOD. After secondary treatment, some of the substances or contaminants cannot be eliminated, which undergo tertiary treatment. This tertiary process eliminates nearly 99% of impurities from the wastewater. This step aims to enhance the water quality by removing bacteria and other pathogens to properly use wastewater for agriculture, street cleaning, or discharge to water sources [25].

Some other wastewater treatment involves adsorption, advanced oxidation process, activated carbon, ozonation, reverse osmosis, UV irradiation, and filtration methods [26]. In the adsorption process, fly ash acts as an adsorbent to remove organic material and toxic compounds in wastewater treatment. Pollutants such as phenols and aromatic compounds can be degraded by the ozonation process [27, 28]. Ozone gas (O_3) is used in ozonation to disinfect, decolorize, and sterilize materials. Ozonation removed 90% of the trace organics more effectively than the trickling filter, although it was still inferior to granular activated carbon. This method is not suited for longer duration because ozone has decomposed in 20 min a short and more expensive [29].

Activated carbon was used in the physical way of treatment, a true technique for removing organic pollutants. This activated carbon will be available in the form of powder or granules. The subsequent filters are filled with activated carbon in particulate form to eliminate cloudiness. On the contrary, powder-activated carbon is introduced into a clarifier since it possesses the maximum adsorption area owing to the small particles and is, therefore, more potent than granular carbon. Additionally, compared with ozonization, activated carbon efficiently eliminated an extensive range of organic waste [30]. Recently eco-friendly sewage treatment has been initiated to remove the phenolic compounds, COD, and turbidity from sewage water. In this process, plants and microbes are potential sources for eliminating COD, turbidity, and toxic compounds. However, it has some demerits, such as high operating and maintenance costs, suitable only for bench-scale processes [31, 32].

7.3 Occurrence of Cellulose and Extracellular Polymeric Substances from Sewage Sludge

Growing industrialization releases enormous amounts of sewage sludge, potentially threatening the environment. Large quantities of wastewater are generated from various industries, mainly the paper and pulp textile industry. This industrial wastewater constitutes a rapid amount of organic and inorganic compounds. This sludge was investigated to meet the regulatory standards and subjected to various treatments to recover value-added products before being released into the environment, as indicated in Figure 7.2. On the other hand, the valorization of the sewage sludge biomass is economically beneficial in sustainable waste management. This valorization's central attractive feature is converting sewage sludge into value-added compounds [33]. The major organic component in municipal sewage sludge is cellulose, the most recoverable product. Recovery of cellulose from sludge waste is essential in sewage management. Various techniques were involved in dissolving and utilizing this cellulose in many applications [34].

Some of the techniques are thermal treatment, autoclaving, and ultrasonication. These above-specialized treatments can reduce the volume of sludge. This cellulose recovery depends on plant and microbial sources. One of the bottlenecks of plant cellulose recovery contains two more polymers: hemicellulose and lignin, limiting cellulose properties [35]. On the other hand, hemicellulose has less economic value than cellulose and other organic matter. Hemicellulose and lignin also affect cellulose extraction since they form strong hydrogen bonds with cellulose. The decomposition of hemicellulose and lignin was performed by acid hydrolysis, enzymatic treatment, and steam explosion, resulting in cellulose alone. The cellulose is recovered from sewage sludge in the paper and pulp industry through conventional methods such as kraft, organosolv, and viscose [36].

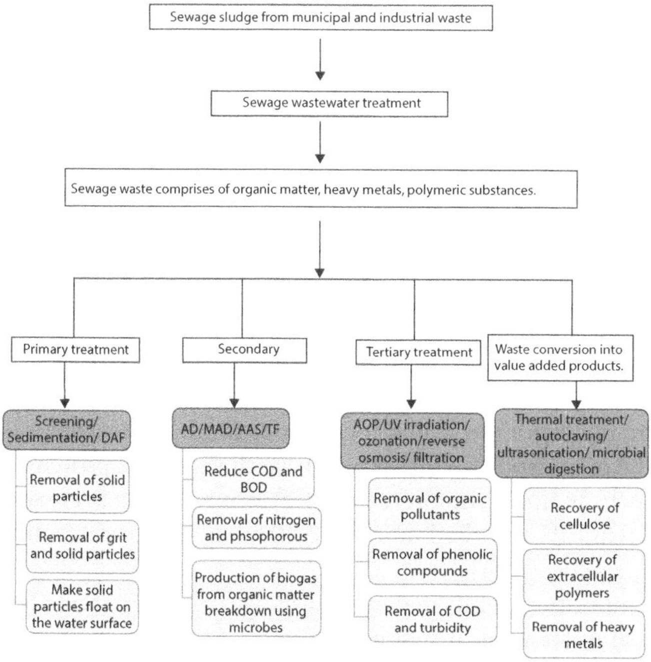

Figure 7.2 Steps in sewage wastewater treatment.

These processes performed well but have some bottlenecks and toxic health effects due to the harmful solvents, pungent smell, and pollutants. In addition, it cannot be easily converted into valuable products and is difficult to recover and reuse [37, 38].

The lignin removal from the lignocellulosic biomass was done using the kraft process. In this process, chemicals, namely, aqueous sodium hydroxide mixed with sulfides, are used to remove the lignin under optimum conditions. However, lignin synthesis has a low heating value and releases more toxic gasses. On the other hand, lignin and hemicellulose were removed using various solvents, acetone, ethanol, and esters through the organosol process. This process was carried out with and without a catalyst over a temperature range of 373 to 473 K. The other methods, viscose and lyocell, were also performed using organic solvents, mainly aqueous sodium hydroxide, and *N*-methylmorpholine-*N*-oxide to recover the cellulose [36].

The processing of plant cellulose also releases toxic components, which cause environmental pollution. Considering this drawback, choosing microbial cellulose extraction is preferable to a plant source. The organic part in the sewage sludge could be utilized for microbial growth, which is helpful for synthesizing biopolymers, mainly cellulose. The recovery of cellulose from microorganisms could be found in pure form and more rapid than that of plant cellulose. Since there is no need for chemical treatments to purify bacterial cellulose and render it void of lignin and hemicellulose, the microbial source of cellulose production is environmentally friendly technologies that lower adverse ecological impacts [39]. Nearly 99% cellulose was obtained through this process, which underwent a bleaching process for purification. The sludge's ash content was removed in the bleaching, and pure cellulose was extracted. Furthermore, levulinic acid was obtained from pure cellulose via hydrothermal liquefaction [40].

Another important compound present in microbial mixtures is EPS. It is a type of organic matter and complex polymer containing huge molecular weight, the major metabolic compounds produced from sewage sludge. In EPS, the main components were polysaccharides and proteins and other details such as lipids, humic acid, and uronic acids. Previous research reported that EPS absorbed more than 60% benzene due to solid intermolecular interactions [41].

The composition of EPS depends on sludge origin, method of extraction, and type of wastewater. It is similar to that of the bacterial growth phase, which can increase the growth of cells concerning incubation time. It was also proved that the EPS can absorb and degrade specific compounds. EPS can also be used as an absorbent to remove and recover metal from sewage sludge [42].

Since it contains a more significant number of negative charge functional groups to absorb the organic pollutant, some of the functional groups of EPS are carboxyl, sulfuryl, hydroxyl, phosphoric, and phenolic groups. The extracted EPS from sludge was rapidly used in many cosmetics, pharmaceutical, and food industries. By considering the economic aspects, the extraction of EPS is very much required. The EPS extract methods are physical, chemical, aerobic, and anaerobic activated sludge. Biological methods require the external forces produced by heating, centrifugation, and ultrasonic to separate the cells from the microbial sludge. The significant merit of the physical process is that sonication lasts for a few minutes. On the other side, centrifugation is the method that is more efficient for both aerobic and anaerobic sewage waste. Another efficient method is heating which yields the EPS even if it is bound to the flocs [43, 44].

Microbiological flocs include EPS, which binds to cells via hydrogen bonds, electrostatic attraction, and van der Waals force. The stability of the microbial aggregates may be increased by these interactions, demonstrating that a more significant EPS results in more excellent stability. Contrarily, chemical techniques require chemical substances to stifle interactions between the microbial cell and EPS. Consequently, EPS may be extracted more effectively via a chemical process than a physical one. Therefore, the activated sludge technique was used to extract EPS from municipal sewage sludge [45]

However, it has several problems during sludge treatment, such as sedimentation and dewaterability. EPS extraction from activated sludge has a specific role in sludge settling, flocculation, and dewatering. The effluent's high quality and low turbidity were achieved through the flocculation ability of microbial cells. The EPS and protein content increases with decreasing the flocculation ability of microbial cells. Some of the research revealed that EPS concentration plays a significant role in bioflocculation [46, 47].

7.4 Recovery Routes of Cellulose from Sewage Sludge

Cellulose is a homopolysaccharide composed of a linear chain of hundreds to thousands of β (1 → 4) glycosidic linkage of repeated D-glucose units [48]. Because of its high crystallinity, it is challenging to dissolve in typical organic solvents. The hydroxyl groups are joined to each polymer chain in the three-dimensional hydrogen bond (H-bond) to form the cellulose structure. The cellulose polymer structure must therefore be able to be separated by a solvent by disrupting the intermolecular H-bonds. Cellulose has many primary characteristics, including biodegradability, biocompatibility, thermal resistance, and chemical resistance; cellulose has a broad spectrum of applications in industry. It is utilized as a raw material to create paper, therapeutic compounds, cosmetics, and textiles [49]. Table 7.1 demonstrates the recovery methods of cellulose from sewage sludge.

The raw materials utilized for cellulose recovery include toilet paper, sewage water, paper and pulp industrial waste, and municipal sewage as a primary resource. Among these, toilet paper contains more than 85% cellulose content, and the recovery rate is also a maximum of 42 g toilet paper per day. Many markets could use the cellulose that can be recovered from

Table 7.1 Recovery routes of cellulose and its yield.

Type of waste	Treatment method	Cellulose yield	Reference
Sewage water	Rotating belt filters	80%	[34]
Primary sludge	Tetrakis ionic liquid	42%	[40]
Municipal sewage	Ionic liquid $P(CH_2OH_4)Cl$	53%	[33]

wastewater, from co-digestion for further biogas generation to construction materials, including asphalt, biocomposites, and concrete. These markets have positive value for this recovered cellulose. In addition to the advantages of cellulose production, they use CellCap RBF as the first treatment, lowering the aeration tank's power consumption by 10 to 15% and creating secondary sludge by 10 to 15%. The total climate effect of the wastewater restoration plant, ignoring the subsequent treatment of the cellulosic screens, is 360 tonnes CO_2-eq/year if the cellulosic screenings are distinguished from the RBF and the cellulose is capitalized on as a resource [34].

Cellulose was recovered from dry sludge instead of wet sludge. Tetrakis ionic liquid was used in the ratio of 1 gDS:10 ml_{IL} heated up to 100°C in an oil bath for 24 hours. The mixture was then cooled to room temperature. Cellulose was recovered using a rotary evaporator at 80°C and 300 mbar [33]. Tetrakis(hydroxymethyl)phosphonium chloride is an ionic liquid utilized to recover cellulose from paper sludge. Using ionic liquid produces the maximum amount of cellulose compared to the conventional method. The less ionic liquid is needed to recover cellulose, but the yield was high [33].

Rotating belt filtration was employed in experimental testing at the wastewater treatment facilities in Falconara and Carbonera to increase the effectiveness of removing particles and recovering cellulose fibers from toilet paper. The removal efficiency of all suspended debris varied between 11 and 74%, while the surface loading rate ranged from 6 to 108 kg of total solids $m^{-2}hour^{-1}$. The average quantity of cellulosic substance recovery was 26.59 g cellulosic material m^{-3}. The findings showed that these fibers enhance the composites' hygrometric properties, promoting human comfort and health. Additionally, including recovered cellulose fibers decreased the mortar's compressive strength while increasing its flexural strength due to the fibers' improved adhesion to the binder paste [50].

Recently, the idea of a "Water Resource Recovery Facility" (WRRF) has replaced the concept of "waste treatment" in WWTPs. The transition to a circular economy is centered on this shift away from removing pollutants and toward resources with value-added (water, energy, and materials). Utilizing cutting-edge techniques is crucial to recovering energy-efficient materials from wastewater. The Geestmerambacht WWTP (CirTec, Netherlands) prototype system is designed to show that sewage may be an economically viable commodity from which assets can be collected and used. This is the world's initial large-scale factory that utilizes sewage to produce 150 to 300 kg/day of high-grade cellulose, which may be reused into manufacturing materials (such as asphalt). For initial treatment, a Salsnes Filter system has been erected at Geestmerambacht. It extracts cellulose fibers from wastewater to produce a highly concentrated sludge.

The sludge is then carried into the treatment facility for further processing. Cellulose is cleaned, dried, and disinfected for marketing purposes.

7.5 Recovery Routes of Extracellular Polymers from Sewage Sludge

Activated sludge is an EPS that plays a significant role in wastewater treatment. In biofilms, granules, and sludge flocs, organisms can be found as microbial floccules. Microbes and extracellular components constitute a large portion of microbial clusters. EPS combines proteins, polysaccharides, and nucleic acids produced after cell lysis or microbial digestion [51]. The quantities and contents of EPS in slurry flocs are affected by several factors, including procedures for extraction, sewage categories, slurry sources, and operational situations. EPS concentrations may be significantly associated with the bacteria's stage of development and increase as incubation time rises throughout the rapid expansion phase. Nevertheless, as an incubating period in the degree of stationary growth grows, EPS content might decrease [52].

Biofilm is a significant component of EPS, accounting for 50 to 90% of the organic carbon in the polymers [53]. They sustain the electrically neutral or negatively charged properties of the biofilms. It has been determined that various bacteria develop varying types and quantities of EPS and that the maturity of the biofilm influences EPS synthesis. Since EPSs are generally water-resistant, significant amounts of water are trapped in particles simultaneously via hydrogen bonding [54]. PHA (polyhydroxyalkanoates) constituents of EPSs are also essential in cell attachment and cohesiveness, influencing the functional integrity, endurance, dewaterability, and biological degradation of waste. They can be further subdivided into soluble EPSs (S-EPSs), firmly bound EPSs (TB-EPSs), and loosely bound EPSs (LB-EPSs). TB-EPSs concentrate both microbiological clusters and the outer layer of cells. S-EPSs are equitably diffused in the water phase, while LB-EPSs form an exceedingly permeable slime layer. Through determining a biofilm, EPSs act as shields from demanding exterior disruptions and carbon storage throughout the biological process [55].

The EPS distribution in the sludge is highly diverse. EPSs are polymers that can interact with other EPSs and contaminants through hydrogen bonds, electrostatic forces, attractive ionic forces, or biological reactions to produce hydrogels [56]. Activated sludge's flocculation, settling, and dewatering characteristics have recently been shown to be affected by the spatial distribution of EPS [36]. Wastewater sludge composed of biofilm and aerobic or anaerobic activated sludge can extract EPS in various ways. For soluble EPS, centrifugation is the most popular extraction technique. Numerous bound EPS extraction techniques have been created, and fresh ideas are constantly being developed. Table 7.2 demonstrates the recovery routes of EPS from sewage sludge.

Table 7.2 Recovery routes of extracellular polymers from sewage sludge.

Extraction method	Treatment parameters			Description	References
	T (°C)	rpm	Time (min)		
Ultrasonic	4	20	10	Compare the EPS content in H-WWTP with M-WWTP in terms of salinity	[66]
Sequencing batch reactor	15	2100	10	Lowering the temperature enhances the protein and EPS content while increasing the temperature increases the polysaccharide content.	[67]
Fermenter		550	15	LB-EPS and TB-EPS were obtained in the aerated fermented A/O_2 process.	[35]
Chemical-EDTA				The extraction is carried out by EDTA –chelating agent between bio-flocculent and bacterial cells, which yields higher nucleic acid.	[68]

7.5.1 Composition Evaluation of Extracted EPS

After the EPS extraction process, its composition has to be analyzed using various techniques. The analysis and identification of EPS compounds in biofilms include a visual inspection and using ^{1}H and ^{13}C NMR, FTIR, Raman, and X-ray photoelectron spectroscopy (XPS). It is possible to examine the composition of the EPS using the three-dimensional excitation-emission matrix (3D-EEM) [57]. Along with the preceding approaches for identifying EPS, spectral analysis methods for *in situ* characterization of EPS components and functional groups have recently been developed. The rate of new EPS generation and the number of production cycles of distinct carbon structures in EPS were determined utilizing a mixture of ^{13}C cross-polarization magic angle spinning (CPMAS) NMR and stable isotope assessments [58]. X-ray absorption fine structure (XAFS) spectroscopy was used to examine the adsorption process of EPS on goethite [59, 60]. Arbuscular mycorrhizal EPS interactions with Cr ions are studied using scanning transmission X-ray microscopy (STXM) and XAFS. EPS structure–correlated spectra can be obtained using STXM. On the surface of EPS, STXM can be used to track the distribution of metal ions in various valence states, such as iron ions. C-NEXAFS analysis using STXM can detect EPS at the cell-pyrite interface. By rupturing the cell, EPS was extracted using an alkaline technique [61]. Mass spectrometry was used to characterize the isolated EPS, which revealed a significant number of beta sheets. FTIR and excitation-emission spectra were used to confirm the presence of beta sheets [62].

7.6 Constraints and Future Opportunities

Sludge production will almost certainly be affected by rapid population increase and rising cultural expectations. Due to fewer landfill sites and higher disposal constraints, it is required to transition from traditional disposal to enhanced valorization options. The prospect of producing energy from sludge is now being researched using various methods. The complexity of the sludge's composition, including high levels of moisture, bacterial components (proteins, lipids, etc.), HMs, and other contaminants [such as polycyclic aromatic hydrocarbons (PAHs), dioxins, and furans], reduces the effectiveness and safety of these processes and raises their overall costs. It is also necessary to design one-of-a-kind machinery or facilities to satisfy this need [63].

Due to the complexity of EPS elements and the variety of EPS operations, additional investigation into the contribution of EPS in the corrosion process is recommended. A more thorough understanding of the relationship between biofilms/EPS and corrosion of metals can provide fresh perspectives into the process of MIC as well as ways to avoid it [64]. Due to EPS's extreme complexity, knowledge of the EPS creation process is far from complete, and additional study is required to fully comprehend the specific functions of EPS in biological treatment processes. More research needs to be done to determine how EPS affects mass transfer among cells and the environment. Additionally, little study has been done on the effects of different EPS components on their functioning. Numerous studies on EPS are presently being conducted. However, more information on EPS is needed to ensure its commercial success for real-world applications.

This study has shown that there is still much to learn about EPS, and it is essential to characterize different components to use them most effectively for removing micropollutants. Proteins and carbohydrates, which comprise the majority of EPS and are crucial in determining how they operate, are the main building blocks of EPS. Furthermore, the presence of HAS and nucleic acid after cellular lysis might enhance the functional properties of EPS. Several growth factors influence EPS's composition, including pH, temperature, carbon substrate, and cell growth. Cutting-edge techniques, including scanning electron microscopy with traditional microbiologist methods, FTIR, XRF, and NMR, and integrating the findings of various disciplines, are crucial for advancing the understanding of EPS composition, operation, organization, and uses in multiple fields. Sludge flocculation, settling down, dehydration, hydrophobicity/hydrophilicity, biosorption metals enforceable, and hazardous organic matter removal are only a few of the various applications of EPS in bio WWT systems. Considering all attempts at comprehending EPS and how they could be employed for eliminating tiny pollutants, substantial holes may be filled by further research. It must be determined how to fully characterize the parts of EPS, how they interact, and how sludge EPS's molecular makeup and purposes work.

Multidisciplinary approaches must be employed to get around characterization constraints, increase EPS functional recognition, and clarify synthetic routes. Suitable methods are required to analyze the interaction between micropollutant-aggregate complexes formed due to hydrophobic, hydrogen bonding, or electrostatic forces. According to a literature review, the study of EPS utilization is still in its early phases of research under controlled laboratory settings. Therefore, full-scale and pilot studies are required to address the shortcomings of nonuniform EPS characteristics in a range of operating scenarios [65].

7.7 Conclusion

The sewage sludge from various municipal and industrial waste is a significant environmental and public health concern. The treatment of sludge was essential for the economic cycle's benefits. The optimization of wastewater treatment processes will be performed on both a lab scale and a pilot scale. The main aim of this wastewater treatment is to provide the best solution for removing organic pollutants, effluent quality, and cost-inexpensive methods. Various methods are performed, and the purified water is passed into the river or pond to meet the water demand. Some wastewater treatment methods were physical, chemical, and biological. The aerobic and anaerobic activated sludge through natural methods gains more attention toward sewage treatment. However, valorizing waste into value-added products is a potential idea to remediate sewage sludge. There was a possibility to recover the bio compounds from the sewage waste produced from the wastewater treatment. Because the sewage sludge comprises a significant amount of organic matter, HMs, and polymeric substances. Anaerobic digestion is the most powerful method in sewage waste treatment because it recovers the biogas. Many other procedures are undertaken to generate biofuels and other products from the wastewater. Until now, a limited study was focused on healing and removing the organic matter and polymers from sewage sludge. Since cellulose and EPSs are significant in sewage sludge, the proper treatment has been performed to recover cellulose and EPS. The recovered cellulose and EPS from sewage have been further utilized to produce bioplastics, bio blocs, flocculants, and insulators. It can also be used as a stabilizer in the additives or binders. Another potential use of recovered cellulose is activated carbon production through hydrothermal carbonization. The toxic substances, mainly HMs or dyes, have been effectively removed using EPS.

References

1. Rajesh Banu, J., Merrylin, J., Kavitha, S., Yukesh Kannah, R., Selvakumar, P., Gopikumar, S., Sivashanmugam, P., Do, K.U., Kumar, G., Trends in biological nutrient removal for the treatment of low strength organic wastewaters. *Curr. Pollut. Rep.*, 7, 1, 1–30, 2021.
2. Asaithambi, P., Govindarajan, R., Yesuf, M.B., Selvakumar, P., Alemayehu, E., Investigation of direct and alternating current–electrocoagulation process for the treatment of distillery industrial effluent: Studies on operating parameters. *J. Environ. Chem. Eng.*, 9, 2, 104811, 2021.

3. Asaithambi, P., Govindarajan, R., Yesuf, M.B., Selvakumar, P., Alemayehu, E., Enhanced treatment of landfill leachate wastewater using sono (US)-ozone (O3)–electrocoagulation (EC) process: role of process parameters on color, COD and electrical energy consumption. *Process Saf. Environ. Prot.*, 142, 212–218, 2020.
4. Periyasamy, S., Temesgen, T., Karthik, V., Isabel, J.B., Kavitha, S., Banu, J.R., Sivashanmugam, P., Wastewater to biogas recovery, in: *Clean energy and resource recovery*, pp. 301–314, Elsevier, 2022.
5. Selvakumar, P., Arunagiri, A., Sivashanmugam, P., Thermo-sonic assisted enzymatic pre-treatment of sludge biomass as potential feedstock for oleaginous yeast cultivation to produce biodiesel. *Renew. Energy*, 139, 1400– 1411, 2019.
6. Periyasamy, S., Temesgen, T., Restoration of contaminated agricultural soils, in: *Biochar and its Application in Bioremediation*, pp. 381–401, Springer, Singapore, 2021.
7. Selvakumar, P., Sivashanmugam, P., Thermo-chemo-sonic pre-digestion of waste activated sludge for yeast cultivation to extract lipids for biodiesel production. *J. Environ. Manage.*, 198, 90–98, 2017.
8. Karthik, V., Senthil Kumar, P., Vo, D.V.N., Selvakumar, P., Gokulakrishnan, M., Keerthana, P., Audilakshmi, V., Jeyanthi, J., Enzyme-loaded nanoparticles for the degradation of wastewater contaminants: a review. *Environ. Chem. Lett.*, 19, 2331–2350, 2021.
9. Periyasamy, S., Karthik, V., Senthil Kumar, P., Isabel, J.B., Temesgen, T., Hunegnaw, B.M., Melese, B.B., Mohamed, B.A., Vo, D.V.N., Chemical, physical and biological methods to convert lignocellulosic waste into value-added products. A review. *Environ. Chem. Lett.*, 20, 2, 1129–1152, 2022.
10. Velusamy, K., Chellam, P., Kumar, P.S., Venkatachalam, J., Periyasamy, S., Saravanan, R., Functionalization of MXene-based nanomaterials for the treatment of micropollutants in aquatic system: A review. *Environ. Pollut.*, 301, 119034, 2022.
11. Selvakumar, P., Sivashanmugam, P., Study on lipid accumulation in novel oleaginous yeast Naganishia liquefaciens NITTS2 utilizing pre-digested municipal waste activated sludge: a low-cost feedstock for biodiesel production. *Appl. Biochem. Biotechnol.*, 186, 731–749, 2018.
12. Selvakumar, P., Sivashanmugam, P., Multi-hydrolytic biocatalyst from organic solid waste and its application in municipal waste activated sludge pre-treatment towards energy recovery. *Process Saf. Environ. Prot.*, 117, 1–10, 2018.
13. Kirchmann, H., Börjesson, G., Kätterer, T., Cohen, Y., From agricultural use of sewage sludge to nutrient extraction: A soil science outlook. *Ambio*, 46, 143–154, 2017.
14. Faria, W., Figueiredo, C., Is sewage sludge biochar capable of replacing inorganic fertilizers for corn production? Evidence from a two-year field experiment. *Arch. Agron. Soil Sci.*, 64, 505–519, 2017.

15. Lasaridi, K.E., Manios, T., Stamatiadis, S., Chroni, C., Kyriacou, A., The evaluation of Hazards to man and the environment during the composting of sewage sludge. *Sustainability (Switzerland)*, 10, 8, 2018.
16. Mehmood, A., Mirza., M.A., Choudhary, M.A., Kim, K.H., Raza, W., Raza, N., Soo Lee, S., Zhang, M., Lee, J.H., Sarfraz, M., Spatial distribution of heavy metals in crops in a wastewater irrigated zone and health risk assessment. *Environ. Res.*, 168, 382–388, 2019.
17. Rosiek, K., Directions and Challenges in the Management of Municipal Sewage Sludge in Poland in the Context of the Circular Economy. *Sustainability*, 12, 3686, 2020.
18. Gottardo, M., Micolucci, F., Mattioli, A., Faggian, S., Cavinato, C., Pavan, P., Hydrogen and methane production from biowaste and sewage sludge by two phases anaerobic codigestion. *Chem. Eng. Trans.*, 43, 379– 384, 2015.
19. Silva, M.I., Mackowiak, C.L., Minogue, P.J., Reis, A.F., Moline, E.F., Potential impacts of using sewage sludge biochar on the growth of plant forest seedlings. *Cienc. Rural*, 47, 1–5, 2017.
20. Moran, S., *An applied guide to water and effluent treatment plant design*, pp. 255–263, Butterworth-Heinemann, 2018.
21. Johnson, B.R., Phillips, J., Bauer, T., Smith, G., Smith, G., Sherlock, J., Startup and Performance of the World's first Large Scale Primary Dissolved Air Floatation Clarifier. *Proceedings of the Water Environment Federation*, pp. 712–721, 2014.
22. Gerba, C.P., Pepper, I.L., Municipal wastewater treatment, in: *Environmental and pollution science*, pp. 393–418, Academic Press, 2019.
23. Shammas, N.K., Wang, L.K., Aerobic Digestion, in: *Biological Treatment Processes*, pp. 635–667, 2009.
24. Jaibiba, P., Vignesh, S.N., Hariharan, S., Working principle of typical bioreactors, in: *Bioreactors*, pp. 145–173, 2020.
25. Mehr, A.S., MosayebNezhad, M., Lanzini, A., Yari, M., Mahmoudi, S.M.S., Santarelli, M., Thermodynamic assessment of a novel SOFC based CCHP system in a wastewater treatment plant. *Energy*, 150, 299–309, 2018.
26. Burch, K.D., Han, B., Pichtel, J., Zubkov, T., Removal efficiency of commonly prescribed antibiotics via tertiary wastewater treatment. *Environ. Sci. Pollut. Res.*, 26, 7, 6301–6310, 2019.
27. Mathon, B., Coquery, M., Miege, C., Penru, Y., Choubert, J.M., Removal efficiencies and kinetic rate constants of xenobiotics by ozonation in tertiary treatment. *Water Sci. Technol.*, 75, 12, 2017.
28. Na, W., Impact of ultrasonication-ozonation pretreatment on anaerobic digestion of sewage sludge. *Nat. Environ. Pollut. Technol.*, 17, 2, 2018.
29. Wei, C., Zhang, F., Hu, Y., Feng, C., Wu, H., Ozonation in water treatment: the generation, basic properties of ozone and its practical application. *Rev. Chem. Eng.*, 33, 1, 2017.
30. Boehler, M., Zwickenpflug, B., Hollender, J., Ternes, T., Joss, A., Siegrist, H., Removal of micropollutants in municipal wastewater treatment plants by powder-activated carbon. *Water Sci. Technol.*, 66, 2115–2121, 2012.

31. Kaushal, J., Mahajan, P., Design and evaluation of hydroponic system for tertiary treatment of STP wastewater: An eco-friendly approach. *Mater. Today: Proc.*, 45, 4914–4918, 2021.
32. Lata, S., Sustainable and eco-friendly approach for controlling industrial wastewater quality imparting succour in water-energy nexus system. *Energy Nexus*, 3, 100020, 2021.
33. Glińska, K., Glińska, G., Aqlan, M., Giralt, J., Torrens, E., Fortuny, A., Montané, D., Stüber, F., Fabregat, A., Font, J., Olkiewicz, M., Bengoa, C., Separation of cellulose from industrial paper mill wastewater dried sludge using a commercial and cheap ionic liquid. *Water Sci. Technol.*, 79, 10, 1897–1904, 2019.
34. Marcelis, P., Wessels, C., Recovery and valorisation of cellulose from waste water, 2019.
35. Liu, Q., Luo, L., Zheng, L., Lignins: biosynthesis and biological functions in plants. *Int. J. Mol. Sci.*, 19, 2, 2018.
36. Zhang, S., Chen, C., Duan, C., Hu, H., Li, H., Li, J., Liu, Y., Ma, X., Stavik, J., Ni, Y., Regenerated cellulose by the lyocell process, a brief review of the process and properties. *BioResources*, 13, 2, 4577–4592, 2018.
37. Chundawat, S.P.S., Balan, V., Da Costa Sousa, L., Dale, B.E., Thermochemical pretreatment of lignocellulosic biomass, in: *Bioalcohol production*, pp. 24–72, Woodhead Publishing, 2010.
38. Molina-Sánchez, E., Leyva-Díaz, J.C., Cortés-García, F.J., Molina-Moreno, V., Proposal of sustainability indicators for the waste management from the paper industry within the circular economy model. *Water, 10*, 8, 1014, 2018.
39. Keshk, J., Bacterial Cellulose Production and Its Industrial Applications. *Bioprocess. Biotechniques*, 4, 150, 2014.
40. Glińska, K., Lerigoleur, C., Giralt, J., Torrens, E., Bengoa, C., Valorization of cellulose recovered from WWTP sludge to added value levulinic acid with a Brønsted acidic ionic liquid. *Catalysts*, 10, 9, 1004, 2020.
41. Qiao, L., Cheng, J., Meng, N., Qian, Y., Yuan, Y., Li, Y., Chen, T., Zhang, Y., Ding, C., An isothermal adsorption model for adsorption of substrates by anammox extracellular polymeric substance proteins: Establishment, verification, and determination of adsorption capacity. *J. Water Process Eng., 43*, 102233, 2021.
42. Mohite, B.V., Koli, S.H., Patil, S.V., Heavy Metal Stress and Its Consequences on Exopolysaccharide (EPS)-Producing Pantoea agglomerans. *Appl. Biochem. Biotechnol.*, 186, 1, 199–216, 2018.
43. Yuan, D., Wang, Y., Qian, X., Variations of internal structure and moisture distribution in activated sludge with stratified extracellular polymeric substances extraction. *Int. Biodeterior. Biodegrad.*, 116, 1–9, 2017.
44. Nouha, K., Kumar, R.S., Balasubramanian, S., Tyagi, R.D., Critical review of EPS production, synthesis and composition for sludge flocculation. *J. Environ. Sci.*, 66, 225–245, 2018.

45. Zhao, P., Zhao, S., Wang, H.G., Lu, M., Li, Z.H., Encapsulation of bacteria in different stratified extracellular polymeric substances and its implications for performance enhancement and resource recovery. *Water Res.*, 220, 118684, 2022.
46. Lai, H., Fang, H., Huang, L., He, G., Reible, D., A review on sediment bioflocculation: Dynamics, influencing factors and modeling. *Sci. Total Environ.*, 642, 1184–1200, 2018.
47. Xing, L., Yang, J., Ni, B.J., Yang, C., Yuan, C., Li, A., Insight into the generation and consumption mechanism of tightly bound and loosely bound extracellular polymeric substances by mathematical modeling. *Sci. Total Environ.*, 811, 152359, 2022.
48. Gherghel, A., Teodosiu, C., De Gisi, S., A review on wastewater sludge valorisation and its challenges in the context of circular economy. *J. Clean. Prod.*, 228, 244–263, 2019.
49. Marquez-Bravo, S., Doench, I., Molina, P., Bentley, F.E., Tamo, A.K., Passieux, R., Lossada, F., David, L., Osorio-Madrazo, A., Functional bionanocomposite fibers of chitosan filled with cellulose nanofibers obtained by gel spinning. *Polymers*, 13, 10, 1563, 2021.
50. Palmieri, S., Cipolletta, G., Pastore, C., Giosuè, C., Akyol, Ç., Eusebi, A.L., Frison, N., Tittarelli, F., Fatone, F., Pilot scale cellulose recovery from sewage sludge and reuse in building and construction material. *Waste Manage., 100,* 208–218, 2019.
51. Yang, X., Xu, T., Cao, P., Qiao, K., Wang, L., Zhao, T., Zhu, J., The viscosity behaviors of bacterial suspensions or extracellular polymeric substances and their effects on aerobic granular sludge. *Environ. Sci. Pollut. Res.*, 26, 29, 30087–30097, 2019.
52. Liu, Y., Lv, W., Zhang, Z., Xia, S., Influencing characteristics of short-time aerobic digestion on spatial distribution and adsorption capacity of extracellular polymeric substances in waste activated sludge. *RSC Adv.*, 8, 56, 32172–32177, 2018.
53. Tsuneda, S., Aikawa, H., Hayashi, H., Yuasa, A., Hirata, A., Extracellular polymeric substances responsible for bacterial adhesion onto solid surface. *FEMS Microbiol. Lett.*, 223, 2, 287–292, 2003.
54. Lee, K.S., Kwon, T.H., Park, T., Jeong, M.S., *Theory and Practice in Microbial Enhanced Oil Recovery*, Gulf Professional Publishing, 2020.
55. Guangyin, Z., Youcai, Z., Harvest of bioenergy from sewage sludge by anaerobic digestion, in: *Pollution control and resource recovery for sewage sludge*, vol. 4, pp. 181–273, 2017.
56. Felz, S., Al-Zuhairy, S., Aarstad, O.A., van Loosdrecht, M.C., Lin, Y.M., Extraction of structural extracellular polymeric substances from aerobic granular sludge. *J. Visualized Exp.*, 115, 54534, 2016.
57. Zeng, J., Gao, J.M., Chen, Y.P., Yan, P., Dong, Y., Shen, Y., Guo, J.S., Zeng, N., Zhang, P., Composition and aggregation of extracellular polymeric

substances (EPS) in hyperhaline and municipal wastewater treatment plants. *Sci. Rep.*, 6, 1, 26721, 2016.

58. Sun, H., Cai, C., Chen, J., Liu, C., Wang, G., Li, X., Zhao, H., Effect of temperatures and alternating anoxic/oxic sequencing batch reactor (SBR) operating modes on extracellular polymeric substances in activated sludge. *Water Sci. Technol.*, 82, 1, 120–130, 2020.
59. Nguyen, V.H., Klai, N., Nguyen, T.D., Tyagi, R.D., Impact of extraction methods on bio-flocculants recovered from backwashed sludge of bio-filtration unit. *J. Environ. Manage.*, 180, 344–350, 2016.
60. Pronk, M., Neu, T.R., Van Loosdrecht, M.C.M., Lin, Y.M., The acid soluble extracellular polymeric substance of aerobic granular sludge dominated by *Defluviicoccus* sp. *Water Res.*, 122, 148–158, 2017.
61. Maqbool, T., Cho, J., Shin, K.H., Hur, J., Using stable isotope labeling approach and two dimensional correlation spectroscopy to explore the turnover cycles of different carbon structures in extracellular polymeric substances. *Water Res.*, 170, 115355, 2020.
62. Lin, H., Wang, C., Zhao, H., Chen, G., Chen, X., A subcellular level study of copper speciation reveals the synergistic mechanism of microbial cells and EPS involved in copper binding in bacterial biofilms. *Environ. Pollut.*, 263, 114485, 2020.
63. Fang, L., Cao, Y., Huang, Q., Walker, S.L., Cai, P., Reactions between bacterial exopolymers and goethite: A combined macroscopic and spectroscopic investigation. *Water Res.*, 46, 17, 5613–5620, 2012.
64. Gao, L., Lu, X., Liu, H., Li, J., Li, W., Song, R., Wang, R., Zhang, D., Zhu, J., Mediation of extracellular polymeric substances in microbial reduction of hematite by Shewanella oneidensis MR-1. *Front. Microbiol.*, 10, 575, 2019.
65. Boleij, M., Seviour, T., Wong, L.L., van Loosdrecht, M.C., Lin, Y., Solubilization and characterization of extracellular proteins from anammox granular sludge. *Water Res.*, 164, 114952, 2019.
66. Raheem, A., Sikarwar, V.S., He, J., Dastyar, W., Dionysiou, D.D., Wang, W., Zhao, M., Opportunities and challenges in sustainable treatment and resource reuse of sewage sludge: A review. *Chem. Eng. J.*, 337, 616–641, 2018.
67. Wang, Y., Zhang, R., Duan, J., Shi, X., Zhang, Y., Guan, F., Sand, W., Hou, B., Extracellular polymeric substances and biocorrosion/biofouling: recent advances and future perspectives. *Int. J. Mol. Sci.*, 23, 10, 5566, 2022.
68. Melo, A., Quintelas, C., Ferreira, E.C., Mesquita, D.P., The role of extracellular polymeric substances in micropollutant removal. *Front. Chem. Eng.*, 4, 778469, 2022.

8

Wastewater-Derived Biomass for Energy

Surbhi Sharma, Ridhika Bangotra, Bisma Habib, Muskaan Chib, Arpana Thakur, Ritu Mahajan and Bijender Kumar Bajaj*

School of Biotechnology, University of Jammu, Jammu, Jammu and Kashmir, India

Abstract

The swift depletion of fossil fuel reserves has not only threatened world energy security but has also raised grave environmental issues. Therefore, the exploration of green energy resources has been emphasized in view of increasing recognition of the circular economy. Wastewater generated from various sources such as domestic, agricultural, and industrial may pose various health hazards and environmental concerns, therefore necessitating its appropriate management, i.e., treatment and disposal. Thus, applying innovative technologies that can lower energy input and/or produce energy from wastewater treatment processes offer promising solutions for wastewater management. As wastewater contains organic and inorganic nutrients, it can be utilized for biomass production for energy generation. Additionally, the organic compounds already present in wastewater can be used for energy generation via biochemical, thermochemical, or microbial fuel cell (MFC) routes. Although this approach is very lucrative for generating renewable energy from wastewater, it faces several challenges. The current chapter aims to review the potential of various wastewater-derived biomasses for energy generation. The chapter entails a brief account of composition, sources, environmental hazards, treatment approaches, and various means for generating energy from wastewater.

Keywords: Wastewater, biomass, algae, treatment, management, renewable energy

**Corresponding author*: bajajbijenderk@gmail.com

Inamuddin, Tariq Altalhi, Mohammad Luqman and Joseph Kapuku (eds.) Sewage and Biomass from Wastewater to Energy, (195–224) © 2024 Scrivener Publishing LLC

8.1 Introduction

Biomass represented by plant residues from agriculture, forest, agro-industry, algae (fresh water and marine), and municipal wastes is considered a sustainable and renewable energy source that has the potential to meet the ever-growing world's energy needs [1]. Biomass energy is an eco-benign green energy source in contrast to fossil fuels which are harmful to the environment in multiple ways in terms of greenhouse gas (GHG) emissions, nitrogen oxide emissions, and sulfur oxide (SOx) emissions [2]. Furthermore, the usage of fossil fuels at breakneck speed may endanger world energy security as these are nonrenewable energy sources and are available in finite reserves. Therefore, the development of alternate and safe energy resources for the sake of the environment and a sustainable energy source is inevitable. The concept of "circular bio-economy", i.e., production of value-added products from renewable resources, is gaining momentum the world over [3–5]. Among various biomass resources (aquatic macrophytes, algae, biofilms, sewage sludge, and others), microalgae represent an important and environmentally friendly biological resource that is available in plenteous and can be used to produce biofuels and other value-added products on an industrial scale [3]. However, the price of producing and processing algal biomass raises concerns about the process's viability from an industrial viewpoint. However, microalgae cultivation in wastewater presents an economically viable alternative to cut down the high production cost of algal biomass and, at the same time, offers cost-effective treatment of wastewater [6]. Thus, wastewater not only offers an inexpensive growth medium for the cultivation of algae but also offers financial advantages related to the wastewater treatment process and may be an effective way of producing renewable energy.

Since wastewater is a major issue worldwide, especially in densely populated regions, therefore, it needs to be recycled or treated before being disposed off. However, wastewater is a rich source of nutrients, minerals, and other growth components and may be exploited for the production of microalgae which not only ensures efficient phycoremediation but increases the productivity of biomass as well as other high-value products (lipids, pigments, proteins, carbohydrates, and others) [7]. Thus, wastewater-integrated closed-biorefineries promise improved process economy and sustainability by recovering resources and at the same time minimizing the ill impact on the environment. Algae grown from wastewater are a rich source of primary (carbohydrates, proteins, and lipids) and secondary (pigments and anti-oxidants) metabolites that can be used to make a wide variety of products such as biofuels, biopolymers, biofertilizers, nutraceuticals, food/health grade compounds, enzymes, feed supplements, and others [5, 8, 9].

For the generation of bioethanol, various algal species such as *Chlamydomonas*, *Spirulina*, *Euglena*, *Chlorella*, *Scenedesmus*, and *Dunaliella* have been thoroughly investigated [10]. The ability of wastewater-grown algae to accumulate carbohydrates has been studied by different researchers. The rich carbohydrate content of the algal biomass may successfully be extracted and subjected to microbial fermentation for the production of bioethanol fuel. Furthermore, extraction of carbohydrates from algae unlike plant biomass does not require the tedious pretreatment processes. Additionally, carbohydrates present in algae may be used for the production of biodiesel, biobutanol, biomethane, biogas, and syngas [10, 11]. Keeping in view the remarkable environmental adaptation ability, high photosynthetic activity, high biomass productivity, efficient CO_2 sequestration potential, and cost-effective process, microalgal biomass may be considered as an excellent feedstock for the production of biofuels.

The wastewater sludge is a serious problem in view of the increasing global population, urbanization, industrialization, and lack of adequate sanitation facilities for large-scale wastewater treatment plants [12]. A viable management solution for wastewater sludge is to use it as a source of energy rather than consider it a waste [13]. The useful components recovered from the sludge, such as carbon and other nutrients, may be utilized for the sustainable production of renewable energy which may not only replace fossil fuel–based energy and help reduce GHG emissions but may also contribute to meeting the future energy demand [12, 13]. Various technologies available for managing sludge are anaerobic digestion (AD), thermochemical processes such as super and subcritical water (hydrothermal) processes, pyrolysis, incineration, and gasification [14]. Biogas is produced via AD, which can be used to generate other useful energy sources including heat, natural gas, and electricity. Biocrude or biofuels and other value-added bioproducts are produced by thermochemical processes, whereas ash and electricity produced from incineration can be used in the construction industries. AD is most frequently preferred among these because of its feasibility and economics [14].

Furthermore, the emergence of MFC technology leads to the generation of bioelectricity by utilizing different wastewater biomasses as feedstock [15]. However, the complexity of biomass requires numerous pretreatment and extraction approaches for its microbial/enzymatic degradation [15]. Due to its reusability and carbon-neutral nature, biomass has gained significant interest in recent years for the production of energy.

Plant-based biomass can be transformed into gaseous, liquid, or solid fuels using a variety of thermochemical, chemical, or biological methods. Biomass-derived fuels, e.g., syngas, biogas, bio-oil, bioethanol, and biochar can be utilized in furnaces, engines, turbines, and fuel cells [16]. Utilizing algal biomass directly for energy generators could be a promising strategy in MFCs which can produce clean bioelectricity through a biocatalytic reaction carried out by electroactive microorganisms like bacteria or yeast. The current chapter encompasses the application potential of various wastewater-derived biomass for the effective production of sustainable energy resources.

8.2 Wastewater Analysis

Poor management of discharged wastewater from various industrial sources leads to the accumulation of harmful pollutants in the water bodies which may impose hazardous impact on the environment. These organic or inorganic pollutants are released directly or indirectly into sewage systems, dumps, and reservoirs and may endanger the environment as well as human/animal/aquatic life. The presence of significant amounts of potentially harmful compounds, such as dissolved salts and heavy metals including Fe^{2+}, Cu^{2+}, Zn^{2+}, Mn^{2+}, Ni^{2+}, and Pb^{2+} makes the wastewater unsafe even for irrigation purposes as it may significantly contaminate the soil in agricultural fields. Therefore, wastewater needs to be assessed for pollution load especially the pollutants of an organic nature on the basis of certain parameters such as biochemical oxygen demand (BOD), chemical oxygen demand (COD), presence of solids, nitrogen, phosphorous, and many others [17].

8.2.1 Biochemical Oxygen Demand

Biochemical oxygen demand (BOD) refers to the amount of oxygen needed to biochemically oxidize organic molecules in water [18]. As a source of carbon, microorganisms utilize the ambient oxygen that has been dissolved in the water to biochemically oxidize organic materials. The BOD value is frequently used as an indicator for the level of organic pollution in water and is mostly measured in milligrams of oxygen used per liter of sample during five days of incubation at 20°C [19]. The amount of biomass produced is regarded as negligible in the laboratory process for calculating the BOD concentration and only the two events (substrate depletion and oxygen consumption) are consequently taken into account. The actual determined concentration of BOD will vary depending on the biomass yield (milligrams of biomass generated per milligrams of substrate utilized) and the experimental conditions. In order to get accurate and repeatable findings from the BOD test, multiple measurements of the same sample (at the same dilution) are typically executed.

8.2.2 Chemical Oxygen Demand

Chemical oxygen demand (COD) is an important analytical variable for determining the quality of water. The level of organic contamination in water bodies is indicated by COD [20]. In the COD test, the oxygen demand is measured based on the chemical oxidation of carbonaceous/organic compounds in water unlike biological oxidation in the case of BOD. Similar to BOD, COD is measured in milligrams of oxygen per liter (mg L^{-1}). The COD test has the benefit of being quick and readily repeatable; however, the drawback is that not all of the organic pollutants detected by COD can be biologically broken down. Since the effectiveness of biological wastewater treatment systems (such as activated sludge, trickling filters, anaerobic digesters, rotating biological contactors, oxidation ponds, and lagoons) is assessed based on BOD and COD levels, it is of utmost importance to determine the amount of oxygen demand corresponding to biodegradation of organic pollutants.

8.2.3 Solids

Wastewater samples must be examined for the presence of solids as they represent the pollutant load of the water (effluent), which is crucial for determining the environmental quality of wastewater generated as an industrial effluent. Both a pollutant load and a load on a receiving stream are represented by solids in wastewater effluent. Solids can be divided into two primary groups: suspended and dissolved. The distinction between the two is frequently quite arbitrary. The suspended solids in a sample do not pass through a 0.45- to 1.2-μm filter, whereas constituents including minerals, salts, and charged ions that pass through the filter are referred to as dissolved solids. The dissolved and suspended solids can be further divided into volatile or fixed solids [21]. At 550°C temperature, fixed solids will not ignite whereas volatile solids would. The assumption is that the organic component of the solids is represented by the volatile solids. The volatile solids in a biological treatment system are frequently linked to biomass and can be used as a measure of the microbial population. The problem with this assumption is that the total particles retained in the system frequently include the organic compounds present in the wastewater.

The solids may not actually represent the component of the actively degrading microbial population, even if the volatile content of these solids is known. So, it is important to be cautious when relating the microbial population and volatile solids concentration. The dissolved solids in the sample comprise organic contaminants in the solution such as carbohydrates, proteins, fats, oils, surfactants, volatile acids, urea, ammonia, and trace pollutants, and also some inorganic substances [21].

8.2.4 Nitrogen

Nitrogen (N) is a crucial component of biological macromolecules such as protein, chlorophyll, and nucleic acids. N-polluted water can have a significant negative impact on human health and ecosystems, leading to issues like eutrophication and hypoxia, and, in extreme conditions, it may even cause the mortality of aquatic animals [22]. In order to minimize the toxic effects of nitrogen, the concept of recovery of nitrogen and application of nitrogen for the production of more sustainable products has been an area of continuous investigation [23]. Nitrogen recovery methods aim to use the ammonia from wastewater for other purposes such as for the production of fertilizers which may not only help reduce energy and petrochemical consumption for fertilizer production but also contribute significantly toward environmental health [24].

8.2.5 Phosphorus

All forms of living organisms require elemental phosphorus (P) as a vital component of nutrition. Considering its importance for plant growth and development, phosphorus constitutes an integral component of fertilizers and, thus, is instrumental for agriculture. As a result of sewage discharges, and from other sources, phosphorus concentration may rise in aquatic ecosystems and lead to eutrophication and related issues. Phosphorus can be found in both organic and inorganic forms. Technology development for the recovery of phosphorus from wastewater and its reusage for the production of sustainable products represent emerging areas of research that may potentially contribute not only to a safer environment but also to economic gains [25].

8.3 Composition of Wastewater

Wastewater is the water that has been contaminated due to domestic, industrial, and commercial activities. The composition of all wastewater from diverse sources keeps on constantly changing and is highly variable [26]. Water makes up 99.9% of wastewater, and the remaining 0.1% is represented by organic materials, bacteria, and inorganic components. Organic matter in wastewater includes human and animal waste, food residues, and other organic detritus. The effluent water carries nutrients, such as nitrogen and phosphorus, which are essential for the growth of plants, but they cause serious environmental concerns, when present in excess. Wastewater also contains a variety of inorganic compounds, such as heavy metals, salts, and minerals [27]. Wastewater can be divided into several categories, including domestic sewage, industrial wastewater, agricultural run-off, and stormwater.

Domestic wastewater includes a wide range of substances such as human and animal waste, food waste, detergents, and other household chemicals/materials. The organic content of domestic wastewater can be assessed using BOD and COD, which represent the measures of the quantity of oxygen required to decompose organic materials by biological means (microorganisms) or by chemical means (chemical oxidizing agents). The BOD and COD levels of domestic wastewater can vary greatly depending on several factors such as population density and water consumption habits. However, generally, the BOD and COD of the domestic water range between 150 and 250 mg/L and 300 and 600 mg/L, respectively [28]. The water that is used for agricultural practices also adds tons of wastewater to the environment and that consists of various agricultural wastes such as fertilizers, pesticides, animal/human waste, crop residues, and runoff water [29].

With the rapid economic developments such as the growth of industries, population explosion, and other factors, there has been tremendous depletion in the environmental quality in terms of increased air, water, and soil pollution. One of the major concerns is the pollution of water bodies due to the discharge of large quantities of sewage/industrial water which is contaminated with heavy metals. The highly soluble nature of heavy metals pollutes the aquatic ecosystems/environment and poses serious environmental health issues. The most common heavy metal contaminants of the environment from diverse sources include zinc (Zn), lead (Pb), cadmium (Cd), nickel (Ni), copper (Cu), chromium (Cr), vanadium (V), silver (Ag), platinum (Pt), tin (Sn), and titanium (Ti) [30].

8.4 Sources of Wastewater

Water is a renewable resource that is vital for life and has several sources like lakes, rivers, streams, and oceans. However, due to the massive increase in human population and environmentally destructive activities

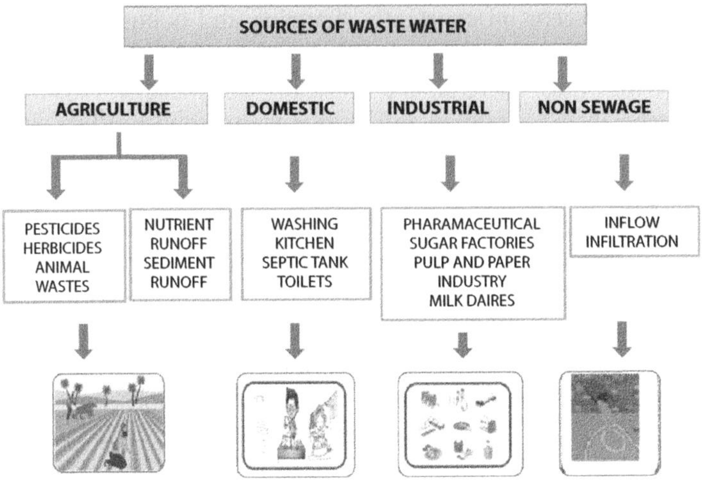

Figure 8.1 Various sources of wastewater.

(industrialization, urbanization, and population explosion) during the past century, fresh water supplies have significantly decreased. The environment is severely threatened by the disposal of human garbage/industrial wastes in water bodies, which impacts all types of living forms/creatures that exist. Therefore, the origins of wastewater and the pollutants present in the water must be comprehended in order to safeguard water bodies and the environment as such [31]. A few of the major sources of wastewater have been discussed in Figure 8.1.

8.4.1 Domestic or Municipal Wastewater

It is the liquid waste released from sanitary facilities inside a building, such as a home, office, or institute. About 80% of domestic sewage is constituted by water from households, and it consists of toilet, bathroom, laundry, and kitchen sink wastes. In the majority of developing nations, domestic wastewater produced in different areas is frequently discharged into rivers and streams without prior treatment. This poses a major threat to the water quality and continued usability and may contaminate downstream water bodies utilized as a source of drinking water, irrigation water, fisheries, and recreational purposes [32]. The public health departments use microbiological treatments of wastewater wherein organic constituents (pollutants) of domestic wastewater are utilized by a wide range of microorganisms (bacteria, fungi, protozoa, algae, and others) [33], and the pollution load is drastically reduced.

8.4.2 Industrial Wastewater

It is the hydrological waste produced when substances are dissolved or suspended in water, usually during the utilization of water in an industrial production process or cleaning activities that occur alongside the process. It frequently contains elements (toxic and hazardous substances, dyes, and metal ions) that need to be cleaned up before being released into a biological treatment facility and subsequently into water bodies [34]. Effluents from industries may include high concentrations of oils, solvents, volatile chemicals, and fats that are hazardous to both humans as well as the environment [35]. Therefore, advanced and effective methods/technologies are being developed for appropriate treatment of industrial wastewater [36]. The final discharge of treated wastewater needs to conform to the applicable national and international requirements and standards in terms of pollution load and other parameters.

8.4.3 Agricultural Wastewater

It includes waste from a variety of farm activities such as irrigation, operations of feeding livestock, growing and handling of agricultural goods, and others. The waste majorly constitutes animal wastes and large quantities of chemical pesticides and herbicides deteriorating the quality of water. Surface runoff during rains is the major cause of agricultural wastewater through which chemical fertilizers are mixed with the nutrients of soil leading to chemical pollution of water [37].

8.4.4 Storm Sewage

It is the untreated surface wastewater from rain and snowmelt, which washes away a significant amount of pollutants from the area around which it flows. It causes dilution of raw sewage with precipitation and conveys the suspended particulates in the sewage system. Pollutants of storm sewage comprise sediments, microorganisms, phosphorus, nitrogen, toxic chemicals, and other materials that are necessary to be treated.

8.4.5 Infiltration

It occurs when groundwater seeps through damaged pipes, joints, connections, pit walls, and other structures into the sewer line and service connections. Urban areas' socioeconomic and environmental elements can suffer greatly from unchecked infiltration and inflow in sanitary sewers.

The ability of the current sewer systems to deal with growing populations can be severely compromised by excessive infiltrations.

8.4.6 Inflow

It refers to wastewater entering the sewage system immediately following storm occurrences via roof and basement drains, as well as a consequence of defective or ruptured connections and damaged sewer holes [38].

8.5 Types of Wastewater

There are several types of wastewater, which are categorized based on their source and composition such as domestic wastewater, industrial wastewater, and agricultural wastewater. Domestic wastewater refers to the water that is discharged from the households. This water is also known as sewage or sanitary wastewater. Domestic wastewater contains liquid and solid discharges from the human population and animals, which add tons of pollutants and microorganisms to the water [39]. But the household wastewater is now being considered as a resource rather than waste for plant growth-promoting nutrients such as nitrogen and phosphorus [40].

The water that is used for agricultural purposes gets contaminated by agricultural waste, livestock waste, irrigation runoff, fertilizers, and pesticides, and it produces large quantities of wastewater in the surrounding environment [41]. This unwanted and polluted water carries large amounts of dissolved nutrients and runoff domestic and agricultural biomass. The organic matter present in agricultural wastewater can be a vital and prominent source of energy. The water from the fields can be used to generate energy, via, a process known as anaerobic digestion (AD). AD is a biological process in which microorganisms break down organic waste in the absence of oxygen, resulting in the production of biogas as a byproduct. Biogas is a gas mixture predominantly composed of carbon dioxide and methane that can be used as fuel for heating, power generation, and transportation [29].

Large-scale/progressive industrialization in the past few decades has elevated the discharge of pollutants into the environment. One of the essential environmental components that have been majorly damaged by industrialization is the hydrosphere [42]. Industries like chemical, distillery, paper and pulp, textile bleaching and dyeing, mining, petroleum and petrochemical, pharmaceutical, metal processing, and electroplating industries are considered to be the main producers of wastewater [43]. The effluent water discharged from industries can be channelized for energy production through MFCs. The MFCs utilize the microorganisms, to generate electricity from the organic matter present in the water discharged from the industries. Inside the MFC, the microbes disintegrate the organic wastes, releasing electrons in the process. The electrons produced are transferred to the electrode, which in turn generates electricity [44].

8.6 Environmental Hazards of Wastewater

The wastewater is an intricate mixture of high-risk pollutants (HRPs) such as pharmaceutical products (PPCPs), endocrine-disrupting chemicals (EDCs), heavy metals, and organic pesticides. Even if the occurrence of these pollutants is low in water, it is concerning as they tend to be highly bioaccumulative and toxic. Once these materials find their way into the environment, they will establish their effect on the growth, reproduction, and overall health of the organisms and may pose potential ecological damage to the ecosystem [45]. Untreated wastewater is typically high in nutrients (nitrogen and phosphorus), heavy metals, herbicides, and other contaminants. The discharge of untreated sewage wastewater into bodies increases the nutrient load and causes heavy pollution. High nutrients in water bodies promote excessive development of algae and aquatic plants and cause oxygen depletion which may be harmful or fatal to the aquatic life. This is referred to as eutrophication [45]. Thus, untreated wastewater has a very detrimental effect both on aquatic and terrestrial ecosystems.

The aquatic primary producers, i.e., the phytoplankton community, plays an essential role in the aquatic environment, but certain waste materials from the industries pose a serious threat to these photosynthetic microorganisms. For instance, when antibiotic residues from pharmaceutical industries enter the aquatic environment, the phytoplankton are the first to be affected among other aquatic life forms. Toxicological research shows that antibiotics have detrimental effects on algae. Different classes of antibiotic residues have a wide range of effects on the aquatic algae, for instance, β-lactam antibiotics mostly impede algal cell wall formation, whereas aminoglycoside antibiotics primarily inhibit algal sugar metabolism. Moreover, tetracycline antibiotics deter algal protein synthesis, and quinolones primarily inhibit algal chlorophyll synthesis and DNA replication [46]. Moreover, wastewater irrigation has been considered as an economic practice in both developed and developing countries, but using wastewater for irrigation purposes may significantly increase heavy metal concentrations in the soil. The occurrence of heavy metals in the edible parts of the crop, which are grown in heavy metal-polluted soil, may be detrimental as they can reach the human body via the food chain. Since heavy metals are persistent in nature, they have the tendency to accumulate in the body organs even at lower concentrations. The accumulation of heavy metals such as Cd, Cu, Ni, Zn, and Pb has antagonistic effects on various essential nutrients in the body, which could lead to immune system failure, psychological dysfunctioning intrauterine growth retardation, gastrointestinal neoplasm, malnutrition, and severe/serious disorders [45].

8.7 Wastewater Treatment Methods

Wastewater treatment is the process of eliminating impurities and pollutants from water used for domestic, commercial, or industrial purposes before it is released back into the environment or utilized again. The primary objective of wastewater treatment is to protect human health and the environment by lowering pollutant concentration to a safe level for disposal or reuse. The basic techniques used for treating wastewater are illustrated in Figure 8.2.

8.7.1 Collection

The planning of wastewater collection is the most crucial step. Based on the source of wastewater generation, the pattern of collection varies. For industries and populated areas where a huge amount of wastewater is

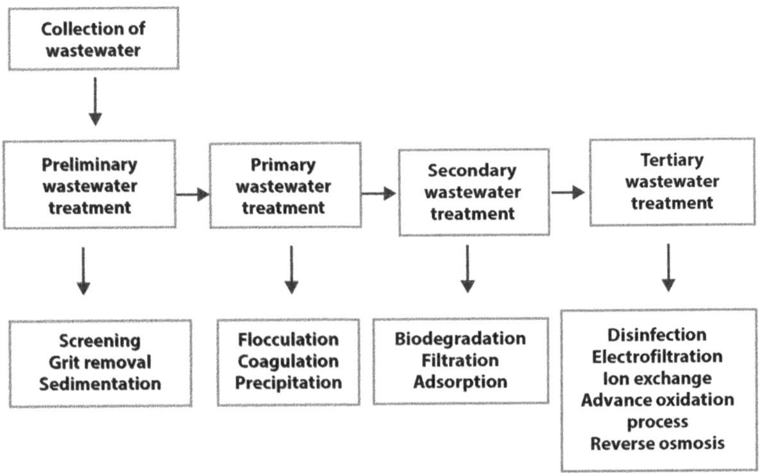

Figure 8.2 Various steps of wastewater treatment process.

generated, the centralized mode is preferred, while in sparsely populated areas and rural areas, the decentralized mode of the wastewater collection system is preferred [47–49].

8.7.2 Preliminary Wastewater Treatment Method

Contaminated water coming from different sources does not have a uniform composition. Furthermore, the wastewater is a mixture of a variety of waste materials such as plastic bags, wood logs, sand, pebbles, rags, sachets, and other residual solids which can clog and destroy the wastewater treatment facility if allowed to pass further, so their removal is utmost important at the very first step. Different types of screens (coarse, medium, and fine screen) are used based on the size of impurities in the wastewater. Grit chambers are for the removal of grit. Grit consists of small particles such as sand, eggshell, bone chips, and other inert inorganic matter which may cause the abrasion, wear, and tear of the equipment [50].

8.7.3 Primary Wastewater Treatment Method

It allows the separation of solids from the liquid. Sedimentation, flocculation, and coagulation are the common processes that take place in this stage. In the process of sedimentation, the floating solids get settled at the bottom by the influence of gravity, while in flocculation, the solids form the flocs or the larger aggregates which can easily be removed. However, in the process of coagulation, the solids are brought together by alteration in their surface charge by using salts or other coagulating agents. In the textile wastewater treatment, ultrasonic-assisted coagulation and flocculation are used extensively [18].

8.7.4 Secondary Wastewater Treatment

Secondary wastewater treatment is often called as the biological treatment. It facilitates the removal of organic matter and residual suspended matter. Both aerobic and anaerobic microbes are being used for the degradation of the organic matter. Biomass may be attached to the inert material (in case of contact bed, intermittent filter, trickling bed, and rotating biological contactor) or may be suspended directly into the secondary tank and aerated well to form the activated sludge. Activated sludge is a flock-like substance containing organic matter from the wastewater and consortia of bacteria and other living organisms, especially microorganisms. Organic matter is being degraded by the microbes and this can be further used as manure. Apart from this, aerated lagoons, disk contractors, contact media, and extended aeration are other methods used in the secondary treatment stage [51].

- Aerobic Treatment - In aerobic treatment, oxygen is consumed by the microorganisms to degrade the pollutants. The organic pollutants are converted into CO_2, H_2O, and biomass. In the aerobic process, aerated lagoons, oxidation ponds, trickling filters, activated sludge processes, and sequencing batch reactors are generally used [36].
- Anaerobic Treatment - In this treatment process, oxygen is not used by the microorganisms to degrade the pollutants. Here, the organic pollutant is converted into CO_2, methane, and biomass. It involves mainly the following three steps: a) hydrolysis- conversion of large inorganic matter into smaller ones in the presence of water; b) acidogenesis- acetic acid, CO_2, and hydrogen are produced as the end products; and c) methane formation- conversion of hydrolysis/acidogenesis products into methane and carbon dioxide. In anaerobic treatment, anaerobic lagoon, anaerobic oxidation ponds, upflow anaerobic sludge blanket, anaerobic baffled reactor, and expanded granular sludge blanket are used [52].

8.7.5 Tertiary Wastewater Treatment

It is the final stage in the wastewater treatment. It allows the removal of leftover contaminating pollutants in water and makes it fit for dispersion into the environment. It allows the removal of leftover antibiotics, heavy metals, dyes, and other pollutants.

Various methods for tertiary treatment are briefly given below:

- Disinfection - It is commonly done using UV radiations and the application of ozone or chlorine. It allows the removal of all the microorganisms. Ozonation is considered better than chlorination because it does not generate disinfection byproducts (DBP) [53].
- Electrofiltration - It is the advanced process of filtration in which the application of an electric field is realized for filtration. Almost all the colloids and microorganisms have charge, so the presence of an electric field allows their separation due to electrostatic interactions. It allows the

improvement in the liquid-solid separation and quality of filtrate, permeates flux, and prevents membrane fouling [54].

- Ion Exchange - This method is used to remove the ions from the wastewater. There are two types of ion exchange resins/membranes based on the ion selectivity. Cation exchange resins contain the negatively charged functional group and selectively bind to cations, while anion exchange resins contain the positively charged functional group and selectively bind to anions. The use of nanofiber in the pore-filling matrix of ion exchange membranes enhances the physical/ionic stability of the membranes and improves the membranes' properties [55].
- Advance Oxidation Process (AOP) - It promotes the oxidation of the persistent organic and inorganic matter into simpler forms by generating the reactive oxygen species using methods such as ozonation, peroxidation, and photocatalysis [56].
- Activated Carbon Adsorption - It allows the removal of organic and inorganic pollutants by the process of adsorption. Various functional groups can be incorporated on the adsorbents' surface to effectively allow the removal of target pollutants. It is used for the removal of pesticides, pharmaceuticals, heavy metals, and other hazardous materials [57].
- Reverse Osmosis - It is commonly used for the desalination of water. It is highly energy efficient and has low operating costs as compared to other desalination methods. A thin film composite (TFC) polyamide membrane is a semipermeable membrane that is responsible for water salt selectivity [58].
- Phytoremediation - It is a method in which plants are used for the wastewater treatment and involves the process like phytoaccumulation, phytodegradation, phytofiltration, phytostimulation, phytostabilization, and phytovolatilization. It is used for the removal of heavy metal and organic pollutants including polycyclic aromatic hydrocarbon. Plants such as *Eichhornia crassipes*, *Typha latifolia*, and water hyacinth [59] have been used for phytoremediation.

8.8 Biomass Derived from Wastewater

Wastewater-derived biomass is an unexploited resource that has enormous potential to be utilized as an energy source. Wastewater is generated as a consequence of industrial and domestic processes that contain organic material, nutrients, and toxins. The release of untreated wastewater into the environment can be hazardous to both human/animal health and the ecosystem. Thus, wastewater treatment is essential to get rid of the contaminants and to make it safe for reuse or disposal. Wastewater treatment can produce different types of biomass which can be subsequently utilized for generating sustainable energy. An account of several of the common biomass types from wastewater is given below:

8.8.1 Sewage Sludge

Sewage sludge is a semi-solid residual matter generated after the treatment (biological, mechanical, and physicochemical) of wastewater [60]. It consists of both organic and inorganic components, such as nutrients, organic materials, and microbes. However, its composition may vary depending on the source and treatment process of wastewater [61]. Despite being a residue, wastewater sludge can be processed into a source of energy or resources, which can eliminate the need to create an equivalent amount of materials or energy from nonrenewable sources with significant environmental consequences [12]. Various forms of energy that can be recovered from sewage sludge, specifically from its organic matter component are a) biofuels production, b) electricity production in MFCs, c) production of biochar, syngas, and other valuable byproducts by gasification and pyrolysis, and d) production of energy and desired chemicals by supercritical wet oxidation, hydrothermal treatment, incineration, and co-incineration of sewage sludge [62].

8.8.2 Wastewater Algae

Algae are mostly aquatic, photosynthetic autotrophic organisms that have the ability to flourish in eco-niches that have abundant organic matter and nutrients. They can be generally categorized into microalgae (unicellular organisms), and macroalgae (multicellular organisms) [63]. Wastewater serves as a cost-effective, readily available, and suitable medium for the growth of algae. It has both macro- and micronutrients that help microalgae flourish. The main nutrients found in wastewater include phosphate, ammonia, urea, nitrate, and other trace elements [64]. Algal biomass generated and collected from wastewater treatment systems can be processed into biofuels such as biogas by anaerobic degradation, biodiesel by transesterification of lipids present in algal biomass, bioethanol by fermentation of carbohydrates, and bio-crude oil by pyrolysis or hydrothermal liquefaction [65]. Utilizing algae as a biorefinery feedstock has various benefits, including quick growth, easy handling, better biomass output, and cultivation on nonarable ground [63].

8.8.3 Plant Biomass

Some wastewater treatment methods involve phytoremediation methods or artificial wetlands, which use particular plants called 'aquatic macrophytes' (emergent, floating, or submerged) to filter out pollutants from the water [66]. These plants produce higher quantities of biomass, which can be harvested and utilized to produce energy (bioethanol, biodiesel, biohydrogen, and biogas), compost, or valuable compounds (biochar, fertilizers, animal feedstock, and adsorbent) [67]. Valorization of aquatic biomass into value-added products is generally carried out by various thermochemical (incineration, pyrolysis, and pyro-gasification), and biochemical conversion (anaerobic digestion) [68].

8.8.4 Biofilms

A biofilm is defined as a group of complex microbial populations that adheres to the wet surface and proliferates in sewage treatments. Various microorganisms, including bacteria, fungi, algae, and others, can be found in these biofilms [69]. A biofilm is formed naturally, with no chemicals employed or added [70]. The microbial communities produce extracellular polymeric substances (EPS) that keep the biofilm intact [71]. The EPS are composed of proteins, phospholipids, nucleic acids, or polysaccharides [69, 72]. EPS facilitate the adhesion of microorganism cells to a surface, forming a diffusive barrier that shields them from the outside world. This barrier shields the cells from harmful substances, phagocytes, and bactericides; however, it can also block the cells from receiving nutrients necessary for cell growth [73]. Biofilm-derived biomass can be used for a variety of purposes, including wastewater treatment, enzyme production, bioremediation, green energy systems, and the generation of microbial products [70].

8.9 Conversion of Wastewater-Derived Biomass into Energy

A sustainable strategy for waste management and energy generation is the conversion of wastewater-derived biomass into various types of energy resources. Considering the escalating global challenges such as meeting

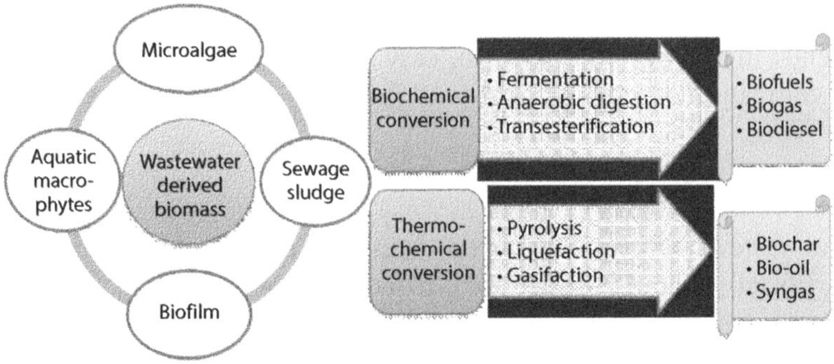

Figure 8.3 Conversion of wastewater-derived biomass into various forms of bioenergy.

ever-growing energy demand, emphasis on renewable energy generation, growing environmental issues, and emphasis on developing environmentally sustainable processes/technologies, it becomes ardently important and critical to find new/novel ways to transform trash into energy or useful resources [12, 74]. Wastewater, a consequence of numerous anthropogenic activities, includes significant amounts of organic materials that can be used to produce various energy types. Biofuels are the most promising replacement for finite, nonrenewable, and nongreen fossil fuels [75]. Biofuels, which are made from nonfossilized biomass or other organic solid wastes, offer the advantage of lowering net carbon emissions, primarily GHGs, and also reduce reliance on crude oil. Renewable biomass is used to produce a variety of biofuels, including biodiesel, biogas, bioethanol, bio-hydrogen, and biomethane [70]. The algal biomass generated and collected during the wastewater treatment process offers numerous opportunities for the production of renewable fuels and value-added goods [76]. An overview of the processing of various wastewater biomasses into biofuels and other energy types via different conversion routes is represented in Figure 8.3.

8.9.1 Biodiesel

Recently, biodiesel has drawn a lot of interest as a sustainable, biodegradable, and environmentally friendly fuel that produces less harmful emissions than normal/petrochemical diesel. In comparison to petro-diesel fuel, biodiesel offers superior physical and chemical qualities, including a greater cetane number, a lower sulfur percentage, a higher point of ignition, and enhanced greasing efficiency because of the oxygen it contains [77]. Triglyceride-enriched biomasses such as microalgae and aquatic plant biomass waste are prominent raw materials that can be used for the production of biodiesel. Fatty acid methyl ester (FAME), often known as biodiesel, can be produced by the transesterification reaction of biomass with alcohol, either in the presence or absence of a homogeneous/heterogeneous catalyst [75]. Various kinds of alcohol such as amyl alcohol, butanol, ethanol, and methanol can be used for transesterification reactions. However, methanol is frequently employed since it is a less expensive, short-chain alcohol that reacts quickly and dissolves easily in the reaction medium [78]. Other renewable or green diesel products like bio-jet fuel, bio-gasoline, and green diesel can also be produced from the lipids obtained from microalgae growing on organic-rich wastewater via a catalytic hydro-processing method [76].

8.9.2 Bioalcohol

The most common types of bioalcohols produced from biomass are biomethanol, bioethanol, biopropanol, and biobutanol. However, to meet the projected increase in energy demand, bioethanol is primarily the most recently developed biofuel [75, 79]. Bioethanol is believed to be better than petroleum fuels due to a variety of characteristics such as high octane number, greater flammability, larger heat of vaporization, clean fuel, rapid combustion, high combustion efficiency, and less hazardous nature [80]. Microalgal biomass rich in carbohydrates, i.e., sugars and/or starches, can be efficiently employed as a feedstock for fermentation-based bioethanol production [76]. Bioconversion of algal biomass to bioethanol involves the following process steps: a) appropriate pretreatment of biomass, b) enzymatic hydrolysis of pretreated biomass with suitable enzymes, and c) fermentation of monomeric sugars via yeast cells [81]. Additionally, the CO_2 generated as a byproduct of this process can be utilized as a source of carbon for microalgae during the growth phase, hence reducing the emission of GHGs [75]. The ethanol produced is separated with the help of a distillation column. Finally, liquefaction of the extremely concentrated ethanol stream is carried out for its usage as a petrol substitute in vehicles [77].

8.9.3 Biogas

One of the most energy-efficient methods of producing bioenergy from biomass is the production of biogas, specifically bio-hydrogen and biomethane. It is an environmentally friendly energy resource that can be utilized immediately as electricity, fuel, fuel cells, and microturbines. It can be obtained effectively from aquatic plant and microbial biomass which are rich in carbohydrates, proteins, and fats [66]. Microalgae serve as the carbon supply for methanogens and bacteria with the hydrogenase enzyme during the production of bio-hydrogen and biomethane [70]. AD represents one of the most promising strategies for bioprocessing of domestic and industrial wastewater sludge into biogas which mainly comprises 50 to 75% methane, 25 to 50% CO_2, and 1 to 2% trace gases [82]. The method of AD involves a series of biochemical reactions: a) hydrolysis of complex organic compounds (lipids, carbohydrates, and proteins) into simpler molecules (alcohols, long-chain fatty acids, sugars, and amino acids) with the help of various microorganisms in the absence of oxygen; b) fermentation of hydrolyzed monomers into volatile fatty acids (propionic, butyric, and valeric acids) by acidogenesis and acetic acid by acetogenesis; and c) acetoclastic methanogenesis of acetate and hydrogenotrophic methanogenesis of hydrogen into biomethane. The process is carried out in anaerobic digesters under ambient process conditions [70].

Bio-hydrogen has been considered as a potential alternative to gasoline due to its limitless potential, sustainability, low energy consumption, and high efficacy. Since only water is produced during the combustion of hydrogen, it is a greener fuel than fossil fuels [66]. There are various methods for producing bio-hydrogen: a) direct biophotolysis by cyanobacteria and green algae to split water into hydrogen and oxygen; b) indirect biophotolysis, which occurs when photosynthetic bacteria photoferment accumulated biomass that has been supplemented with carbohydrates; and c) dark fermentation of organic matter [75]. Among all the processes (biophotolysis, photofermentation, and microbial electrolysis), dark fermentation is favored due to its ability to produce repeated hydrogen, lack of sunlight dependence, and economical operation [77].

8.9.4 Microbial Fuel Cell

The production of electricity from wastewater with the help of a microalgae-based MFC has attracted the attention of several scientists recently. MFCs are bioelectrochemical devices that produce bioelectricity through biocatalytic reactions catalyzed by electroactive microorganisms (yeast or bacteria) [83]. Microalgae can be employed in MFCs as photosynthetic microalgae MFCs (PAMFC), microbial carbon capture (MCC) cells, sediment MFCs (SMFC), or algae-aided cathode systems. These MFCs have been demonstrated to be effective at removing N and P from wastewater through symbiotic interactions between microalgae and bacterial consortia, capturing CO_2, and for power generation with little risk of carbon emissions [84]. An MFC typically comprises a cathode chamber and an anode chamber with an ion-exchange membrane separator. The basic working principle under an algal fuel cell is that it generates electrons by the oxidation of a substrate at the anode, which is then transported through a connecting wire and reduced at the cathode to generate energy. Algae may exist in the algal fuel cell either exclusively in the cathode chamber or equally in the anode and cathode chambers as an anodic half-cell of the fuel cell. Algae become a component of the wastewater's microbiome and aid in oxidizing the substrate when utilized with activated sludge or wastewater. Thus, by using photosynthetic organisms like algae in the cathode chamber, biomass, dissolved oxygen, and electron acceptors from photosynthesis are obtained, which facilitate the reduction of electrons at the cathode [85].

8.10 Challenges and Future Prospects

The utilization of wastewater released from various industries, households, sewage, etc. for energy generation represents the application of the circular economy. There are a number of challenges at every step from the collection of wastewater to converting it into energy. The safe disposal of wastewater from different sources to the wastewater treatment plant is not always very feasible. Leakage in wastewater collection pipelines can cause menace. Therefore, automatic and intelligent fault detection, fault diagnosis, and fault evolution are a must for the safe and efficient operation of wastewater treatment systems [51]. Apart from this, wastewater treatment processes may also have some limitations; these limitations need to be overcome. For example, although the AOPs wastewater treatment is very efficient, it is limited to pilot scale in batch mode. The focus should be given to implementing it on a large scale in real conditions in continuous flow mode [56]. The use of disinfectant in wastewater treatment leads to the generation of DBP which may have ill effects on human health and cause cytotoxicity or developmental toxicity [86]. Membrane fouling is also a big challenge in the treatment of wastewater. Focus should be given to designing antifouling membranes, cleaning monitoring, and optimizing the operating conditions [87]. Phytoremediation is a slow process and the rate varies with the species of plant used and is limited only to the top layers of the soil [59].

In algal biomass production, harvesting of algal biomass involves about 20 to 30% cost, and focus should be given to decreasing its cost by bringing advanced harvesting techniques. While growing the algal biomass in the wastewater, there are chances of contamination with the other strains of the algae. Also, as the wastewater from different sources may have different compositions, the presence of some pollutants/toxins may behave as an inhibitor for algal growth. Efforts can be put to develop genetically engineered strains of algae that are more robust and tolerate fluctuating conditions up to a high extent. The downstream process of algal biomass into biofuel is also complicated and expensive [5].

Although MFC has the potential to generate energy from wastewater, this method is still facing serious challenges. The cost of platinum rods is very high, and there may be development of biofilms on the surface of the cathode, which may reduce the process efficacy. Apart from this, varying properties/composition of wastewater, e.g., different pH, nutrient levels, and organic content, may affect the efficiency of MFC as the chemicals and other pollutants may inhibit the activity of MFC. Monitoring and harvesting of the power generated in the system is also a challenge [88, 89].

8.11 Conclusion

In this chapter, the energy-generating potential of wastewater has been reviewed along with a brief account of composition, sources, and environmental toxicity, and the treatment of wastewater has been presented. Given the significant impact of untreated effluents on the surrounding environment, the necessity for creating resource-efficient wastewater treatment systems cannot be underestimated. Despite intensive research, the various combinations of wastewater treatment technologies have not yet been fully optimized to the best of process configurations. Furthermore, current investigations are concentrating on integrating wastewater treatment plants with bioenergy manufacturing systems with continuous energy supply at every level, establishing a multichannel system that could not only save a significant amount of energy needed for waste treatment but also result in the production of cleaner fuels in the form of bioenergy. MFCs and biofuel generation using algal biomass are renowned examples of using wastewater biomass for bioenergy production. A diverse range of wastewater can be utilized in this way for bioenergy production, the efficiency of which depends on various parameters such as pH, temperature, mode of operation, and other factors. Although a substantial quantity of energy can be recovered using this approach, there are challenges in each step from collection to energy production, and further research and developmental studies are needed. Overall, the strategy is anticipated to become more effective in the future with the transition of wastewater treatment from pilot scale in batch mode to large scale in continuous flow mode, use of antifouling membranes, and improved/innovative technologies for wastewater-derived biomass conversion processes to bioenergy.

Acknowledgements

Prof. B.K. Bajaj (BKB) thankfully acknowledges the Govt. of India funding agencies (DST, DBT, CSIR, ICMR, and UGC) for granting financial aid in the form of Research Projects; BKB acknowledges international agencies for overseas fellowships/research stays: IUSSTF Research Fellowship at Ohio State University, USA; Commonwealth Fellowship at Institute of Biological, Environmental and Rural Sciences (IBERS), Aberystwyth University, Aberystwyth, UK; VLIR-UOS-Short Research Stay Scholarship at Antwerp University, Antwerp, Belgium; ERASMUS MUNDUS Invited Professorship at University of Naples, Italy; and COFUND-International Senior Research Fellowship at Durham University, Durham, UK. Miss Surbhi Sharma, Miss Ridhika Bangotra, and Miss Bisma Habib acknowledge the University of Jammu for providing a University Research Fellowship for Ph.D. research. Authors gratefully acknowledge the School of Biotechnology, University of Jammu for research infrastructure and facilities.

References

1. Lee, J., Kim, S., You, S., Park, Y. K., Bioenergy generation from thermochemical conversion of lignocellulosic biomass-based integrated renewable energy systems. *Renewable Sustainable Energy Rev.*, 178, 113240, 2023.
2. Motevali, A., Hooshmandzadeh, N., Fayyazi, E., Valipour, M., Yue, J., Environmental impacts of biodiesel production cycle from farm to manufactory: An application of sustainable systems engineering. *Atmosphere*, 14, 399, 2023.
3. Chan, Y. H., Lock, S. S. M., Chin, B. L. F., Wong, M. K., Loy, A. C. M., Foong, S. Y., Lam, S. S., Progress in thermochemical co-processing of biomass and sludge for sustainable energy, value-added products and circular economy. *Bioresour. Technol.*, 129061, 2023.
4. Mohan, S. V., Dahiya, S., Amulya, K., Katakojwala, R., Vanitha, T. K., Can circular bioeconomy be fueled by waste biorefineries—A closer look. *Bioresour. Technol. Rep.*, 7, 100277, 2019.
5. Shahid, A., Malik, S., Zhu, H., Xu, J., Nawaz, M. Z., Nawaz, S., Mehmood, M. A., Cultivating microalgae in wastewater for biomass production, pollutant removal, and atmospheric carbon mitigation: A review. *Sci. Total Environ.*, 704, 135303, 2020.
6. Nazloo, E. K., Moheimani, N. R., Ennaceri, H., Biodiesel production from wet microalgae: Progress and challenges. *Algal Res.*, 102902, 2022.

7. Gupta, S., Pawar, S. B., An integrated approach for microalgae cultivation using raw and anaerobic digested wastewaters from food processing industry. *Bioresour. Technol.*, 269, 571-576, 2018.
8. Ghaffar, I., Deepanraj, B., Sundar, L. S., Vo, D. V. N., Saikumar, A., Hussain, A., A review on the sustainable procurement of microalgal biomass from wastewaters for the production of biofuels. *Chemosphere*, 137094, 2022.
9. Martins, M.G., da Silva Arouche, T., Neto, A.F.G., da Cruz, J.N., da Costa, F.L.P., Fernandes, L.L., de Carvalho Junior, R.N., da Silva Costa, J.F., de Jesus Chaves Neto, A.M., Density functional theory for the thermodynamic gas-phase investigation of butanol biofuel and its isomers mixed with gasoline and ethanol. *J. Mol. Model.* 27, 80, 2021.
10. Mehar, J., Shekh, A., Uthaiah Malchira, N., Mudliar, S., Potential of microalgae for integrated biomass production utilizing CO2 and food industry wastewater. In: *Application of Microalgae in Wastewater Treatment*, Gupta, S.K. and Bux, F. (eds.) pp. 41–67. Springer International Publishing, Cham, 2019.
11. Shahid, A., Ishfaq, M., Ahmad, M. S., Malik, S., Farooq, M., Hui, Z., Mehmood, M. A., Bioenergy potential of the residual microalgal biomass produced in city wastewater assessed through pyrolysis, kinetics and thermodynamics study to design algal biorefinery. *Bioresour. Technol.*, 289, 121701, 2019.
12. Gherghel, A., Teodosiu, C., De Gisi, S., A review on wastewater sludge valorisation and its challenges in the context of circular economy. *J. Clean. Prod.*, 228, 244-263, 2019.
13. Tyagi, V. K., Lo, S. L., Sludge: A waste or renewable source for energy and resources recovery?. *Renewable Sustainable Energy Rev.*, 25, 708-728, 2013.
14. Moraes, E.S., Reis, G.F., Cruz, J., Cozzolino, K., Neto, A.F.G., Andrade-Filho, T., Neto, A.M.J.C., Thermodynamics study of biokerosene from coconut and palm kernel oils and JP-8 aircraft fuels in the gas phase by the DFT method. *J. Mol. Model.* 26, 79, 2020.
15. Savla, N., Pandit, S., Mathuriya, A. S., Gupta, P. K., Khanna, N., Babu, R. P., Kumar, S., Recent advances in bioelectricity generation through the simultaneous valorization of lignocellulosic biomass and wastewater treatment in microbial fuel cell. *Sustainable Energy Technol. Assess.*, 48, 101572, 2021.
16. Wang, F., Ouyang, D., Zhou, Z., Page, S. J., Liu, D., Zhao, X., Lignocellulosic biomass as sustainable feedstock and materials for power generation and energy storage. *J. Energy Chem.*, 57, 247-280, 2021.
17. Ma, J., Wu, S., Shekhar, N. V., Biswas, S., Sahu, A. K., Determination of physicochemical parameters and levels of heavy metals in food waste water with environmental effects. *Bioinorg. Chem. Appl.*, 2020.
18. Azanaw, A., Birlie, B., Teshome, B., Jemberie, M., Textile effluent treatment methods and eco-friendly resolution of textile wastewater. *Case Stud. Chem. Environ. Eng.*, 100230, 2022.

19. Some, S., Mondal, R., Mitra, D., Jain, D., Verma, D., Das, S., Microbial pollution of water with special reference to coliform bacteria and their nexus with environment. *Energy Nexus*, 1, 100008, 2021.
20. Li, J., Luo, G., He, L., Xu, J., Lyu, J., Analytical approaches for determining chemical oxygen demand in water bodies: A review. *Crit. Rev. Anal. Chem.*, 48(1), 47-65, 2018.
21. Aniyikaiye, T. E., Oluseyi, T., Odiyo, J. O., Edokpayi, J. N., Physico-chemical analysis of wastewater discharge from selected paint industries in Lagos, Nigeria. *Int. J. Environ. Res. Public Health.*, 16(7), 1235, 2019.
22. Wang, Y., Bai, Y., Su, J., Ali, A., Gao, Z., Huang, T., Ren, M., Advances in microbially mediated manganese redox cycling coupled with nitrogen removal in wastewater treatment: A critical review and bibliometric analysis. *J. Chem. Eng.*, 461, 141878, 2023.
23. Shao, Q., Zhang, Y., Liu, Z., Long, L., Liu, Z., Chen, Y., Huang, L. Z., Phosphorus and nitrogen recovery from wastewater by ceramsite: Adsorption mechanism, plant cultivation and sustainability analysis. *Sci. Total Environ.*, 805, 150288, 2022.
24. Beckinghausen, A., Odlare, M., Thorin, E., Schwede, S., From removal to recovery: An evaluation of nitrogen recovery techniques from wastewater. *Appl. Energy*, 263, 114616, 2020.
25. Nagtode, V.S., Cardoza, C., Yasin, H.K.A., Mali, S.N., Tambe, S.M., Roy, P., Singh, K., Goel, A., Amin, P.D., Thorat, B.R., Cruz, J.N., Pratap, A.P., Green Surfactants (Biosurfactants): A petroleum-free substitute for sustainability—Comparison, applications, market, and future prospects. *ACS Omega*. 8, 11674, 2023.
26. Ezugbe, E., Rathilal, S., Membrane technologies in wastewater treatment: A review. *Membranes*, 10(5), 89, 2020.
27. Bhateria, R., Jain, D., Water quality assessment of lake water: A review. *Sustain. Water Resour. Manage.*, 2, 161-173, 2016.
28. Manyuchi, M. M., Mbohwa, C., Muzenda, E., Municipal sewage wastewater treatment, in: *Resource Recovery from Municipal Sewage Plants,* M.M. Manyuchi, C. Mbohwa, E. Muzenda (Ed.), pp. 5-14, CRC Press, 2018.
29. Prajapati, K. B., Singh, R., Enhancement of biogas production in bio-electrochemical digester from agricultural waste mixed with wastewater. *Renewable Energy*, 146, 460-468, 2020.
30. Shrestha, R., Ban, S., Devkota, S., Sharma, S., Joshi, R., Tiwari, A. P., Joshi, M. K., Technological trends in heavy metals removal from industrial wastewater: A review. *J. Environ. Chem. Eng.*, 9, 105688, 2021.
31. Kesari, K. K., Soni, R., Jamal, Q. M. S., Tripathi, P., Lal, J. A., Jha, N. K., Ruokolainen, J., Wastewater treatment and reuse: A review of its applications and health implications. *Water Air Soil Pollut.*, 232, 1-28, 2021.
32. Qin, Y., Wang, K., Xia, Q., Yu, S., Zhang, M., An, Y., Zhou, Z., Up-concentration of nitrogen from domestic wastewater: A sustainable strategy from removal to recovery. *Chem. Eng. J.*, 451, 138789, 2023.

33. Devi, N. D., Sun, X., Hu, B., Goud, V. V., Bioremediation of domestic wastewater with microalgae-cyanobacteria co-culture by nutritional balance approach and its feasibility for biodiesel and animal feed production. *Chem. Eng. J.*, 454, 140197, 2023.
34. Tawfik, A., Bakr, M. H., Nasr, M., Haider, J., Lim, H., Qyyum, M. A., Lam, S. S., Economic and environmental sustainability for anaerobic biological treatment of wastewater from paper and cardboard manufacturing industry. *Chemosphere*, 289, 133166, 2022.
35. Gudkov, A. G., Pavlov, M. V., Karpov, D. F., *IOP Conf. Ser.: Earth Environ. Sci.*, 1154 012041, 2023.
36. Sathya, R., Arasu, M. V., Al-Dhabi, N. A., Vijayaraghavan, P., Ilavenil, S., Rejiniemon, T. S., Towards sustainable wastewater treatment by biological methods–A challenges and advantages of recent technologies. *Urban Clim.*, 47, 101378, 2023.
37. Buta, M., Hubeny, J., Zieliński, W., Harnisz, M., Korzeniewska, E., Sewage sludge in agriculture–the effects of selected chemical pollutants and emerging genetic resistance determinants on the quality of soil and crops–A review. *Ecotoxicol. Environ. Saf.*, 214, 112070, 2021.
38. Piri, J., Pirzadeh, B., Keshtegar, B., Givehchi, M., A hybrid statistical regression technical for prediction wastewater inflow. *Comput. Electron. Agric.*, 184, 106115, 2021.
39. Rout, P. R., Shahid, M. K., Dash, R. R., Bhunia, P., Liu, D., Varjani, S., Surampalli, R. Y., Nutrient removal from domestic wastewater: A comprehensive review on conventional and advanced technologies. *J. Environ. Manage.*, 296, 113246, 2021.
40. McCarty, P. L., Bae, J., Kim, J., Domestic wastewater treatment as a net energy producer–can this be achieved? *Environ. Sci. Technol.*, 45, 7100-7106, 2011.
41. Akhtar, N., Syakir Ishak, M. I., Bhawani, S. A., Umar, K., Various natural and anthropogenic factors responsible for water quality degradation: A review. *Water*, 13(19), 2660, 2021.
42. Ajiboye, T. O., Oyewo, O. A., Onwudiwe, D. C., Simultaneous removal of organics and heavy metals from industrial wastewater: A review. *Chemosphere*, 262, 128379, 2021.
43. Dutta, D., Arya, S., Kumar, S., Industrial wastewater treatment: Current trends, bottlenecks, and best practices. *Chemosphere*, 285, 131245, 2021.
44. Ucar, D., Zhang, Y., Angelidaki, I., An overview of electron acceptors in microbial fuel cells. *Front. Microbiol.*, 8, 643, 2017.
45. Jiang, X and Li M., Ecological safety hazards of wastewater, in: *High-Risk Pollutants in Wastewater*, pp.101-123, 2020.
46. Fu, L., Huang, T., Wang, S., Wang, X., Su, L., Li, C., Zhao, Y., Toxicity of 13 different antibiotics towards freshwater green algae *Pseudokirchneriella subcapitata* and their modes of action. *Chemosphere*, 168, 217-222, 2017.
47. Asano, T., Burton, F., Leverenz, H., *Water reuse: issues, technologies, and applications. McGraw-Hill Education,* 2007.

48. Duan, N., Xiong, J., Feng, Q., Wang, L., Yang, F., Ding, H., Special rural sewage treatment plan in Jiangxia district, Wuhan City, China. *Sustainability*, 15(3), 1764, 2023.
49. Estévez, S., García, S., Feijoo, G., Moreira, M. T., How decentralized treatment can contribute to the symbiosis between environmental protection and resource recovery. *Sci. Total Environ.*, 812, 151485, 2022.
50. Sathya, K., Nagarajan, K., Malar, G., Rajalakshmi, S., Lakshmi, P., A comprehensive review on comparison among effluent treatment methods and modern methods of treatment of industrial wastewater effluent from different sources. *Appl. Water Sci.*, 12(4), 70, 2022.
51. Liu, Y., Ramin, P., Alsina, X., Gernaey, K. V., Transforming data into actionable knowledge for fault detection, diagnosis and prognosis in urban wastewater systems with AI techniques: A mini-review. *Process Saf. Environ. Prot.*, 2023.
52. Sangamnere, R., Misra, T., Bherwani, H., Kapley, A., Kumar, R., A critical review of conventional and emerging wastewater treatment technologies. *Sustain. Water Resour. Manage.*, 9(2), 58, 2023.
53. Saravanan, A., Kumar, P. S., Jeevanantham, S., Karishma, S., Tajsabreen, B., Yaashikaa, P. R., Reshma, B., Effective water/wastewater treatment methodologies for toxic pollutants removal: Processes and applications towards sustainable development. *Chemosphere*, 280, 130595, 2021.
54. Mostafazadeh, A. K., Zolfaghari, M., Drogui, P., Electrofiltration technique for water and wastewater treatment and bio-products management: A review. *J. Water Process. Eng.*, 14, 28-40, 2016.
55. Swanckaert, B., Geltmeyer, J., Rabaey, K., De Buysser, K., Bonin, L., Clerck, K., A review on ion-exchange nanofiber membranes: properties, structure and application in electrochemical (waste) water treatment. *Sep. Purif. Technol.*, 120529, 2022.
56. Cardenas, J. A., García, B., Agüera, A., Pérez, J. A., Agugliaro, F., Wastewater treatment by advanced oxidation process and their worldwide research trends. *Int. J. Environ. Res. Public Health.*, 17(1), 170, 2020.
57. Sellaoui, L., Aviles, A., Dhaouadi, F., Bedia, J., Petriciolet, A., Rtimi, S., Belver, C. Adsorption of emerging pollutants on lignin-based activated carbon: Analysis of adsorption mechanism via characterization, kinetics and equilibrium studies. *Chem. Eng. J.*, 452, 139399, 2023.
58. Wang, L., He, J., Heiranian, M., Fan, H., Song, L., Li, Y., Elimelech, M., Water transport in reverse osmosis membranes is governed by pore flow, not a solution-diffusion mechanism. *Sci. Adv.*, 9(15), eadf8488, 2023.
59. Lakshmi, K. S., Sailaja, V. H., Reddy, M. A. Phytoremediation-a promising technique in waste water treatment. *Int. J. Sci. Res.*, 5(06), 5480-5489, 2017.
60. Sugurbekova, G., Nagyzbekkyzy, E., Sarsenova, A., Danlybayeva, G., Anuarbekova, S., Kudaibergenova, R., Moldagulova, N., Sewage sludge management and application in the form of sustainable fertilizer. *Sustainability*, 15(7), 6112, 2023.

61. Collivignarelli, M. C., Abbà, A., Carnevale Miino, M., Torretta, V., What advanced treatments can be used to minimize the production of sewage sludge in WWTPs? *Appl. Sci.*, 9(13), 2650, 2019.
62. Rulkens, W., Sewage sludge as a biomass resource for the production of energy: Overview and assessment of the various options. *Energy Fuel*, 22(1), 9-15, 2008.
63. Bhatt, P., Bhandari, G., Turco, R. F., Aminikhoei, Z., Bhatt, K., Simsek, H., Algae in wastewater treatment, mechanism, and application of biomass for production of value-added product. *Environ. Pollut.*, 309, 119688, 2022.
64. Salama, E. S., Kurade, M. B., Abou-Shanab, R. A., El-Dalatony, M. M., Yang, I. S., Min, B., Jeon, B. H., Recent progress in microalgal biomass production coupled with wastewater treatment for biofuel generation. *Renewable Sustainable Energy Rev.*, 79, 1189-1211, 2017.
65. Park, J. B. K., Craggs, R. J., Shilton, A. N., Wastewater treatment high rate algal ponds for biofuel production. *Bioresour. Technol.*, 102(1), 35-42, 2011.
66. Kurniawan, S. B., Ahmad, A., Said, N. S. M., Imron, M. F., Abdullah, S. R. S., Othman, A. R., Hasan, H. A., Macrophytes as wastewater treatment agents: Nutrient uptake and potential of produced biomass utilization toward circular economy initiatives. *Sci. Total Environ.*, 790, 148219, 2021.
67. Kinidi, L., Salleh, S., Phytoremediation of nitrogen as green chemistry for wastewater treatment system. *Int. J. Chem. Eng.*, 2017.
68. Edgar, V. N., Fabián, F. L., Mario, P. C. J., Ileana, V. R., Coupling plant biomass derived from phytoremediation of potential toxic-metal-polluted soils to bioenergy production and high-value by-products-A review. *Appl. Sci.*, 11(7), 2982, 2021.
69. Saini, S., Tewari, S., Dwivedi, J., & Sharma, V., Biofilm mediated wastewater treatment: A comprehensive review. *Mater. Adv.*, 2023.
70. Udaiyappan, A. F. M., Hasan, H. A., Takriff, M. S., Abdullah, S. R. S., A review of the potentials, challenges and current status of microalgae biomass applications in industrial wastewater treatment. *J. Water Process. Eng.*, 20, 8-21, 2017.
71. Bhat, B., Bajaj, B. K., Hypocholesterolemic and bioactive potential of exopolysaccharide from a probiotic *Enterococcus faecium* K1 isolated from kalarei. *Bioresour. Technol.*, 254, 264-267, 2018.
72. Bhat, B., Bajaj, B. K., Hypocholesterolemic potential and bioactivity spectrum of an exopolysaccharide from a probiotic isolate *Lactobacillus paracasei* M7. *Bioact. Carbohydr. Diet. Fibre*, 19, 100191, 2019.
73. Roeselers, G., Loosdrecht, M. V., Muyzer, G., Phototrophic biofilms and their potential applications. *J. Appl. Phycol.*, 20, 227-235, 2008.
74. Sharma, S., Nargotra, P., Sharma, V., Bangotra, R., Kaur, M., Kapoor, N., Bajaj, B. K., Nanobiocatalysts for efficacious bioconversion of ionic liquid pretreated sugarcane tops biomass to biofuel. *Bioresour. Technol.*, 333, 125191, 2021.

75. Voloshin, R. A., Rodionova, M. V., Zharmukhamedov, S. K., Veziroglu, T. N., & Allakhverdiev, S. I., Biofuel production from plant and algal biomass. *Int. J. Hydrog. Energy*, 41, 17257-17273, 2016.
76. Zhou, W., Chen, P., Min, M., Ma, X., Wang, J., Griffith, R., Ruan, R., Environment-enhancing algal biofuel production using wastewaters. *Renewable Sustainable Energy Rev.*, *36*, 256-269, 2014.
77. Sarwer, A., Hamed, S. M., Osman, A. I., Jamil, F., Al-Muhtaseb, A. A. H., Alhajeri, N. S., Rooney, D. W., Algal biomass valorization for biofuel production and carbon sequestration: A review. *Environ. Chem. Lett.*, 20, 2797-2851, 2022.
78. Abdullah, S. H. Y. S., Hanapi, N. H. M., Azid, A., Umar, R., Juahir, H., Khatoon, H., Endut, A., A review of biomass-derived heterogeneous catalyst for a sustainable biodiesel production. *Renewable Sustainable Energy Rev.*, 70, 1040-1051, 2017.
79. Sharma, S., Tsai, M. L., Sharma, V., Sun, P. P., Nargotra, P., Bajaj, B. K., Dong, C. D., Environment friendly pretreatment approaches for the bioconversion of lignocellulosic biomass into biofuels and value-added products. *Environments*, 10, 6, 2022.
80. Toor, M., Kumar, S. S., Malyan, S. K., Bishnoi, N. R., Mathimani, T., Rajendran, K., Pugazhendhi, A. An overview on bioethanol production from lignocellulosic feedstocks. *Chemosphere*, 242, 125080, 2020.
81. Sirajunnisa, A. R., Surendhiran, D., Algae–A quintessential and positive resource of bioethanol production: A comprehensive review. *Renewable Sustainable Energy Rev.*, 66, 248-267, 2016.
82. Sharma, I., Rackemann, D., Ramirez, J., Cronin, D. J., Moghaddam, L., Beltramini, J. N., Doherty, W. O., Exploring the potential for biomethane production by the hybrid anaerobic digestion and hydrothermal gasification process: A review. *J. Clean. Prod.*, 132507, 2022.
83. Elshobary, M. E., Zabed, H. M., Yun, J., Zhang, G., Qi, X., Recent insights into microalgae-assisted microbial fuel cells for generating sustainable bioelectricity. *Int. J. Hydrogen Energy*, 46(4), 3135-3159, 2021.
84. Arun, S., Sinharoy, A., Pakshirajan, K., Lens, P. N., Algae based microbial fuel cells for wastewater treatment and recovery of value-added products. *Renewable Sustainable Energy Rev.*, 132, 110041, 2020.
85. Kannan, N., Donnellan, P., Algae-assisted microbial fuel cells: A practical overview. *Bioresour. Technol. Rep.*, 15, 100747, 2021.
86. Xiao, R., Ou, T., Ding, S., Fang, C., Xu, Z., Chu, W. Disinfection by-products as environmental contaminants of emerging concern: a review on their occurrence, fate and removal in the urban water cycle. *Crit. Rev. Environ. Sci. Technol.*, *53*(1), 19-46, 2023.
87. Li, H., Guo, Y., Liu, C., Zhou, Y., Lin, X., Gao, F., Microbial deposition and growth on polyamide reverse osmosis membrane surfaces: Mechanisms, impacts, and potential cures. *Desalination*, 548, 116301, 2023.

88. Elhenawy, S., Khraisheh, M., AlMomani, F., Al-Ghouti, M., Hassan, M. K., From waste to watts: Updates on key applications of microbial fuel cells in wastewater treatment and energy production. *Sustainability*, 14, 955, 2022.
89. Kolajo, O. O., Pandit, C., Thapa, B. S., Pandit, S., Mathuriya, A. S., Gupta, P. K., Upadhye, V. J., Impact of cathode biofouling in microbial fuel cells and mitigation techniques. *Biocatal. Agric. Biotechnol.*, 102408, 2022.

9

Recovery of Value-Added Products from Sewage Sludge: Processes, Life Cycle Assessment, and Costs

Abiola E. Taiwo[1], Olayomi A. Falowo[2], Anthony I. Okoji[3], Lekan M. Latinwo[4] and Eriola Betiku[4*]

[1]*Faculty of Engineering, Mangosuthu University of Technology, Durban, South Africa*
[2]*Department of Chemical Engineering, Landmark University, Omu-Aran, Kwara State, Nigeria*
[3]*Department of Chemical Engineering, Covenant University, Ota, Ogun State, Nigeria*
[4]*Department of Biological Sciences, Florida Agricultural and Mechanical University, Tallahassee, FL, USA*

Abstract

Biological sewage sludge (SS) from wastewater treatment is generating environmental concerns due to the amount generated worldwide. In the past, sludge disposal was considered a nuisance associated with wastewater treatment. However, a growing body of evidence suggests that emerging pollutants have contributed to significant limitations and higher costs associated with the conventional disposal of SS. As the water industry has been pursuing sustainability and circular economy principles in recent years, the focus of investigations has shifted to sludge-embedded resource recovery. Sludge from sewage treatment plants has been used to process energy, nutrients, and other platform chemicals for use as resource recovery products. However, to identify an environment-friendly process route for optimum SS utilization at minimum cost, there is a need to evaluate the environmental impacts of sludge treatment alternatives using life cycle assessment (LCA) and economic indicators. This chapter seeks to integrate the products recovery of value-added products from SS, considering available product production processes, economic impact, and environmental sustainability through LCA for a circular economy.

**Corresponding author*: eriola.betiku@famu.edu

Inamuddin, Tariq Altalhi, Mohammad Luqman and Joseph Kapuku (eds.) Sewage and Biomass from Wastewater to Energy, (225–258) © 2024 Scrivener Publishing LLC

Keywords: Economics, life cycle assessment, recovery, sewage sludge, value-added products, waste management, wastewater

9.1 Introduction

Disposing of organic waste products obtained from wastewater plants is a major environmental concern owing to the vast amounts of this waste generated globally. Sewage sludge (SS) is a semisolid wastewater treatment process byproduct in the primary or secondary sedimentation tank and other water treatment facilities [1]. The high concentration of organics in sludge and the possibility of recovering components make it an important renewable resource with potential economic benefits for energy, nutrients, or other platform chemicals [2]. There are many techniques for recovering value-added materials from SS, including chemical, biological, physical, thermochemical, and electrochemical processes. Studies have shown the recovery and potential availability of many value-added products from SS, with application for energy use, nutrients, and other valuable resources [1, 3, 4]. This evolution in the bioprocess recovery of value-added products from SS could improve the energy mix by providing an alternative to fossil fuel through its use as a substrate in biogas, bioethanol, and bio-hydrogen production [5]. In the sustainability context, wastewater should be acknowledged as a renewable resource since it offers opportunities from which essential resources, such as chemicals, bioplastics, fertilizer, water, and energy, are recovered [6]. SS is projected to ensure social, economic, and environmental sustainability if guided by the appropriate treatment technologies and infrastructure since it is considered an asset instead of a liability [7]. Likewise, the concerted effort on sludge management in wastewater industries would be reduced if recovery of value-added products is integrated into the production system.

Rapid urbanization has resulted in the continuous SS generation. Analyzing the impact of various technical strategies on the environment and assessing their sustainability is essential to minimizing resource input and waste output. In the 21st century, life cycle assessment (LCA) and costing can be valuable tools for SS environmental management worldwide [8, 9]. A slight deviation from critical process parameters can cause an environmental gain or loss. Based on key parameters, sludge treatment technologies must be evaluated regarding their environmental benefits and drawbacks. Sensitivity analysis and uncertainty analysis are both elements of LCA that can help with this type of identification. Hence, the primary focus of this chapter is to highlight the products recovery of value-added products from SS considering production processes, economic impact, and LCA for a circular economy. As a result of this phase of technological advancement and upscaling processes of SS, it is expected that leading industries and companies will be able to see value-added materials from wastewater as part of the circular bioeconomy.

9.2 Methods and Product Recovery Processes

Resource recovery techniques for SS involve biochemical, physical, and physicochemical processes or a combination of two or more processes [10]. Due to increased population, industrialization, and urbanization, the volume of SS generated is expected to be on the increase [11]. This poses a challenge as ensuring the supply of fresh water and wastewater treatment becomes vital to the growing population's public safety. Recently, sludge from wastewater treatment plants (WWTPs) has been considered a biomass and a promising renewable energy source with relatively high energy content [12]. Likewise, it is claimed to be a viable source of recovering chemicals, fuels, nutrients, and energy [13]. Much attention is paid to the way the world manages resources, materials, and products. Our environment's sustainability is related to how resources and waste are managed. For a long time, water and wastewater system policies have been primarily driven on account of function, human/environmental protection, and cost-benefit analysis [6]. Going by the depletion of available resources and the consideration for the environment, resources in wastewater should be recovered from sludge for reutilization. Developing infrastructures, technologies, research, and policies promoting cost-effective recovery will have broad geopolitical impacts, particularly large-scale economic applications [6]. The recovery pathways, technologies, and resources from SS are depicted in Figure 9.1.

Reutilization of wastewater provides a unique opportunity to solve water scarcity challenges and offers the possibility for nutrients, chemicals, and energy recovery, further offsetting part of the exploitation of valuable resources and treatment costs [15].

A new start of the innovation economy involves generating value-added products from wastewater systems, from which energy, organic products, nutrients, and other resources are recovered as valuable products rather than wastefully destroyed [16]. A concerted approach is required

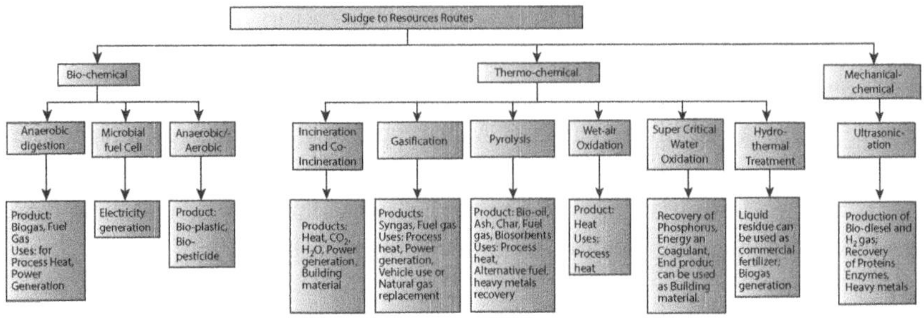

Figure 9.1 Processing techniques from sludge to products (adapted from Tyagi and Lo [14] with permission from Elsevier).

to maximize resource recovery from SS so that each industrial sector concerned can integrate its waste management methodologies with others. The technological solutions provided by the expert may not be sufficient in achieving sustainable resource recovery without the collaborative input of stakeholders which is central to a sustainable planning and design paradigm [6]. The conventional disposal of SS is through landfilling or land spreading until stricter regulations restrict such practices. Improvements in treatment technologies allow sludge to be used as feedstocks or substrates in anaerobic digestion, composting, incineration, wet air oxidation, gasification, and pyrolysis processes. Figure 9.2 depicts recovery technologies and the value-added products generated from SS.

9.2.1 Conventional Method

The conventional method is mainly the traditional technique of managing SS. The approach favors waste management or environmental protection rather than resource recovery. However, this method, for a long time, has been applied to benefit agricultural practices. The most conventional method of handling SS is landfilling.

9.2.2 Landfilling

Conventionally, sludge is disposed of through landfilling or land-spreading operations. The method of landfilling SS is designed to manage solid waste rather than resource recovery. The current practice of landfilling sludge involves mixing concentrated sludge with other municipal solid wastes (MSWs) [17]. Across all levels, there is a gradual shift and usage in landfilling techniques of SS as modern techniques shift toward efficient technologies (thermal and electrochemical) for resource recovery [18].

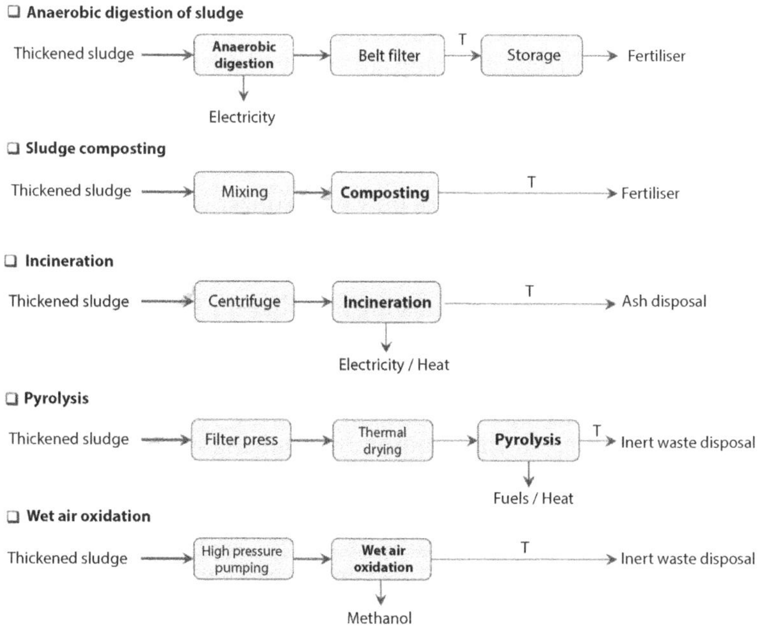

Figure 9.2 Overview of sludge treatment and resources recovery routes (adapted from Tarpani and Azapagic [13] with permission from Elsevier).

Before implementing landfill techniques in handling sludge, certain engineering guidelines must be followed in the design and construction of the site. Generally, landfill disposal of SS affects the leachate production and the CO_2 emissions directly into the air in landfilled sites [19]. Recovering materials and resources via landfilling is being avoided because it does not represent an efficient environmental approach.

Besides, landfills generally involve using vast large areas [20], thereby taking up some land that should have been available for other purposes. Without pathogens or toxic substances, land application of SS is considered a great incentive due to its fertilizer and soil conditioning properties [21]. It is claimed that SS compost could serve as a soil conditioner in excavated land due to landfill operation and could be a cost-effective strategy to restore the fertility of reclaimed soil [20]. The practice of claiming valuable material through landfilling causes ecological disturbances. A landfill, where most SS is reclaimed, usually applies mined soil for reclamation. After landfill closure, a huge amount of organic fertilizer is needed to induce revegetation and fertility since most reclaimed soil contains very little organic matter.

9.2.3 Composting

Compost produced from mechanically and biologically treated processes is believed to be a good substitute for inorganic fertilizer. Composting is a cost-effective technology to manage SS and eliminate pathogens [22]. It is an ideal way of converting sludge into a stable organic matter which could serve as a soil fertilizer. Not only biofertilizer can be generated from composting methods but also the extraction of heavy metals during composting [23]. Moreover, important nutrients such as nitrogen can be recovered via composting [24]. Composting of SS has the following advantages: stabilization of sludge, destruction of pathogenic microorganisms, and reduction of mass and water content [25]. Sludge composting involves an aerobic process in a batch composter (inclined rotating cylinder). Before the main composting in the windrow, the bulking agent is added to the composting mixture to adjust the carbon-to-nitrogen ratio and create an ideal aeration environment. The bulking agents include agricultural, industrial, green, food, and wood wastes [22]. The composting materials are then transferred to windrows to decompose biodegradable organic compounds. They can then be left for a few weeks to several months to complete the composting process. The stable compost fraction, mainly humus and macrofauna, is generated [25]. The finished compost can be applied as a soil conditioner [13]. The commonly used methods in composting sludge are reactor composting and vermicomposting (lab-scale), windrowing composting, and pile composting (plant-scale) [22]. Handling SS can be problematic; hence, constructing facilities where the sludge can be stored for some time offers a promising solution. The stored sludge can be composted with other organic wastes such as sawdust, mushroom residue, straw, fly ash, and green waste to generate matured compost. Adding another organic matter to the stored sludge is known as co-composting. This modification can improve its quality, overcoming some sludge-associated problems such as high water content, contamination with pathogenic microorganisms, and lack of stability [26].

Unlike the landfilling technique, sludge composting offers promising results, and it is adopted across all levels for agricultural operations. The agricultural wastes compost produced from the olive grove wastes and from sheep manure generated on the farm, as well as meat flour, is used as biofertilizer. The finished compost is usually applied on the same farm as an organic amendment of the soil [27]. The compost is produced from agricultural wastes and sludge in a composting plant designed to collect and temporarily store SS and agricultural wastes from the local producers. The compost materials are usually formulated using 30% SS, 25% agricultural wastes, and 45% woody materials, resulting mixture being allowed to stabilize by composting [27].

9.3 Biochemical Methods

Many new technologies are aiding the recovery of value-added products from wastewater; however, biological methods provide the most robust outcomes in recovering valuable resources from sludge mixtures. Biomass involves the biological recovery process including chemotrophic, microalgae, photosynthetic bacteria, phototrophic, and terrestrial plants for organics recovery. The biological method can facilitate the recovery of bioenergy (biogas, biodiesel, bio-hydrogen, and bioethanol), nutrients (single-cell protein, carbon, organic fertilizer, and organic acids), hormones, enzymes, radioactive materials, and biopolymer (bioplastics) from sludge. The technology can also synthesize highly specialized metal oxidizing and reducing organisms for heavy metal recovery [16]. Some of the components or forms of biochemical sludge utilization are discussed in this section.

9.3.1 Fermentation

Fermentation of waste sludge involves the biodegradation of soluble organic substances in sludge by the microbial community. The sludge fermentation process is an economical approach to recovering organic carbon and phosphorus [28]. Besides, photofermentation and dark fermentation processes are potential routes for energy recovery (hydrogen production) from waste sludge due to their high organic matter [24].

9.3.2 Anaerobic Digestion

This method involves degrading organic materials in sludge by microbial communities under anaerobic conditions, followed by dewatering in a belt filter. This technique is the most popular sludge stabilization technology, transforming organic solids into biogas (a mixture of CH_4 and CO_2). The process involves four biochemical reaction phases: hydrolysis, acidogenesis, acetogenesis, and methanogenesis [14]. To aid biogas production, sludge pretreatment is proposed to accelerate the hydrolytic step. Applying microwave, ultrasonic, and thermal pretreatment of sludge has led to an incremental biogas yield [29]. After digestion, the leftover sludge rich in nutrients can be applied on a farm as a biofertilizer. Biogas produced in the digester can be used to supply thermal energy to the biodigester, and the excess biogas can be upgraded to generate electricity [30]. Anaerobic digestion of sludge offers several advantages: effective pathogen removal, less biomass sludge production, minimal odor emissions, improved digestate as biofertilizer, and carbon-neutral energy [31]. The energy generated from the anaerobic digestion of sludge can meet a portion of the energy requirements for the waste treatment facility. To ensure satisfactory management of the anaerobic digestion process from sludge, kinetic models were proposed to improve process performance [32]. Before digestion, sludge undergoes some treatment to reduce its volume and improve its characteristics by transforming the organic matter into stable organic and inorganic residues. The biogas produced from this technique has a high calorific value and factors like alkalinity, temperature, pH, and retention time that can affect the rate of the digestion process [33]. Apart from energy recovery, other nutrients, such as struvite, can be recovered from domestic waste sludge. Phosphorus recovery can then be initiated from the phosphorus-containing compound (struvite) [34].

9.4 Thermal Techniques

The application of thermal energy in resource recovery is increasingly applied though on a laboratory scale. Certain thermal energy is applied to improve resource recovery from SS. Sometimes pretreatment of sludge is carried out to enhance the possibility of recovering value-added products. In this context, the microwave technique is applied in the pretreatment of sludge [35], taking advantage of microwave energy. The common methods under the thermochemical technique of recovery of value-added products from SS are incineration, pyrolysis, gasification, supercritical water oxidation, wet air oxidation, and hydrothermal technology. All these technologies involved heat treatment of sludge at elevated temperatures before resources could be recovered.

9.4.1 Incineration

The goal of sludge incineration is to ensure complete oxidation of the organic matter at high temperatures. The incineration of organic compounds takes place in a combustion chamber supplied with oxygen to form CO_2, water, and inert material (ash) [14]. SS incineration is a popular treatment technique in the European Union, Japan, the United States, and Hong Kong [36]. Several valuable resources can be recovered from this method. Heat and electricity can be recovered by combusting sludge.

Hence, incineration with a power generation unit is implemented simultaneously. Besides, ash leftover after incineration is a potential raw material for building [14]. The sludge is mixed with a dewatering agent to reduce sludge volume and dewatered in centrifuges before incinerating to boost the efficacy of energy recovery from sludge incineration. The typical composition of primary and secondary SS has 1 to 4 wt% solids, and the sludge mixture must be pumped into tanks for further pretreatment. At a thickening stage, the sludge settles so that the supernatant is removed, and the solids content is increased to 3 to 8 wt% solids. The thickened sludge is then dewatered, typically using plate or belt presses. At this stage, organic or inorganic additives may be added to improve dewatering sludge [36]. Figure 9.3 shows the flow and recovery of resources from a typical incinerated system.

A vital advantage of sludge incineration is the control level provided to the operation managers. Incinerated sludge ash is used as an alternative to clay in sintered bricks, pavers, and tiles, as well as a raw material for making lightweight aggregate and in the manufacture of high-density glass-ceramics [36]. Interestingly, the sludge incineration method can also be applied to recover nutrients. This method recovers phosphorus via sludge incineration, and the residual ashes were then treated with subsequent acid

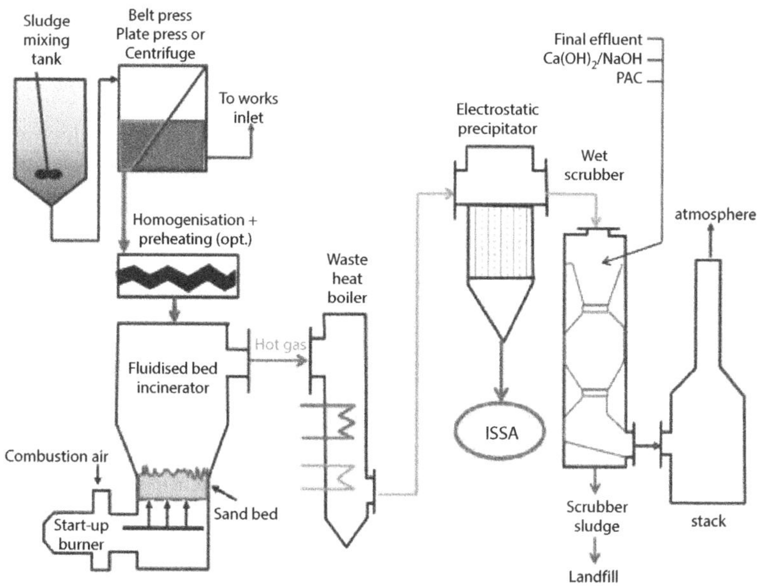

Figure 9.3 Resource recovery from the incineration facility (adapted from Donatello and Cheeseman [36] with permission from Elsevier).

extraction and adsorption process, which recovered phosphate from the ashes [10].

9.4.2 Pyrolysis

Sludge pyrolysis is a creative technique to manage sludge and recover energy. The sludge is thermally treated between 350°C and 500°C under pressure in the absence of oxygen [37]. The pyrolysis method begins with dewatering sludge in a filter press to reduce sludge volume, followed by thermal drying. Value-added products from sludge pyrolysis include combustible gasses, biochar, and bio-oil, which can substitute natural gas heating, charcoal, and diesel oil, respectively [13]. Like in the gasification method, sludge pyrolysis produces synthesis gasses, destroys organic compounds, provides stable residue, provides heat, lowers dioxins/furans, and lowers the volume of flue gas [14]. The pyrolysis product yields are affected by reaction time, process temperature, raw materials characteristics, and operating pressure. Pyrolysis of oily sludge is one of the prevalent practices in the oil industry for the recovery of chemicals and fuels. When pyrolyzed at 700°C, dewatered cake sludge is reported to be a good source of inorganic phosphorus. Hydrolysis treatment recovers nutrients such as phosphorus from SS, and the biochar obtained is considered promising for soil amendment or fertility [38].

9.4.3 Gasification

Gasification is a thermal-chemical complex treatment involving decomposing dried sludge at high temperatures into ash and combustible gasses in a system with a limited amount of oxygen [39]. The thermal decomposition of sludge takes place in a gasifier after the removal of water. Gasification products include heat, char, and syngas. The chemical characteristics of the gasification products and the energy value from the process are affected by the gasification agents, the operating temperature and pressure, and feed characteristics [14]. There are five zones in gasification thermal technology, namely, drying, pyrolysis, reduction, oxidation, and gasification zone. Oxidation of volatile materials from the pyrolysis zone is usually from 1000°C to 1100°C [40]. The temperature in the gasification zone is usually from 1400°C to 1700°C and pressure from 0.6 to 2.6 MPa leads to the formation of producer gasses (CO, H, and CH_4). At high temperatures, the ash from the sludge produces a molten slag, which is cooled and removed from the bottom of the gasifier, forming fine-grained slag particles. The raw gasses from the gasifier are scrubbed to remove CN, NH_3, and H_2S to produce high-quality wet syngas. Vitrified slag, fly ash, and tar can be used in concrete mixes [14].

9.4.4 Hydrothermal

The hydrothermal method of treating sludge involves a chemical digestion process in an aqueous solvent at elevated temperature and pressure. This can convert SS into nontoxic components and potentially produce valuable industrial byproducts [41]. Hydrothermal processing in sludge treatment can be broadly categorized into two main types, oxidative and non-oxidative techniques. Although chemicals or catalysts may or may not be applied in both cases, one major difference between the two techniques is the presence of oxidants and, subsequently, the end product obtained [42]. The non-oxidative thermal hydrolysis of sludge is a promising technique to increase methane production and simultaneously decrease the volume of solid materials. Hydrothermal treatments involve the fractionation of municipal SS by water, and the process is highly suitable for the recovery of resources from municipal wastewater since dewatering is avoided [3]. There are two basic types of hydrothermal treatment, namely, hydrothermal gasification (HTG) and hydrothermal liquefaction (HTL). Each methodology is considered promising in handling SS disposal while recovering fuels in the process. Production of syngas (H_2 and CH_4) is via HTG, while liquid fuel is through HTL and hydrochar by hydrothermal carbonization [12].

9.4.5 Wet Air Oxidation

This technique involves the oxidation of SS in a reactor where the sludge is mixed with air at high temperatures and pressure [13]. It is representative of the oxidative techniques of hydrothermal technologies, and it is usually carried out at temperatures greater than 200°C in the presence of an oxidant [41]. This method can treat high concentrations of toxic and hazardous industrial effluents. Hence, it is suitable for eliminating complex contaminants [43]. Wet air oxidation uses both heat and oxidants to degrade organic compounds [42]. A typical condition for wet oxidation of SS is the temperature (180°C to 350°C), pressure (2 to 15 MPa), and residence times in the range of 15 to 120 min, as well as the presence of oxidant mainly active oxygen species [44]. The technology adopted in this technique is flexible to treat different kinds of effluent containing high organic matter. For instance, a two-stage wet air oxidation was introduced to recover 93.1% of oil from oily sludge and reduce the volume of this type of sludge by 85.4% [43]. Mineralization into inorganic compounds in the form of inorganic salts, CO_2, and H_2O occurs when organic contaminants are partially degraded by oxidizing agents into biodegradable intermediates. The elemental species are in the process converted to inorganic salts: sulfur to sulfate, phosphorus to phosphates, halogens to halides, organic nitrogen to nitrate, and ammonia [44]. However, the wet air oxidation method's major weakness is its inability to achieve complete mineralization of organics since some oxygenated compounds are resistant to further transformation to CO_2.

9.5 Mechanical-Chemical Technique

Mechanical methods such as screening, grit removal, and sedimentation are usually applied to generate primary sludge from wastewater treatment. The mechanical sludge disintegration technique is based on cell disruption by shear stresses. Examples of mechanical treatment applications involve the use of stirred ball mills, high-pressure homogenizers, ultrasonic, and mechanical jet smash [29]. Recovery of value-added products using this technique is uncommon; however, a mechanical sludge pretreatment process is being applied to aid resource recovery. For example, mechanical pretreatment of sludge using a high-pressure homogenizer increases the rate of extracellular and intracellular substances released in the process and consequently increases methane yield [35].

9.5.1 Ultrasonication Method

The ultrasonic technology of resource recovery is the most common mechanical-chemical technique for sludge pretreatment. The method is economical, environmentally friendly, and efficient in higher product quality. The application of the ultrasonic method in waste recovery is to aid sludge disintegration. The cellular matter of the SS is disrupted, solubilizing the organic matter, and consequently improving biogas production. Ultrasonic wave–induced liquid can activate the organic solid and increase the movement of targeted nutrients or metal from the surface of waste sludge into the liquid phase. The sonication time and intensity are important parameters influencing the effectiveness of the technique [35]. The ultrasonic resource recovery method efficiently recycles heavy metals from waste sludge, such as in the case of Cu and Fe recovery from electronic waste sludge [45].

9.5.2 Electrochemical Technology

Electrochemical technology is an extensively exploited technique to recover resources from wastewater. This method works based on the foundation of an electrochemical reaction without adding a chemical agent. Electrochemical technologies can recover valuable resources such as Ag, Cu, Zn, Ni, and Cr, inorganic nutrients (N, P, and S), hydrogen (H_2), and value-added organic substances [46]. For example, Ag^+ ions are recovered from wastewater using electrochemical ion exchange technology [47]. Also, tellurium (Te) and copper (Cu) are electrochemically extracted from wastewater with a high extraction ratio [48]. P, NH_3, and N recovery from wastewater or sludge-containing solution is achieved through electrochemical stripping, electrochemically induced precipitation, and membrane-based electrochemical technique [49, 50]. The recovery of elemental sulfur via bioelectrochemical cell coupling with an electrochemical cell (EC) is achieved from wastewater through electrochemical oxidation. At operating conditions of 30 Am^{-2} and the EC of 7.5 Am^{-2}, 92.9% of elemental sulfur of all sulfate is recovered [51].

This technology is believed to be promising for resource recovery from sludge due to low maintenance and chemical requirements. The technology can rapidly uptake selective ions and release target ions by modulating the redox states without the side effects of secondary pollutants [47]. In addition, it possesses a unique advantage since the electrode reaction can be adjusted by tunning current and voltage, thereby minimizing side reactions while enhancing the purity of the products [46]. The electrochemical technologies of recovery value-added resources work on the fundamentals of electrochemical oxidation and reduction, electrodialysis, electrocoagulation, and capacitive deionization. However, the application of this methodology depends on the chemical features of metals, the concentration of the wastewater, and the targeted goal of the process [46].

9.6 Recoverable Products from SS

9.6.1 Energy Product

The previously mentioned techniques, including incineration, aerobic composting, anaerobic decomposition, pyrolysis, and hydrothermal carbonization, can help with the energy or recovering resources and are gaining much attention. The key trade-off between waste utilization and resource or energy recovery is an issue of debate among the aforementioned treatment procedures. Several studies have been performed to extend these boundaries so that targeted energy recovery systems are as environmentally friendly as possible [1, 2, 52]. As a viable energy source, sludge supports using clean energy while lowering greenhouse gas (GHG) emissions by reducing fossil fuel imports. The energy produced from sludge can provide numerous environmental benefits, such as decreased emissions of GHGs, acids, and gasses, an alternative to the exhaustion of natural fossil fuel reserves, and management of water and soil pollution [53]. As compared to landfills with energy recovery, MSW landfills were found to negatively impact ecosystems, conservation efforts, human health, and resource damage more significantly and outperform other schemes due to the disorderly gas released from landfills and the drawbacks of refraining from producing substances that cause acidification [53]. The incineration process can turn wastewater sludge into electricity. Power plants, incineration facilities, and other industrial sectors can work effectively and productively with WWTPs. However, depending on how much energy is recovered, it may or may not perform well in terms of the environment. Xu *et al.* [54] concluded that incineration causes the least environmental damage compared to landfills and anaerobic digestion of SS. Liu *et al.* [55] found that refined oil, municipal sludge, and coal were co-combusted to reduce global warming potential. Chen *et al.* [56] observed that the co-incineration of MSW and SS has complementary synergistic effects that resulted in value-added products, energy efficiency, and profitability of the process [56]. The co-incineration technique has the potential to recover a significant amount of heat and electricity while consuming less nonrenewable energy and completely utilizing the excess heat to dry the sludge. Therefore, using the bottom recycled slag for building material production has proven advantageous [56]. Nakakubo *et al.* [57] compared the different treatments of MSW and anaerobically digested sludge to co-incineration. The findings confirm that the co-incineration plant reduced CO_2 emissions by 18% compared to the separate treatment plant, owing primarily to eliminating some sludge pretreatment technologies. The result is identical to the conclusion reached by Nakatsuka *et al.* [58] who discovered a 35% reduction in CO_2 emissions by integrating WWTPs and incineration plants. These researchers also discovered that the amount of MSWs disposed of, and the efficiency of electricity generated play critical roles in the long-term utilization of municipal biosolids.

9.6.2 Nutrients and Vitamin Products

Studies have also been conducted that extend their focus to consider the environmental effects of using value-added products from sludge products as a substitute for chemical fertilizers. One method for closing the cycling of carbon and nutrients on farmland is to apply SS to it. Open dumping, pyrolysis, agricultural use, and replacing fuel with industrial processes have been the most popular sewage treatment handling methods over the past ten years; however, agricultural applications are preferred due to the waste hierarchy and the valuable N, P, and organic matter contained in sludge [59]. According to studies by Jeffery *et al.* [60], by improving nutrient contents, water holding capacity, and utilizing the lime effect, biosolids are considered a practical disposal method due to their ability to recycle materials and increase crop productivity. The long-term development of agriculture will be enhanced by recycling organic waste residues and incorporating them into farmland [61]. As a result of their complementary strengths, composting SS from other industries has a wide range of benefits and long-term development prospects. The composition of organic carbon in SS may not biodegrade and therefore detach from the soil, thereby assisting in the reduction of global warming. Moreover, obtaining nutrients from SS enhances soil quality and reduces environmental impact, suggesting clear economic advantages [38].

According to Yoshida *et al.* [62], site topography and climatic conditions significantly influence sludge assessment results. As a result, sludge characteristics and local environmental conditions affect the emission coefficient. Despite its economic and biological advantages, sludge is still utilized on the land, so it is essential to evaluate and analyze the lasting influence of sludge on the soil before applying it [63]. Products made from sludge are created utilizing a variety of treatment procedures.

(a) Phosphorus

In developed nations with wastewater and sewage treatment facilities, sludge treatment has replaced other phosphorus (P) recovery techniques [1]. Research has demonstrated that (P) recovery can reduce eutrophication opportunities (reducing mining of phosphate), which reduces pollution from mining phosphate water [64]. It has been reported that the application of agricultural SS minimizes heavy metal pollution [65]. Waste product's potential as mineral fertilizers is largely determined by the percentage composition of (P) present in soil [66]. There was little difference in (P) fate between fertilizers applied to low P soils [1]. Studies based on land use change analysis have shown that agricultural waste applications significantly impact categories such as groundwater nutrient enrichment, climate change, and abiotic resource depletion [67]. Amann *et al.* [68] assessed the key technologies for extracting P from wastewater sludge and sewage ash. According to their findings, extraction from sewage soot has the most promising environmental prospects since it is capable of significant recovery rates, eliminates organic micropollutants that do not exist, minimizes gas emissions, and reduces energy consumption.

(b) Nitrogen
It has been stated that the release of industrial use of N into the ecosystem exceeds Earth's ecological limits. This can abruptly disrupt the natural N cycle, causing environmental damage [69]. It may be possible to alleviate the redundant N fixation by reusing fixed nitrogen that has been reactivated and imported into the biological N cycle [70]. A certain amount of nutrients is removed from SS by thermal drying and incineration, accelerating the eutrophication of water. Deviatkin *et al.* [70] recommended reclaiming N separated during the sewage treatment, resulting in a 28% reduction in global warming potential. N mineralization in the soil can affect both plant absorption and hydrosphere leaching.

(c) Other Products (Chemical Fertilizer Substitution)
A sensitivity analysis was conducted by Vadenbo *et al.* [71], which showed that replacing surrogate products with recovered byproducts was essential for improving results. As chemical fertilizer manufacturing expands, pollutants of nitrate and phosphate sources are released into the atmosphere, adversely impacting climate change, acidification, and eutrophication of the environment. Extraction of organic substances in sludge provides favorable advantages that exceed the challenges linked to sludge management, in addition to avoiding some products. Using dry processed sludge as fertilizer reduces the diffusion of active nitrogen into the atmosphere, mitigating freshwater eutrophication produced by sludge use in the soil [72]. Sludge or digestate is an environmentally friendly alternative to chemical fertilizer because it avoids the rising emissions from synthetic chemicals [62]. It is commonly believed that triple superphosphates or urea can be substituted for phosphate fertilizers, whereas the nitrate of ammonium can be substituted for nitrogen fertilizers [66]. In composting materials, nitrogen content determines the extent of the avoidance effect, and N content also determines how much synthetic fertilizer is substituted. The less fertilizer used, the less pollution produced [73]. The eco-friendly sludge P extraction technique can be commercially realized if source P prices increase and energy utilization rises [65]. Using the final products benefits the environment by reducing emissions from chemical fertilizer manufacture, shipping, and application [74]. Using mineral fertilizer and avoiding CO_2 in mineral fertilizer manufacture, N_2O emissions are reduced, resulting in the lowest level of eutrophication and the highest level of resource

Table 9.1 Product category, techniques, recoverable products, and efficiency techniques.

Product category	Recovery technique/ process	Specific recoverable product	Percentage efficiency/yield	References
Nutrient	Hydrothermal carbonization and filtration	Phosphorus	13–17% (concentration)	Oliver-Tomas *et al.* [67]
	Electro-concentration	Nitrogen	≤87 ± 8.5%	Koskue *et al.* [77]
Energy	Upflow anaerobic sludge blanket reactor	Electricity and biogas	611 ± 275 Nm^3 per day (average biogas production), 534.1 MWh per year (electricity)	Arthur *et al.* [78]
	Fermentation	Sugar extraction and bioethanol	75.5% and 1.8 g/L (concentration)	Supaporn and Yeom [5]
	Thermochemical	Biochar	72.3% (yield)	Gopinath *et al.* [79]
	Microbial fuel cell	Bioelectricity	The current, voltage, and power obtained in 17.28 mA, 815.32 mV,	Sivakumar [80]
	Transesterification	Biodiesel	98–99% (yield)	Usman *et al.* [81]

recovery in digested sludge [62]. The amount of methane and N_2O emissions produced by artificial fertilizers and biosolids is still unclear [75]. Compared to other chemical fertilizers, calcium ammonium nitrate, for example, has little environmental impact (outside of ionizing radiation impacts associated with urea) [76]. Table 9.1 shows some of the techniques used to recover different sewage treatment products and their corresponding efficiency or yield. It can be seen from Table 9.1 that concerted efforts are directed toward energy and nutrient synthesis from sewage treatment.

9.7 LCA for Sludge Treatment

LCA is a useful method for measuring the prospective adverse impacts of a product or service across its entire life span. In general, LCA investigations are divided into three phases: initial material acquisition, production of products, and product utilization and waste [82]. LCA is the most frequently employed approach for calculating environmental impacts since it allows for a holistic view of a process (ISO14040, 2006). Sludge management schemes could be examined and assessed using this methodology. For decades, sewage waste disposal has been a major concern since inadequate control could cause climate change and harm the environment, including unpleasant odors, disease transmission, and heavy metal pollution [56]. Classification and characterization are the commonly used steps in the LCA methodology [83]. These two methods could be used to assess the various impacts and process routes. The guideline for carrying out LCA studies on sludge for product recovery is shown in Figure 9.4, based on

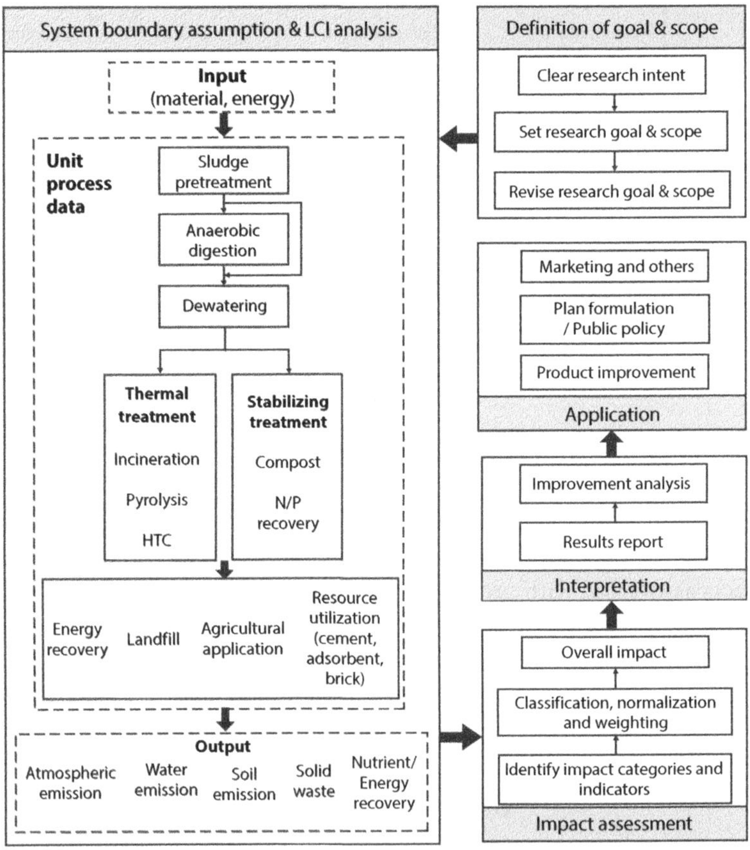

Figure 9.4 Steps and framework of LCA research and main process of sludge treatment and disposal (adapted from Ding *et al.* [1] with permission from Elsevier).

technology and approaches used in past studies. Several methodologies have been employed to conduct detailed LCA analyses, the most notable of which is the ReCiPe Midpoint (H) technique, which is used to examine the impact of global warming, depletion of the ozone layer, soil acidification, fine particulate production, toxic effects on humans, depletion of water resources, and fossil fuels exhaustion. LCA studies also involve various tools to simulate the desired product recovery and assess the life cycle impact. EASETECH, SimaPro, GaBi, PROBIO, Umberto, OpenLCA, Sphera, and other LCA software have been reported [84, 85].

Consequently, the environmental impact and advantages of sludge removal and treatment operations must be examined at the system level [1]. The report of Murray *et al.* [86] on the LCA of sewage shows that anaerobic digestion is the most effective treatment option for SS. Likewise, Xu *et al.* [54], Li *et al.* [87], and Li and Feng [88] concluded that anaerobic decomposition may be the most effective sludge treatment approach in terms of both energy recovery rate and cost-effectiveness measures. Despite its excellent environmental sustainability, heavy metals in anaerobic digestion have a serious ecotoxic effect on freshwater, according to Tarpani *et al.* [72]. Furthermore, Lishan *et al.* [89] discovered that hydrothermal carbonization is more environmentally friendly and adaptable than incineration, landfilling, or composting. Zhou *et al.* [90] developed an LCA-based multi-criteria decision-making model, which found that land application performed best against environmental, economic, and social considerations. Study boundaries differed significantly between the studies mentioned earlier. Comparing and synthesizing the results of sludge treatment and disposal techniques were challenging because each study employed various data sources.

Using LCA, it is possible to measure the environmental effects of various sludge treatment options and discover ecologically favorable process routes [52, 91]. Anaerobic digestion, composting, burning, and landfilling are a few of the more conventional techniques [1, 89]. Researchers have investigated cutting-edge techniques like wet oxidation, sludge cultivation in earthworms, and integrated food and sludge treatment in the past few years. To effectively account for the benefits gained from sludge resource recovery environmentally, LCA must understand the influence of resource recovery across the entire sludge treatment system. The potential for global warming can be reduced by substituting AD energy for coal. A further reduction in environmental impacts could be achieved by collecting biochar after the AD process. A general environmental benefit of reusing sludge's abundant nutrients (nitrogen and phosphorus) is to reduce the use of conventional chemical fertilizers [92] and reduce freshwater eutrophication by spreading sludge on cropland. Moreover, a combination of sludge and agricultural waste contributes to increased resource recovery, which results in greater environmental benefits [93]. Comparing the LCA results between different sludge treatment technologies is generally difficult since the resource recovery process involves several treatment parameters [94]. A study by Yoshida *et al.* [62] considered the effect of parameters on LCA results and initiated a strong correlation between coverage of the initial data used for the study and LCA performance. Currently, most sludge recycling processes are being developed only in the laboratory research stages. In the process integration of sludge recycling to industrial practice, it is crucial to identify the key process parameters that contribute most to environmental gain or loss. Analysis around sensitivity, uncertainty, and cost benefits are components of LCA that can help identify critical parameters in the sludge recycling process [76, 87, 93, 94]. For instance, Li *et al.* [87] examined the uncertainties of biogas production by conducting a comparative LCA of five AD processes. Based on the author's results, the biomethane yield and the organic content of the sludge influence the assessment results. Gourdet *et al.* [76] performed a sensitivity analysis on 17 sludge AD and dewatering process parameters to identify the key life cycle environmental variables. Sludge treatment could improve environmental performance by improving biogas production by increasing the decomposition rate of volatile solids. Table 9.2 presents an overview of LCA cases and results obtained in selected studies regarding SS management.

9.7.1 Economic Indicators

Based on bibliographic information, costs can be divided into operational and capital costs, which are construction, equipment, and maintenance costs. To achieve an accurate cost capture, sludge disposal, electricity, chemical consumption, and staff costs should also be included in the operational costs. Hermelink and de Jager [95] calculated the cost based on Net Present Value (NPV), where n represents the useful life of the asset and i represents the inflation-adjusted discount rate of 5% (Equation 9.1).

$$NPV = CAPEX + \sum_{n} \frac{OPEX}{(1+i)^n} \tag{9.1}$$

Table 9.2 Overview of LCA in selected studies in relation to SS management.

Source/waste	Study area	LCA software/technique/ model used	Scenarios/cases	Optimum case	Inference	Reference
Slaughter house waste (SHW) organic fraction of municipal solid waste	United States	Cradle-To-Grave	Case1: AD of SHW Case 2: Co-anaerobic digestion of SHW +organic solid waste Case 3: Co-anaerobic digestion of SHW with SS	SHW with co-AD outperforms causing by replacing fossil fuels, generating energy instead, and producing nitrogen fertilizers as output results	The findings show that the AD of SHW can both create biofuel that can provide U.S. energy security and significantly lower GHG emissions from U.S. slaughterhouses	Wang *et al.* [82]
SS	China	ReCipe method	Case1: Incineration Case 2: Aerobic composting Case 3: Brick molding Case 4: Anaerobic digestion	Brick production and use as building materials had the least negative effects on the environment and operating costs.	According to an analysis of pollutant sources and composition, the majority of the environmental consequences are caused by treatment and recycling processes	Zhang and Matsumoto [96]
Reject water and condense sate in a municipal wastewater treatment plant	Finland	ISO standards 14040 and 14044 GaBi and ReCiPe method	Case 1: Conventional treatment without nitrogen recovery Case 2: Utilizing air stripping and subsequent scrubbing to recover nitrogen Case 3: Recovery of nitrogen through ammonia adsorption	In three of the five effect categories considered in this analysis, Case 3 performs better	The information in this study lends credence to the notion that integrated resource recovery in wastewater systems may benefit the environment, the economy, and human health	Saud *et al.* [4]

(*Continued*)

Table 9.2 Overview of LCA in selected studies in relation to SS management. (*Continued*)

Source/waste	Study area	LCA software/technique/ model used	Scenarios/cases	Optimum case	Inference	Reference
SS treatment paths	Germany	Gate-to-cradle approach ReCiPe Midpoint 2016 method ISO 14040 and 14,044 openLCA 1.10 ecoinvent database 3.5	The basis scenario was the application of digested sludge to agricultural soils (system: AD + spreading). Case 1: Centralized incineration of sludge (system: AD + incineration) Case 2: Decentralized hydrothermal carbonization followed by centralized incineration (system: AD + HTC + incineration) Case 3: Decentralized pyrolysis of sludge followed by centralized incineration (system: AD + P + I).	Comparative analysis revealed that AD + spreading had the least negative effects on the environment across most categories	Drivers of SS treatment facilities are asked to look for measures to lessen environmental consequences, regardless of the study methodology they choose	Mayer *et al.* [91]
Sludge in either agriculture or forestry	Sweden	GaBi software tool GaBi Professional ecoinvent	Case 1: Thermophilic digestion Case 2: Pasteurization, Case 3: Thermal hydrolysis Case 4: The mono-incineration of urea with the recovery of phosphorus	In all cases, efforts can be made to minimize emissions but no technologies are inherently free of such emissions	The total environ mental impact turned out to be surprisingly similar for the studied technologies	Svanström *et al.* [97]

(*Continued*)

Table 9.2 Overview of LCA in selected studies in relation to SS management. (*Continued*)

Source/waste	Study area	LCA software/technique/ model used	Scenarios/cases	Optimum case	Inference	Reference
UASB or upflow anaerobic sludge blanket	Brazil	PROBIO 1.0 software SimaPro software (model 8.4 version) ReCiPe 2016 Midpoint (H) method	Case 1: The sludge is dried in a rotary dryer with the use of biogas, and the dried sludge is then applied to farming Case 2: The dewatered sludge is burned, and the heat produced is used to dry it while the ashes are used in agriculture Case 3: Similar to Case 2, except that ashes are dispersed on the ground	In comparison to the baseline scenario, all of the effect categories that were analyzed significantly improved in the scenario where heat from biogas combustion is used to dry the sludge	The research results highlight several novel approaches for advancing environmental sustainability linked to UASB reactors' holistic handling of biogas and sludge	do Amaral *et al.* [98]
Food waste, woody biomass, and used oil waste are employed in the sludge treatment process	China	ReCipe method ecoinvent 3.4 database openLCA	Case 1: Co-digestion of food waste and sludge Case 2: Co-gasification of woody waste and sludge Case 3: Combined incineration of used oil and sludge Case 4: Landfilling Case 5: Incineration Case 6: Anaerobic digestion + incineration	Sludge and woody waste co-gasification performed the best in terms of the environment	Adding parametric resource recovery uncertainty to the decision-making process, it can be asserted, makes the evaluation process clearer	Zhou *et al.* [52]

(*Continued*)

Table 9.2 Overview of LCA in selected studies in relation to SS management. (*Continued*)

Source/waste	Study area	LCA software/technique/ model used	Scenarios/cases	Optimum case	Inference	Reference
Wastewater treatment plants and resource recovery facilities	Denmark	EASETECH (v.3.4.1.31) Environmental Footprint (EF) 3.0	Case 1: Direct application of sludge in agriculture Case 2: Incineration of sludge and landfilling of sludge ashes	The study revealed that using biochar instead of sludge directly on the land is preferable	In Denmark, simultaneous sludge drying, and decomposition is becoming more popular since it reduces the water content of the sludge from about 80% to almost none, lowering disposal expenses	Faragò *et al.* [9]

Moreover, Equation 9.2 is also a useful method for evaluating the payback time, whereas *Ms* represents the amount of sludge produced in a year (ton/year), *Cd* represents the final disposal cost in €/ton, *E* represents the change in electricity production in the sludge line (kWh/year); *Ce* represents the electric cost, which is based on the electricity price, and *C* represents the total capital cost.

$$Payback\ time = \frac{Ms\ x\ Cd + \Delta E\ x\ Ce - C}{Investment\ cost\ (t)} \tag{9.2}$$

9.8 Conclusion and Future Perspectives

Sludge management strategies are typically designed based on the nature of the sludge, the ecosystem, the costs associated with the management, and the potential environmental damage to be reduced. Currently, sludge is disposed of primarily in landfills, leading to high transportation costs and a significant loss of nutrients and energy. The issuance of higher landfill taxes, stricter landfilling disposal regulations, and the reluctance of landfill operators to accept sludge are forcing treatment facilities to embrace more efficient strategies. Gasification of waste sludge is expected to produce significant energy recovery. However, the innovation is developing before recovered energy can be maximally utilized. The application of microwave-assisted resource recovery from sludge is mostly on a laboratory scale, though it has the potential for large-scale adoption. Presently, no single technology can perfectly recover resources for all scenarios of sludge management. The complication at this level is due to the unstable characteristics of the SS. Having a perfect outcome for sludge treatment may be far from feasible. However, investment in resource recovery could be a sustainable approach to reducing the cost of SS treatment in wastewater plants. The wastewater contains substantial amounts of nutrients such as nitrogen and phosphorus. As a result, future wastewater treatment plants should be designed to be energy-efficient and to recover every nutrient possible, particularly nitrogen and phosphorus. There is also a growing interest in recovering items from waste and wastewater from various industrial, manufacturing, and mining sources. The chemical obtained from the streams mentioned above is very stable and has a low biodegradability. There are differences in published LCA results, which weakens their comparative validity. Thus, most research focuses on comparison situations. It should be noted that LCA is not a method for determining which procedure is more efficient; instead, it analyzes potential consequences rather than measuring actual environmental impacts. The outcomes may also differ depending on the evaluation methodologies or basic data sources used. However, with increased interest in sludge recovery as a resource, there is an increasing demand for LCA to clarify the environmental impact contribution of resource recovery as it relates to the entire sludge treatment life cycle to account for all the environmental advantages that can be accomplished. Most sludge recycling procedures are still in the laboratory. Before moving on to actual applications, it is crucial to identify important process variables that can result in environmental benefit or loss with only a minor variance. Identifying essential parameters is critical for understanding the environmental consequences of resource recovery and analyzing the environmental benefits and drawbacks of sludge treatment technology. Considering the recovered resources from SS, different technologies and procedures are considered before implementation, and a trade-off is always seen. It is vital to evaluate the process economics that confirms substantial resources in sludge to justify recovery. This would help major drivers of the circular economy, government, environment specialists, and industrialists push to recover and regain all the valuable resources in SS treatments.

References

1. Ding, A., Zhang, R., Ngo, H.H., He, X., Ma, J., Nan, J., Li, G., Life cycle assessment of sewage sludge treatment and disposal based on nutrient and energy recovery: A review. *Sci. Total Environ.*, 769, 144451, 2021.
2. Capodaglio, A.G., Biorefinery of Sewage Sludge: Overview of Possible ValueAdded Products and Applicable Process Technologies. *Water (Switzerland)*, 15, 1195, 2023.
3. Djandja, O.S., Yin, L.X., Wang, Z.C., Duan, P.G., From wastewater treatment to resources recovery through hydrothermal treatments of municipal sewage sludge: A critical review. *Process Saf. Environ. Prot.*, 151, 101, 2021.
4. Saud, A., Havukainen, J., Peltola, P., Horttanainen, M., Environmental Performance of Nitrogen Recovery from Reject Water of Sewage Sludge Treatment Based on Life Cycle Assessment. *Recycling*, 8, 43, 2023.
5. Supaporn, P., Yeom, S.H., Optimized Sugar Extraction and Bioethanol Production from Lipid-extracted Sewage Sludge. *Biotechnol. Bioprocess Eng.*, 27, 119, 2022.
6. Guest, J.S., Skerlos, S.J., Barnard, J.L., Beck, M.B., Daigger, G.T., Hilger, H., Jackson, S.J., Karvazy, K., Kelly, L., Macpherson, L., Mihelcic, J.R., Pramanik, A., Raskin, L., Van Loosdrecht, M.C.M., Yeh, D., Love, N.G., A new planning

and design paradigm to achieve sustainable resource recovery from wastewater. *Environ. Sci. Technol.*, 43, 6126, 2009.
7. Peccia, J., Westerhoff, P., We Should Expect More out of Our Sewage Sludge. *Environ. Sci. Technol.*, 49, 8271, 2015.
8. Teoh, S.K., Li, L.Y., Feasibility of alternative sewage sludge treatment methods from a lifecycle assessment (LCA) perspective. *J. Clean. Prod.*, 247, 119495, 2020.
9. Faragò, M., Damgaard, A., Logar, I., Rygaard, M., Life Cycle Assessment and Cost-Benefit Analysis of Technologies in Water Resource Recovery Facilities: The Case of Sludge Pyrolysis. *Environ. Sci. Technol.*, 56, 17988, 2022.
10. Solon, K., Volcke, E.I.P., Spérandio, M., Van Loosdrecht, M.C.M., Resource recovery and wastewater treatment modelling. *Environ. Sci. Water Res. Technol.*, 5, 631, 2019.
11. Liu, F., Zhou, L., Zhou, J., Song, X., Wang, D., Improvement of sludge dewaterability and removal of sludge-borne metals by bioleaching at optimum pH. *J. Hazard. Mater.*, 221–222, 170, 2012.
12. Chen, S., Sun, Z., Zhang, Q., Hu, J., Xiang, W., Steam gasification of sewage sludge with CaO as CO2 sorbent for hydrogen-rich syngas production. *Biomass Bioenergy*, 107, 52, 2017.
13. Tarpani, R.R.Z., Azapagic, A., Life cycle costs of advanced treatment techniques for wastewater reuse and resource recovery from sewage sludge. *J. Clean. Prod.*, 204, 832, 2018.
14. Tyagi, V.K., Lo, S.L., Sludge: A waste or renewable source for energy and resources recovery? *Renew. Sustain. Energy Rev.*, 25, 708, 2013.
15. Yang, H., Ye, S., Zeng, Z., Zeng, G., Tan, X., Xiao, R., Wang, J., Song, B., Du, L., Qin, M., Yang, Y., Xu, F., Utilization of biochar for resource recovery from water: A review. *Chem. Eng. J.*, 397, 125502, 2020.
16. Puyol, D., Batstone, D.J., Hülsen, T., Astals, S., Peces, M., Krömer, J.O., Resource recovery from wastewater by biological technologies: Opportunities, challenges, and prospects. *Front. Microbiol.*, 7, 2017.
17. O'kelly, B.C., Sewage sludge to landfill: Some pertinent engineering properties. *J. Air Waste Manage. Assoc.*, 55, 765, 2005.
18. Kacprzak, M., Neczaj, E., Fijałkowski, K., Grobelak, A., Grosser, A., Worwag, M., Rorat, A., Brattebo, H., Almås, Å., Singh, B.R., Sewage sludge disposal strategies for sustainable development. *Environ. Res.*, 156, 39, 2017.
19. Kiliç Taşeli, B., Comparative Life Cycle Assessment of Sewage Sludge (Biosolid) Management Options. In: Taşeli, B.K. (ed.) *Sustainable Sewage Sludge Management and Resource Efficiency*. p. Ch. 5. IntechOpen, Rijeka, 2020.
20. Song, U., Lee, E.J., Environmental and economical assessment of sewage sludge compost application on soil and plants in a landfill. *Resour. Conserv. Recycl.*, 54, 1109, 2010.
21. Singh, R.P., Agrawal, M., Potential benefits and risks of land application of sewage sludge. *Waste Manage.*, 28, 347, 2008.

22. Lü, H., Chen, X.H., Mo, C.H., Huang, Y.H., He, M.Y., Li, Y.W., Feng, N.X., Katsoyiannis, A., Cai, Q.Y., Occurrence and dissipation mechanism of organic pollutants during the composting of sewage sludge: A critical review. *Bioresour. Technol.*, 328, 124847, 2021.
23. Amir, S., Hafidi, M., Merlina, G., Revel, J.C., Sequential extraction of heavy metals during composting of sewage sludge. *Chemosphere*, 59, 801, 2005.
24. Wainaina, S., Awasthi, M.K., Sarsaiya, S., Chen, H., Singh, E., Kumar, A., Ravindran, B., Awasthi, S.K., Liu, T., Duan, Y., Kumar, S., Zhang, Z., Taherzadeh, M.J., Resource recovery and circular economy from organic solid waste using aerobic and anaerobic digestion technologies. *Bioresour. Technol.*, 301, 122778, 2020.
25. Kwarciak-Kozłowska, A., Co-composting of sewage sludge and wetland plant material from a constructed wetland treating domestic wastewater. In: *Industrial and Municipal Sludge: Emerging Concerns and Scope for Resource Recovery*, Prasad, M.N.V., de Campos Favas, P.J., Vithanage, M., and Mohan, S.V.B.T.-I. and M.S. (eds.) pp. 337–360. Butterworth-Heinemann, 2019.
26. Smith, S.R., A critical review of the bioavailability and impacts of heavy metals in municipal solid waste composts compared to sewage sludge. *Environ. Int.*, 35, 142, 2009.
27. Alvarenga, P., Mourinha, C., Farto, M., Santos, T., Palma, P., Sengo, J., Morais, M.C., Cunha-Queda, C., Sewage sludge, compost and other representative organic wastes as agricultural soil amendments: Benefits versus limiting factors. *Waste Manage.*, 40, 44, 2015.
28. Lin, L., Li, R. hong, Li, Y., Xu, J., Li, X. yan, Recovery of organic carbon and phosphorus from wastewater by Fe-enhanced primary sedimentation and sludge fermentation. *Process Biochem.*, 54, 135, 2017.
29. Climent, M., Ferrer, I., Baeza, M. del M., Artola, A., Vázquez, F., Font, X., Effects of thermal and mechanical pretreatments of secondary sludge on biogas production under thermophilic conditions. *Chem. Eng. J.*, 133, 335, 2007.
30. Oloko-Oba, M.I., Taiwo, A.E., Ajala, S.O., Solomon, B.O., Betiku, E., Performance evaluation of three different-shaped bio-digesters for biogas production and optimization by artificial neural network integrated with genetic algorithm. *Sustain. Energy Technol. Assess.*, 26, 116, 2018.
31. Ward, A.J., Hobbs, P.J., Holliman, P.J., Jones, D.L., Optimisation of the anaerobic digestion of agricultural resources. *Bioresour. Technol.*, 99, 7928, 2008.
32. Martins, M.G., da Silva Arouche, T., Neto, A.F.G., da Cruz, J.N., da Costa, F.L.P., Fernandes, L.L., de Carvalho Junior, R.N., da Silva Costa, J.F., de Jesus Chaves Neto, A.M., Density functional theory for the thermodynamic gas-phase investigation of butanol biofuel and its isomers mixed with gasoline and ethanol. *J. Mol. Model.*, 27, 80, 2021.
33. Appels, L., Baeyens, J., Degrève, J., Dewil, R., Principles and potential of the anaerobic digestion of waste-activated sludge. *Prog. Energy Combust. Sci.*, 34, 755, 2008.

34. Puyol, D., Flores-Alsina, X., Segura, Y., Molina, R., Padrino, B., Fierro, J.L.G., Gernaey, K. V., Melero, J.A., Martinez, F., Exploring the effects of ZVI addition on resource recovery in the anaerobic digestion process. *Chem. Eng. J.*, 335, 703, 2018.
35. Saha, M., Eskicioglu, C., Marin, J., Microwave, ultrasonic and chemo-mechanical pretreatments for enhancing methane potential of pulp mill wastewater treatment sludge. *Bioresour. Technol.*, 102, 7815, 2011.
36. Donatello, S., Cheeseman, C.R., Recycling and recovery routes for incinerated sewage sludge ash (ISSA): A review. *Waste Manage.*, 33, 2328, 2013.
37. Stolarek, P., Ledakowicz, S., Thermal processing of sewage sludge by drying, pyrolysis, gasification and combustion. *Water Sci. Technol.*, 44, 333, 2001.
38. Zhu, Y., Zhao, Q., Li, D., Li, J., Guo, W., Performance comparison of phosphorus recovery from different sludges in sewage treatment plants through pyrolysis. *J. Clean. Prod.*, 372, 133728, 2022.
39. Jaeger, M., Mayer, M., The Noell Conversion Process - A gasification process for the pollutant-free disposal of sewage sludge and the recovery of energy and materials. *Water Sci. Technol.*, 41, 37, 2000.
40. Dogru, M., Midilli, A., Howarth, C.R., Gasification of sewage sludge using a throated downdraft gasifier and uncertainty analysis. *Fuel Process. Technol.*,75, 55, 2002.
41. Hii, K., Baroutian, S., Parthasarathy, R., Gapes, D.J., Eshtiaghi, N., A review of wet air oxidation and Thermal Hydrolysis technologies in sludge treatment. *Bioresour. Technol.*, 155, 289, 2014.
42. Strong, P.J., McDonald, B., Gapes, D.J., Combined thermochemical and fermentative destruction of municipal biosolids: A comparison between thermal hydrolysis and wet oxidative pre-treatment. *Bioresour. Technol.*, 102, 5520, 2011.
43. Zhao, Y., Yan, X., Zhou, J., Li, R., Yang, S., Wang, B., Deng, R., Treatment of oily sludge by two-stage wet air oxidation. *J. Energy Inst.*, 92, 1451, 2019.
44. Levec, J., Pintar, A., Catalytic wet-air oxidation processes: A review. *Catal. Today*, 124, 172, 2007.
45. Xie, F., Cai, T., Ma, Y., Li, H., Li, C., Huang, Z., Yuan, G., Recovery of Cu and Fe from Printed Circuit Board waste sludge by ultrasound: Evaluation of industrial application. *J. Clean. Prod.*, 17, 1494, 2009.
46. Liu, Y., Deng, Y.Y., Zhang, Q., Liu, H., Overview of recent developments of resource recovery from wastewater via electrochemistry-based technologies. *Sci. Total Environ.*, 757, 143901, 2021.
47. Niu, J., Yan, W., Du, J., Hao, X., Wang, F., Wang, Z., Guan, G., An electrically switched ion exchange film with molecular coupling synergistically-driven ability for recovery of Ag+ ions from wastewater. *Chem. Eng. J.*, 389, 124498, 2020.
48. Jin, W., Hu, M., Hu, J., Selective and Efficient Electrochemical Recovery of Dilute Copper and Tellurium from Acidic Chloride Solutions. *ACS Sustain. Chem. Eng.*, 6, 13378, 2018.

49. Liao, M., Liu, Y., Tian, E., Ma, W., Liu, H., Phosphorous removal and high-purity struvite recovery from hydrolyzed urine with spontaneous electricity production in Mg-air fuel cell. *Chem. Eng. J.*, 391, 123517, 2020.
50. Kuntke, P., Zamora, P., Saakes, M., Buisman, C.J.N., Hamelers, H.V.M., Gas- permeable hydrophobic tubular membranes for ammonia recovery in bio-electrochemical systems. *Environ. Sci. Water Res. Technol.*, 2, 261, 2016.
51. Blázquez, E., Gabriel, D., Baeza, J.A., Guisasola, A., Freguia, S., Ledezma, P., Recovery of elemental sulfur with a novel integrated bioelectrochemical system with an electrochemical cell. *Sci. Total Environ.*, 677, 175, 2019.
52. Zhou, X., Li, J., Zhao, X., Yang, J., Sun, H., Yang, S.S., Bai, S., Resource recovery in life cycle assessment of sludge treatment: Contribution, sensitivity, and uncertainty. *Sci. Total Environ.*, 806, 150409, 2022.
53. Fernández-Nava, Y., Del Río, J., Rodríguez-Iglesias, J., Castrillón, L., Marañón, E., Life cycle assessment of different municipal solid waste management options: A case study of Asturias (Spain). *J. Clean. Prod.*, 81, 178, 2014.
54. Xu, C., Chen, W., Hong, J., Life-cycle environmental and economic assessment of sewage sludge treatment in China. *J. Clean. Prod.*, 67, 79, 2014.
55. Liu, Q., Jiang, P., Zhao, J., Zhang, B., Bian, H., & Qian, G., Life cycle assessment of an industrial symbiosis based on energy recovery from dried sludge and used oil. *J. Clean. Prod.* 19(15), 1700-1708, 2011.
56. Chen, G., Wang, X., Li, J., Yan, B., Wang, Y., Wu, X., Velichkova, R., Cheng, Z., Ma, W., Environmental, energy, and economic analysis of integrated treatment of municipal solid waste and sewage sludge: A case study in China. *Sci. Total Environ.*, 647, 1433, 2019.
57. Nakakubo, T., Yoshida, N., Hattori, Y., Analysis of greenhouse gas emission reductions by collaboratively updating equipment in sewage treatment and municipal solid waste incineration plants. *J. Clean. Prod.*, 168, 803, 2017.
58. Nakatsuka, N., Kishita, Y., Kurafuchi, T., Akamatsu, F., Integrating wastewater treatment and incineration plants for energy-efficient urban biomass utilization: A life cycle analysis. *J. Clean. Prod.*, 243, 118448, 2020.
59. Rovira, J., Mari, M., Nadal, M., Schuhmacher, M., Domingo, J.L., Use of sewage sludge as secondary fuel in a cement plant: Human health risks. *Environ. Int.*, 37, 105, 2011.
60. Jeffery, S., Verheijen, F.G.A., van der Velde, M., Bastos, A.C., A quantitative review of the effects of biochar application to soils on crop productivity using meta-analysis. *Agric. Ecosyst. Environ.*, 144, 175, 2011.
61. Kirchmann, H., Börjesson, G., Kätterer, T., Cohen, Y., From agricultural use of sewage sludge to nutrient extraction: A soil science outlook. *Ambio*, 46, 143, 2017.
62. Yoshida, H., ten Hoeve, M., Christensen, T.H., Bruun, S., Jensen, L.S., Scheutz, C., Life cycle assessment of sewage sludge management options including long-term impacts after land application. *J. Clean. Prod.*, 174, 538, 2018.

63. Bruun, S., Yoshida, H., Nielsen, M.P., Jensen, L.S., Christensen, T.H., Scheutz, C., Estimation of long-term environmental inventory factors associated with land application of sewage sludge. *J. Clean. Prod.*, 126, 440, 2016.
64. Smol, M., Kulczycka, J., Lelek, Ł., Gorazda, K., Wzorek, Z., Life Cycle Assessment (LCA) of the integrated technology for the phosphorus recovery from sewage sludge ash (SSA) and fertilizers production. *Arch. Environ. Prot.*, 46, 42, 2020.
65. Lederer, J., Rechberger, H., Comparative goal-oriented assessment of conventional and alternative sewage sludge treatment options. *Waste Manag.*, 30, 1043, 2010.
66. ten Hoeve, M., Bruun, S., Naroznova, I., Lemming, C., Magid, J., Jensen, L.S., Scheutz, C., Life cycle inventory modeling of phosphorus substitution, losses and crop uptake after land application of organic waste products. *Int. J. Life Cycle Assess.*, 23, 1950, 2018.
67. Oliver-Tomas, B., Hitzl, M., Owsianiak, M., Renz, M., Evaluation of hydrothermal carbonization in urban mining for the recovery of phosphorus from the organic fraction of municipal solid waste. *Resour. Conserv. Recycl.*, 147, 111, 2019.
68. Amann, A., Zoboli, O., Krampe, J., Rechberger, H., Zessner, M., Egle, L., Environmental impacts of phosphorus recovery from municipal wastewater. *Resour. Conserv. Recycl.*, 130, 127, 2018.
69. Steffen, W., Richardson, K., Rockström, J., Cornell, S.E., Fetzer, I., Bennett, E.M., Biggs, R., Carpenter, S.R., De Vries, W., De Wit, C.A., Folke, C., Gerten, D., Heinke, J., Mace, G.M., Persson, L.M., Ramanathan, V., Reyers, B., Sörlin, S., Planetary boundaries: Guiding human development on a changing planet. *Science (80-.)*, 347, 2015.
70. Deviatkin, I., Lyu, L., Chen, S., Havukainen, J., Wang, F., Horttanainen, M., Mänttäri, M., Technical implications and global warming potential of recovering nitrogen released during continuous thermal drying of sewage sludge. *Waste Manage.*, 90, 132, 2019.
71. Vadenbo, C., Hellweg, S., Guillén-Gosálbez, G., Multi-objective optimization of waste and resource management in industrial networks - Part I: Model description. *Resour. Conserv. Recycl.*, 89, 52, 2014.
72. Tarpani, R.R.Z., Alfonsín, C., Hospido, A., Azapagic, A., Life cycle environmental impacts of sewage sludge treatment methods for resource recovery considering ecotoxicity of heavy metals and pharmaceutical and personal care products. *J. Environ. Manage.*, 260, 109643, 2020.
73. Righi, S., Oliviero, L., Pedrini, M., Buscaroli, A., Della Casa, C., Life Cycle Assessment of management systems for sewage sludge and food waste: Centralized and decentralized approaches. *J. Clean. Prod.*, 44, 8, 2013.
74. Thomsen, M., Seghetta, M., Mikkelsen, M.H., Gyldenkærne, S., Becker, T., Caro, D., Frederiksen, P., Comparative life cycle assessment of biowaste to resource management systems – A Danish case study. *J. Clean. Prod.* 142, 4050, 2017.

75. Chiaradia, J.J., Chiba, M.K., de Andrade, C.A., do Carmo, J.B., de Oliveira, C., Lavorenti, A., Fluxos de CO2, CH4 e N2O de um argissolo tratado com lodo de esgoto e cultivado com mamona. *Rev. Bras. Cienc. do Solo*. 33, 1863, 2009.
76. Gourdet, C., Girault, R., Berthault, S., Richard, M., Tosoni, J., Pradel, M., In quest of environmental hotspots of sewage sludge treatment combining anaerobic digestion and mechanical dewatering: A life cycle assessment approach. *J. Clean. Prod.* 143, 1123, 2017.
77. Koskue, V., Freguia, S., Ledezma, P., Kokko, M., Efficient nitrogen removal and recovery from real digested sewage sludge reject water through electroconcentration. *J. Environ. Chem. Eng.* 9, 106286, 2021.
78. Arthur, P.M.A., Konaté, Y., Sawadogo, B., Sagoe, G., Dwumfour-Asare, B., Ahmed, I., Bayitse, R., Ampomah-Benefo, K., Evaluating the Potential of Renewable Energy Sources in a Full-Scale Upflow Anaerobic Sludge Blanket Reactor Treating Municipal Wastewater in Ghana. *Sustainability*, 15, 3743, 2023.
79. Gopinath, A., Divyapriya, G., Srivastava, V., Laiju, A.R., Nidheesh, P. V., Kumar, M.S., Conversion of sewage sludge into biochar: A potential resource in water and wastewater treatment. *Environ. Res.*, 194, 110656, 2021.
80. Sivakumar, D., Wastewater treatment and bioelectricity production in microbial fuel cell: salt bridge configurations. *Int. J. Environ. Sci. Technol.*, 18, 1379, 2021.
81. Usman, M., Cheng, S., Cross, J.S., Biodiesel production from wet sewage sludge and reduced CO2 emissions compared to incineration in Tokyo, Japan. *Fuel*, 341, 127614, 2023.
82. Wang, S., Sahoo, K., Jena, U., Dong, H., Bergman, R., Runge, T., Life-cycle assessment of treating slaughterhouse waste using anaerobic digestion systems. *J. Clean. Prod.*, 292, 126038, 2021.
83. International Organization for Standardization 14044, *Environmental Management. Life Cycle Assessment. Requirements and Guidelines*, http://tienda.icontec.org/brief/NTC-ISO14044.pdf, (2007).
84. Herrmann, I.T., Moltesen, A., Does it matter which Life Cycle Assessment (LCA) tool you choose? - A comparative assessment of SimaPro and GaBi. *J. Clean. Prod.*, 86, 163, 2015.
85. Nagtode, V.S., Cardoza, C., Yasin, H.K.A., Mali, S.N., Tambe, S.M., Roy, P., Singh, K., Goel, A., Amin, P.D., Thorat, B.R., Cruz, J.N., Pratap, A.P., Green Surfactants (Biosurfactants): A Petroleum-Free Substitute for Sustainability-Comparison, Applications, Market, and Future Prospects. *ACS Omega*, 8, 11674, 2023.
86. Murray, A., Horvath, A., Nelson, K.L., Hybrid life-cycle environmental and cost inventory of sewage sludge treatment and end-use scenarios: A case study from China. *Environ. Sci. Technol.*, 42, 3163, 2008.

87. Li, H., Jin, C., Zhang, Z., O'Hara, I., Mundree, S., Environmental and economic life cycle assessment of energy recovery from sewage sludge through different anaerobic digestion pathways. *Energy,* 126, 649, 2017.
88. Li, H., Feng, K., Life cycle assessment of the environmental impacts and energy efficiency of an integration of sludge anaerobic digestion and pyrolysis. *J. Clean. Prod.,* 195, 476, 2018.
89. Lishan, X., Tao, L., Yin, W., Zhilong, Y., Jiangfu, L., Comparative life cycle assessment of sludge management: A case study of Xiamen, China. *J. Clean. Prod.,* 192, 354, 2018.
90. Zhou, G., Gu, Y., Yuan, H., Gong, Y., Wu, Y., Selecting sustainable technologies for disposal of municipal sewage sludge using a multi-criterion decision-making method: A case study from China. *Resour. Conserv. Recycl.,* 161, 104881, 2020.
91. Mayer, F., Bhandari, R., Gäth, S.A., Life cycle assessment of prospective sewage sludge treatment paths in Germany. *J. Environ. Manage.,* 290, 112557, 2021.
92. Improved life cycle modelling of benefits from sewage sludge anaerobic digestion and land application. *Resour. Conserv. Recycl.,* 122, 126, 2017.
93. Albini, E., Bacchi, D., Ferrara, G., Francini, G., Galoppi, G., Lombardi, L., Pecorini, I., Susini, C., Bioenergy recovery from waste: Comparison of different treatment scenarios by LCA. *Energy Procedia,* 148, 34, 2018.
94. Cartes, J., Neumann, P., Hospido, A., Vidal, G., Life cycle assessment of management alternatives for sludge from sewage treatment plants in Chile: does advanced anaerobic digestion improve environmental performance compared to current practices? *J. Mater. Cycles Waste Manage.,* 20, 1530, 2018.
95. Hermelink, A. H., de Jager, D., Evaluating our future. The crucial role of discount rates in European Commission energy system modelling. The European Council for an Energy Efficient Economy (ECEEE) & Ecofys, 2015. http://www.eceee.org/policy-areas/discount-rates/evaluating-our-future-report
96. Zhang, J., Matsumoto, T., Comparative Life Cycle Assessment Analysis of Sewage Sludge Recycling Systems in China. *Nat. Environ. Pollut. Technol.,* 20, 2099, 2021.
97. Svanström, M., Heimersson, S., Peters, G., Harder, R., I'Ons, D., Finnson, A., Olsson, J., Life cycle assessment of sludge management with phosphorus utilisation and improved hygienisation in Sweden. *Water Sci. Technol.,* 75, 2013, 2017.
98. Do Amaral, K.C., Aisse, M.M., Possetti, G.R.C., Prado, M.R., Use of life cycle assessment to evaluate environmental impacts associated with the management of sludge and biogas. *Water Sci. Technol.,* 77, 2292, 2018.

10

Various Biomasses from Wastewater and Possibilities of Conversion to Energy Resources

Neelaambhigai Mayilswamy and Balasubramanian Kandasubramanian*

Nano Surface Texturing Lab, Department of Metallurgical and Materials Engineering, Defence Institute of Advanced Technology (DU), Pune, Maharashtra, India

Abstract

With the aggravating ecological considerations and the surge to fulfill the objectives of sustainable development, it is an arduous mission to seek an individual energy resource that may unequivocally supersede fossil fuels. The energy acquired from the biomass present in effluents serves as an exemplary alternative to the development of valuable bioenergy in the form of oil, gas, heat, or electricity depending on the biomass conversion technique being employed. The bioenergy from sewage sludge, algae, and lignocellulosic biomass from wastewater can permit sustainable waste management. This chapter propounds the composition of various wastewater-extracted biomass (sewage sludge, algae, and lignocellulose), with a comprehensive description of different bioenergy transformation pathways (physical, chemical, thermochemical, biochemical, and physicochemical conversion methods) leading to the formulation of biofuel (bio-oil, biodiesel), biogas, or biochar. Furthermore, a propensity toward future prospects and associated challenges based on the current technology is also presented.

Keywords: Sewage sludge, microalgae, lignocellulosic biomass, fermentation, hydrothermal carbonization, supercritical transformation

**Corresponding author*: meetkbs@gmail.com

Inamuddin, Tariq Altalhi, Mohammad Luqman and Joseph Kapuku (eds.) Sewage and Biomass from Wastewater to Energy, (259–282) © 2024 Scrivener Publishing LLC

10.1 Introduction

The global energy requisites are augmenting incessantly ascribable to technical advancement, and industrial enterprise, and the demographic growth is envisaged to attain the 9.2 billion mark by the year 2050 [1]. Petrochemicals, petroleum derivatives, and coal resources have always served as an entrenched component of humankind's subsistence for almost two centuries, impelling industrialization and other consequential technological advancements. Nevertheless, the apprehensiveness over the climate and environmental crisis has coerced mankind to seek alternative energy sources. The sustainable development goals executed by the UN General Assembly and the Paris Climate Change Accords (2015), constituted an infrastructure for worldwide cooperation in reconnoitering a virtuous, cost-efficient, and eco-friendly renewable energy resource by the year 2030 [2]. Comprehensive investigations towards other credible energy sources to supersede fossil fuels are being carried out owing to the indefiniteness associated with the futuristic availability of fossil fuels as they depreciate [3]. Energy resources contingent on fossil fuels might thereby be providently unsubstantial in the near future, albeit its footprint on the climatic conditions must be considered, and global diminutions in overall greenhouse gas emissions (GHGEs) would be imperative to prevent unfavorable repercussions. GHGEs are the underlying effects of the melting of poles and global warming, also bringing about the depletion of aquatic agrobiodiversity via oceanic acidulation. Oceans have been ascertained to assimilate one-third of the anthropic carbon dioxide exuded annually, which undergoes dissolution, progressively diminishing the water's pH. Therefore, the implementation of renewable sources is crucial to preserve Earth for progeniture [4]. The Energy strategy of the European (EU) Commission (2020) has bequeathed a long-range goal for the EU and other developed nations to bring about around 80 to 95% reduction in GHGE by the year 2050. Furthermore, the UN Paris Agreement purports a global diminution to 40 Gt of carbon dioxide equivalent emissions in lieu of the 55 Gt intended for by the year 2030 [5].

Biomass encompassing organic matter in the nonfossil form emanating from water- and land-based verdure and organic debris is an intricate combination predominantly comprising lipids, proteins, organic acids, and carbohydrates [6]. Biomass as a prospective energy resource possesses significant benefits ascribable to its facile storage and worldwide accessibility. They also serve to promote other commercial energy requisites including transportation fuel, electricity, and heat [7–9]. Biomass can be classified with respect to its utilization and application as follows:

- Biomass waste: Sewage sludge (SS), livestock, forestry, agricultural residues, black liquor, sawdust, etc.
- Traditional biomass sources: Fishery, agriculture, forestry, livestock farming, chips, pulp, etc.
- Plantation biomass: Herbaceous (sorghum, sugarcane, and switchgrass), forestry (willow, eucalyptus, oil palm, and poplar), and aquatic (water hyacinth and giant kelp) biomass [10].

Owing to the multitudinous resources from which biomass is obtained, there exists an inconstancy in the composition, characteristics, and attributes of the biomass, resulting in versatile product quality. Moreover, most of the bioenergy transformation methodologies being fostered are in their incipient phases [11]. Chemical, physical, biochemical, and thermochemical transformation methodologies are being employed nowadays to retrieve energy from various biomass feedstocks.

This chapter delineates the constitution of various wastewater-derived biomass, including SS, lignocellulosic, and algal biomass along with a comprehensive overview of various chemical (hydrolysis, solvent extraction, and supercritical transformation), physical (mechanical extraction), thermochemical [hydrothermal carbonization (HTC), gasification, pyrolysis, and torrefaction], physicochemical (transesterification), and biochemical (anaerobic digestion and fermentation) techniques employed for the conversion of waste biomass to energy leading to the production of biofuel (biodiesel and bio-oil), biogas, or biochar contingent with the transformation techniques being utilized.

10.2 Composition of Different Types of Biomass Procured from Wastewater

10.2.1 Sewage Sludge

SS, an ineludible derivative of water treatment facilities, is often considered a biomass commodity for energy (syngas, biogas, fuel gas, bio-oil, biodiesel, etc.) and resource reclamation ascribable to its greater invaluable inorganic and organic matter content [12, 13].

They are often exemplified by the following six classes of constituents as enlisted below:

(i) Water spanning from a few to greater than 95%.
(ii) Innocuous organic and inorganic contaminants: (a) Pesticides, dioxins, polychlorinated biphenyls (PCBs), linear-alkyl sulfonates, polybrominated flame retardant, nonylphenols, etc. (b) Toxic metal ions like Ni, Hg, Cr, Pb, Cu, Zn, As, and Cd (concentrations spanning from greater than 10^6 ppb to lower than 10^3 ppb).
(iii) Innocuous organic compounds (roughly around 60% on a dry basis).
(iv) Microbiological contaminants and pathogens.
(v) Inorganic components including aluminates, magnesium, and calcium comprising compounds, silicates, etc.
(vi) Phosphorous and nitrogen comprising compounds.

Sustained processing entails the restoration and effectual reutilization of essential elements and the diminishment of the plausible detrimental effects of SS or the remnants from SS handling on the ecosystem and mankind [14].

10.2.2 Algal Biomass

Aquatic biomass be categorized into macroalgae, microalgae, and other incipient plants moderately submersed in swamps or marshes, where the former relates to marine algae or seaweed that are perceivable by the unaided eye. They comprise photosynthetic plants capable of transforming sunlight into nutrients. The composition and distribution of macroalgal species are greatly inconsistent, contingent on the geographical location. They are capable of sequestering carbon dioxide, alleviating climatic variability while rendering sustainable feedstocks for fuel, or bioenergy. On the other hand, microalgae include monocellular photosynthetic microbes that are eukaryotic or prokaryotic, customarily originating from sea and freshwaters, capable of transforming water, sunlight, and carbon dioxide into algal biomass. The major constituents include carbohydrates, lipids, and proteins that are reserved within the cells along with other minor substances including chlorophylls, phycobiliproteins, and carotenoids [15].

Algal biomass comprises phosphorous, nitrogen, and carbon as the major components, besides containing oligo nutrients such as cobalt, zinc,

Table 10.1 Biomass constitution of a few microalgal species [62].

Sr. no.	Microalgal species	Carbohydrate (%)	Protein (%)	Lipid (%)
1.	*Scenedesmus obliquus*	13.41	30.38	4.66
2.	*Chlorella vulgaris*	20.99	41.51	15.67
3.	*Chlamydomonas reinhardtii*	-	45.70	22.40
4.	*Nannochloropsis oculata*	17.00	39.00	20.00
5.	*Dunaliella tertiolecta*	21.69	61.32	2.87

and iron, functioning as stimulants of methanogenic activities. This mineral constitution of algal biomass is adequate for the nutritional requirements of microbes accountable for the degradation phenomenon [16]. Microalgal species rendered a changeable cellular composition commensurate with concentrations of the order of 5 to 23%, 7 to 23%, and 6 to 52% corresponding to carbohydrates, lipids, and proteins, respectively. For various genres, a greater ratio of proteins imparts a lower carbon-to-nitrogen ratio, in contrast with the terrene species [16].

Table 10.1 illustrates the biomass composition of some microalgal species.

10.2.3 Lignocellulose-Derived Biomass

Lignocellulose-derived biomass is a copiously obtainable resource possessing a periodical yield of around 200 billion dry metric tonnes per annum. The fundamental structure of lignocellulose predominately constitutes cellulose (35 to 50%), lignin (10 to 25%), and hemicellulose (20 to 35%), in conjunction with minor amounts of other nonorganic and organic constituents including lipids, proteins, and other extracts [17]. It can exemplify starting materials for bioethanol generation ascribable to its several inherent advantages including cost efficiency, copiousness, renewability, and greater carbohydrate proportion. Moreover, this feedstock can be endowed at a massive level from industrial, municipal wastes, and agrarian refuse. In contrast with sugar and starch-derived raw materials, lignocellulose-based biomass provides a substantial endowment toward ecological preservation, curtailing overall carbon dioxide expulsion by roughly 90% [18].

10.3 Various Techniques for the Transformation of Different Wastewater-Derived Biomass Feedstock into Energy

Various techniques customarily employed for wastewater-derived biomass energy conversion can be categorized as follows (Figure 10.1):

(i) Physical conversion techniques
(ii) Thermochemical conversion techniques
(iii) Physicochemical conversion techniques
(iv) Biochemical conversion techniques
(v) Chemical conversion techniques

10.3.1 Physical Conversion Techniques

10.3.1.1 Mechanical Extraction

Crude oil can be excavated from microalgal species and other aquatic biomass by means of the application of mechanical pressure employing screw gauges [19]. Mechanical pressing can be achieved via pre-pressing or full-pressing methods. The pre-pressing technique is customarily used in combination with the solvent extraction technique, where almost 18 to 20% of oil is withdrawn via pressing, and the remainder of it uses solvent extraction. This technique is often utilized for biomass feedstock possessing a greater amount of oil (around 30 to 40%) [20]. The full pressing method

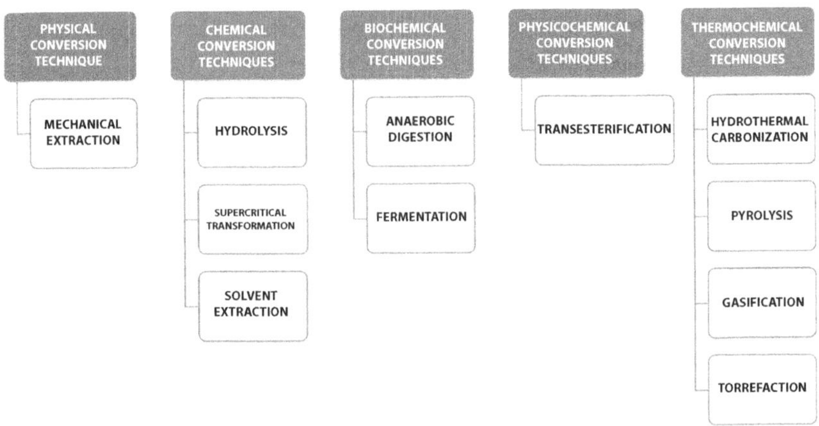

Figure 10.1 Classification of different biomass-to-energy transformation techniques.

utilizes pressures of the order of 95,000 kPa on the biomass feedstock to extricate up to 3 to 5% of the residuum oil. The mechanical extraction process is generally integrated with the transesterification technique to transform the excavated oil into biodiesel along with other value-aggregated derivatives [21].

10.3.2 Chemical Conversion Techniques

10.3.2.1 Hydrolysis

Chemical hydrolysis is associated with the presence of a pre-fermentation stage for transforming extended chains of carbohydrates into facile sugar molecules. Acidolysis can be utilized for microalgal and lignocellulosic biomass employing an extensive array of inorganic acids like phosphoric acid, nitric acid, oxalic acid, acetic acid, sulfuric acid, hydrochloric acid, etc. Some factors including acid concentration, surface-to-volume ratio, temperature, and time are known to have an impact on the hydrolysis mechanism. [22]. Consolidating hydrolysis with microwave or ultrasound methods is currently being explored to enhance the process yield [23, 24]. Microwaving augments the hydrolytic efficacies of biomass by ensuring ameliorated heating and temperature regulation [24]. On the other hand, utilizing the ultrasound method ensues in the shearing of the biomass cell walls due to the cavitation phenomena that amplifies the emittance of low molar mass sugars from polysaccharides [23].

Castro *et al.* carried out the optimization of sugar extrication from wastewater algae employing an acid hydrolysis technique for the consequential fermentation of butanol, acetone, and ethanol via *Clostridium saccharoperbutylacetonicum* N1-4. Pre-treatment of 10% wastewater microalgae (desiccated) via acid hydrolysis employing 1M sulfuric acid at temperatures spanning from 80°C to 90°C, for 2 hours was considered the optimum conditions imparting a sugar yield of around 166.1 g/kg of desiccated microalgae [25].

10.3.2.2 Supercritical Transformation

Supercritical biomass transformation serves as an efficacious substitute for the enzymatic or chemical hydrolysis method. Enzymatic hydrolysis necessitates preliminary treatment of biomass while chemical hydrolysis is a cost-intensive process. As opposed to these methods, supercritical carbon dioxide or water can facilely transform biomass into an admixture of organic acids, oils, alcohol, cellulose, and methane into sugar [26].

Supercritical fluids are substances that can either exist in the gaseous or liquid phase and are commonly employed at states higher than the critical pressure and temperature where liquid and gaseous substances can co-occur. They have the capability to permeate the crystalline framework of lignocellulose-based biomass, surmounting the mass transfer constraints experienced during other pre-treatment processes [27]. Supercritical water gasification serves as a propitious hydrothermal methodology for the conversion of residual biomass to hydrogen-rich synthesis gas at elevated pressures and temperatures. Another significant application of supercritical water is in the gasification of organic matter and residual biomass to yield hydrogen. The product distribution from supercritical water gasification varies substantially commensurate with different processing parameters including pressure, feed concentration, temperature, reactor configuration, biomass particle size, catalyst type, etc. [28].

Wang *et al.* [29] analyzed bio-oil generation from SS via the supercritical transformation method and obtained an utmost bio-oil yield of 39.73% at a 0-min residence time and temperature of 375°C [29].

Table 10.2 represents various fuel characteristics of SS-extracted bio-oil.

Table 10.2 Fuel characteristics of bio-oil obtained from sewage sludge–based biomass [29].

Sr. no.	Fuel characteristics	Bio-oil (procured from SS)
1.	Ash (weight percent)	0.06
2.	Freezing point (°C, maximum)	–45
3.	Flash point (°C)	35
4.	Carbon residue (weight percent, db)	0.46
5.	Density (20°C, g cc^{-1})	1.207
6.	Acid value (mg KOH g^{-1})	4.03
7.	Viscosity (40°C, mm^2 s^{-1}, maximum)	6.83
8.	Higher heating value (MJ kg^{-1})	31.08
9.	Water amount and mechanical impurity (weight percent)	0.3777

10.3.2.3 Solvent Extraction

The traditional solvent extraction method is associated with the excavation of solvents from solid matrices, employing solid-liquid extraction strategies including maceration, distillation, and Soxhlet extraction [30]. Several depuration and chemical extraction processes are contingent on the solvents being employed. Practically, all the biochemical entities can be dissolved in distinct solvents. Therefore, the effective implementation employing solvent extraction necessitates the amendment of concurrent biomolecular excavation [31]. Biofuel generation from microalgal species necessitates energy-intensive methods for the extrication and transformation of triacylglyceride lipids to fuel, congruent with the cogeneration of nutraceuticals and protein feed. The wet solvent extraction method entails mechanical cell breaching, lipid extrication by virtue of solvent contacting, physical phase segregation, thermal solvent retrieval, and transesterification [32]. Some microalgae species such as *Botryococcus braunii* are capable of proactively exuding oil that can be retrieved without harming the cell wall structure by employing solvents such as decane. Moreover, an enhancement in oil extraction was observed from 5.6 to 18.8% on consolidation with the bead milling technique during extraction from *Chlorella protothecoides* species employing hexane as the solvent [33].

10.3.3 Biochemical Conversion Techniques

These methodologies enable the chemical disintegration of the biomass matter into carbohydrates which are further transformed into biogas or liquid fuels as delineated below.

10.3.3.1 Anaerobic Digestion

It is a biochemical conversion technique associated with the conversion of organic substrates to biogas. This method involves the deterioration and stabilization of organic matter under an anaerobic environment implemented via microbes (such as acetogenic bacteria, acidogenic bacteria, and methanogenic archaebacteria) ensuing in the generation of biogas possessing ameliorated energy content. Typically, the constitution of biogas procured from anaerobic digestion includes hydrogen sulfide (1 to 2%), carbon dioxide (25 to 45%), methane (55 to 75%), and relics of hydrogen and ammonia. The calorific values of the engendered biogas span from 10 to 27 MJ Nm^{-3}, based on the amount of methane. The degeneration technique occurs in digesters contemplated to ascertain optimum growth for microbes (temperature, mixing, anaerobic atmosphere, hydraulic residence time, organic load, and pH). Biogas acquired finds various applications including biomethane generation, heat and electricity generation, etc. The conversion of intricate organic biomass into biogas occurs through four stages involving the operation of discrete metabolous groups of microbes which are listed below:

(i) Hydrolysis Stage: Preliminary stage of the technique in which the organic matter (carbohydrates, fats, and proteins) is converted into explicable substances (alcohols, ketones, and inflammable fatty acids), followed by their hydrolysis to facile monomeric molecules employing bacteria via the fermentation process.
(ii) Acidogenesis Stage: The phase post-hydrolysis that is associated with the additional disintegration of compounds generated during the initial phase, via acidogenic microbes, yielding alcohols, hydrogen, ammonia, hydrogen sulfide, carbon dioxide, and other lightweight volatile substances.
(iii) Acetogenesis Stage: Involves the fermentative disintegration of sugars, ammonia, acids, and other volatile fatty acids in butyrate, propionate, hydrogen, acetate, and carbon dioxide. The terminal fermentation derivates serve as predecessors of methane generation occurring in the subsequent stage.
(iv) Methanogenesis Stage: The methodology for generating methane emanating from formic acid or acetic acid through the decarboxylation mechanism or via the reduction of carbon dioxide employing hydrogen as the co-substratum. This stage of the anaerobic digestion methodology is actuated by methanogenic bacterial species [15].

Olsson *et al.* [34] performed co-digestions of various classes of algal biomass and sludges and a statical synergy effect was ascertained with 37% (wt/wt) of wet algae-based slurry and 63% (wt/wt) of indigested SS; the admixture yielded 23% greater amounts of methane compared to the SS digested separately [34].

10.3.3.2 Fermentation

The fermentation technique is inextricably linked with the conversion of biomass into ethanol in congruence with the reaction given below, employing biochemical routes (Equation 10.1):

$$C_6H_{12}O_6 \rightarrow 2C_2H_5OH + 2CO_2(\Delta H = -71.4\ KJ\ mol^{-1}) \quad (10.1)$$

After cultivation and transportation steps in processing plants, the biomass is exposed to pre-treatments to ensure facile disintegration in the ensuing stages by means of enzymes. The customarily employed pre-treatments include a consolidation of acid treatment and steam to foster the hydrolysis of biomass into hemicellulose, lignin, and cellulose. The decrease in the feedstock size influences the efficacy of the fermentation approach. As per Equation 10.1, the fermentation of sugars possessing 6 C atoms involves transformation into 2 mol of carbon dioxide and 2 mol of ethyl alcohol when subjected to heat. In the fermentation approach, the admixture is subjected to inoculation with microorganisms like bacteria, and yeast to assimilate sugars and exude substances that can be utilized as biofuel. The overall transformation methodology generally requires around 3 to 5 days. In instances of ethanol, the liquid is segregated using distillation, owing to the boiling point differences among various constituents. The resultant ethanol is gathered and cleansed for utilization in combination with the fuel [15]. Lately, macro- and microalgal biomasses are being explored as the starting material for the fermentation technique. A diverse spectrum of algae-based biomass comprises different varieties of polysaccharides such as red algae consisting of cellulose, and Floridean starch, and green algae consisting of starch and cellulose. In conjunction with glucans, certain non-glucans inclusive of alginate, carrageenan, and agar also exist. For enhanced ethanol generation, hydrolysis of non-glucans, as well as glucans, is imperative [35].

Fermentation of carbohydrate-abundant fragments of algal biomass after lipid extraction attained butanol yields of around 8.05 g/L [36]. In another study, dry and wet biomass of *Scenedesmus obliquus* was subjected to fermentation for the generation of bio-hydrogen employing *Enterobacter aerogenes* and *Clostridium butyricum* possessing equivalent rates of purity and bio-hydrogen generation irrespective of the water conditions of the biomass [37].

10.3.4 Physicochemical Conversion Technique

Physicochemical transformation methods like transesterification advance into higher-density biofuels including biodiesel [38].

10.3.4.1 Transesterification

Biodiesel an alternative fuel is developed by employing renewable sources and residual oils. Biodiesel generation necessitates reasonably facile chemical reactions, the most prevalent being transesterification and transesterification subjected to acid catalysis. The latter is associated with the reaction of an alcohol (methanol or ethanol) with triglyceride in a catalytic environment [39, 40]. The transformation of triglyceride to fatty acid methyl ester or fatty acid ethyl ester renders glycerol as the side derivative. Catalysts employed in the process can be heterogeneous (varied phases, typically solid) or homogeneous (similar catalyst and reactant phases); moreover, catalysts utilized in biodiesel generation can be categorized into three different classes, namely, enzymes, alkalis, and acids.

Customarily, acid-based catalysts are employed for the transesterification approach of algae-based lipids. Although algae-derived biomass comprises a greater proportion of lipids, only certain components of the overall lipids constitute triglycerides and free fatty acids capable of generating biodiesel (fatty acid methyl ester). Subsequently, fatty acid methyl esters yield a proportion of dry algae-derived biomass and overall lipid differs widely spanning from 0.6 to 33.5% of the dry weight [41]. A species of green algae, *Chlorella vulgaris* possessed an oleic acid ratio of around 76% after 1 week of cultivation, yielding biodiesel of ameliorated quality possessing diminished NO_x exudations [42].

In situ transesterification of SS employing the hydrothermal technique has been utilized to generate biodiesel. Methanol dosage values of 20 g g^{-1} desiccated SS, along with 2% H_2SO_4 at temperature conditions of 160°C for 4 hours duration, yielded 8.63% biodiesel [43].

10.3.5 Thermochemical Conversion Techniques

The thermochemical methodologies are associated with the transformation of biomass into energy via the application of chemical and thermal disintegration methods under variegated temperature conditions and oxygen concentration.

10.3.5.1 Hydrothermal Carbonization

HTC refers to the thermochemical transformation of damp biomass, targeting the transformation of carbohydrate constituents of the biomass feedstock into a carbon-prolific material [44]. Predominantly, the apprehension of the HTC process is similar to wet torrefaction. The HTC technique is extensively utilized to generate coal possessing greater carbon content that can be employed as a fuel, fertilizer, or soil enhancer. Hydrochar, the end product of the HTC process, comprises an aqueous phase and a solid

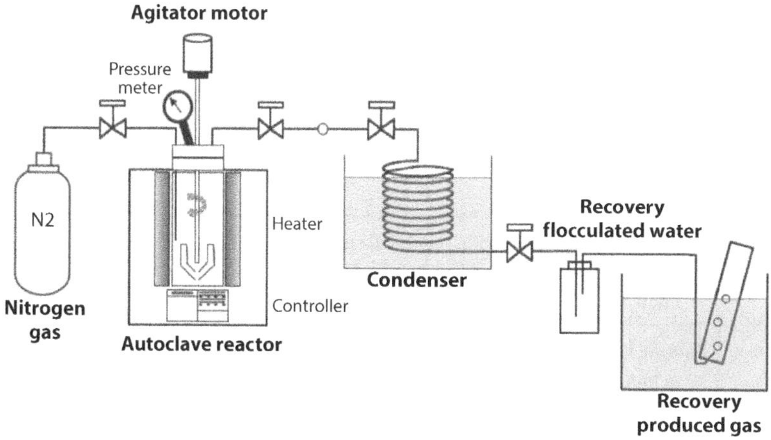

Figure 10.2 Schematic representation of a lab-scale HTC reactor operated in an inert environment. Reprinted with permission from [59]. Copyright 2018, Elsevier.

char phase. In this method, the biomass imparts the medium as well as the reactants for an intricate sequence of reactions during the HTC process, incorporating dehydrogenation, decarboxylation, and demethanation, with decreased H/C and O/C proportions in the raw material ensuing in an energy-efficient suspension [45]. Even though various biomasses have been utilized in the HTC methodology, microalgal species have attained considerable emphasis ascribable to their proficiency to fix carbon, and the capability to remediate effluents for the concurrent nutrient expulsion and biomass generation, thereby substantially lowering the starting material expenditure [46]. Figure 10.2 depicts a laboratory-scale hydrothermal reactor operated in an inert atmosphere.

HTC of blue-green microalgal species performed at durations and temperatures of <1 hour and 200°C, respectively, yielded algal char residue of the bituminous coal grade. Plausible uses of this derivative encompass the generation of syngas and transformation into gasoline and other chemicals and as a carbon-free addendum to natural coal for the production of electric power [47].

Table 10.3 represents the HTC analyses of some artificially cultivated algal biomass species.

10.3.5.2 Pyrolysis

Pyrolysis is associated with the thermochemical transformation of biomass feedstock in oxygen-deficient conditions at elevated temperatures [48].

Table 10.3 HTC analyses of some synthetically cultivated algal biomass.

Sr. no.	Biomass type	Catalyst	Reaction parameters	Yield (%)	HHV (MJ/kg)	Ref.
1.	*Chlamydomonas reinhardtii* - char	Oxalic acid	t: 120 min, T: 200°C	39	31.6	[63]
2.	*Aphanizomenon flos-aquae* - char	Citric acid	t: 120 min, T: 213°C	16	-	[47]
3.	*Nannochloropsis oculate* - char	-	t: 30 min, T: 210°C	41	28.4	[63]
4.	*Chlorella vulgaris* – raw biomass	-	-	-	23.8	[64]
5.	*Chlorella vulgaris* - char		t: 30 min, T: 170°C	51.8	26	[64]
6.	*Spirulina* sp. - char	Oxalic acid	t: 180 min, T: 213°C	27	-	[47]
7.	Chlorella sp. - char	Oxalic acid	t: 180 min, T: 200 °C	39.5	-	[47]

Abbreviations: t: Residence time, **T:** Reaction temperature, **HHV:** Higher heating value

The pyrolysis technique can be categorized into different types, namely, flash, catalytic, slow, and fast pyrolysis based on the reaction time and operating temperatures [49]. Pyrolysis yields an admixture of various components including bio-oil, biochar, and gases (carbon dioxide, hydrogen, methane, and carbon monoxide) [50]. The preliminary stage of the pyrolysis process is biomass pre-treatment at temperatures spanning from 100°C to 120°C. The next stage involves the pyrolysis reaction where the preheated biomass is subjected to the pyrolysis reactor, encompassed by an exterior plenum chamber in which combustion occurs, implicitly heating the biomass comprehended in the reaction container. During this step of the pyrolysis process, thermal degradation of organic matter commences at temperatures of around 350°C to 800°C. Following the pyrolytic decomposition, diminutive entities are obtained in the form of incondensable vapors, gases, and liquid byproducts such as oil, tar, and solid char. Pyrolysis contradicts other elevated temperature conversion techniques like hydrolysis and combustion due to the lack of chemical reactions with water, oxygen, and other reactive substances. Various kinds of reactors customarily employed for the pyrolysis process possessing different practicable functioning and designs are enlisted below [15]:

- Auger reactor
- Fixed bed reactor
- Ablative reactor
- Vacuum pyrolysis reactor
- Fluidized bed reactor
- Rotating cone reactor

Maddi *et al.* carried out pyrolysis studies with lignocellulosic and algal biomass employing analogous reactor conditions for analyzing the constitution and yield of various products (biochar, bio-oil, and gas). The comprehensive bio-oil yield from lignocellulosic biomass and algae were almost identical; albeit the algae-derived bio-oils differed on the basis of composition and possessed multitudinous N- compounds from deterioration of protein structure [51].

10.3.5.3 Gasification

The gasification process primarily endeavors to acquire synthesis gas products that can be utilized explicitly for heating requirements or additional processing into dimethyl ether, methanol, and other chemical substances via the Fischer-Tropsch technique [52]. It is considered to be an extended version of pyrolysis, in which the raw material initially subjected to heating, experiences desiccation and pyrolysis, liberating volatile components, thereby egressing solid residual matter like char, accompanied by gas-gas and gas-solid reactions with the help of gasifying agents [53]. It permits the transformation of aquatic biomass into combustive matter in congruence with the following equation (Equation 10.2):

$$Biomass + O_2 + H_2O \rightarrow CO + CO_2 + H_2 + CH_4 + N_2 + evidence\ of\ other\ matter \tag{10.2}$$

The gas generated is an admixture comprising around 40 to 52% nitrogen, 17 to 24% carbon monoxide, 10 to 12% hydrogen, 9 to 14% carbon dioxide, 1 to 4% methane, and other substances in minor amounts [15].

Gasification reactors can be categorized into two types, updraft and downdraft gasifiers as portrayed in Figure 10.3(a and b), respectively. In an upward gasifier, the gasifying agents (namely, steam, oxygen, or air) travel

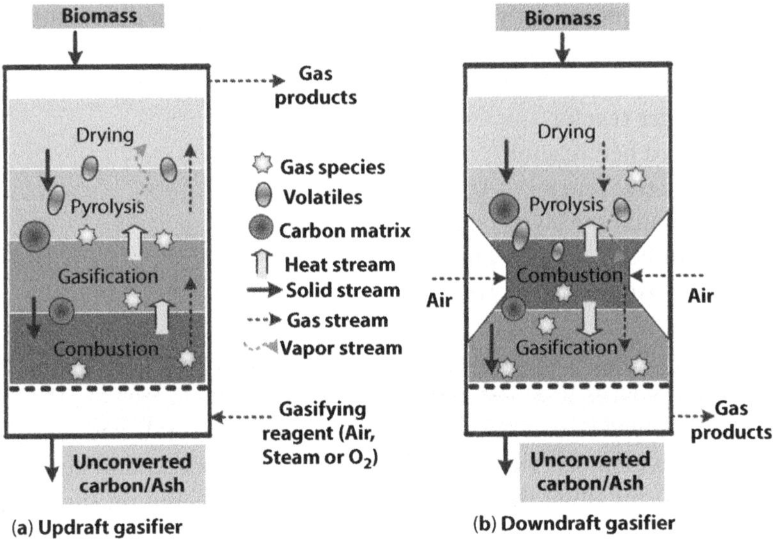

Figure 10.3 Pictorial illustration of the biomass reaction mechanisms occurring in the (a) updraft and (b) downdraft gasifiers. Reprinted with permission from [60]. Copyright 2020, American Chemical Society.

from the underside, and the biomass enters the reactor from the uppermost region along the counter-current direction. Contrarily, both the gasifying agent and the biomass feedstock travel downward in a co-current fashion in the downdraft gasifier.

SS could be gasified to generate substandard-grade combustive gas and it could serve as an auxiliary source against fossil fuels for engendering sustainable energy. It was recommended to function the downdraft gasifier at air fuel ratios, feed rates, turn-down ratios, and specific gasification rates of 2.28 to 2.34 N m^3 kg^{-1} ± 1.84%, 3.69 to 3.71 kg h^{-1} ± 1.43%, 93.64 to 94.15% ± 1.5%, and 497.74 to 514 N m^3 m^{-2} h^{-1} ± 1.5%, respectively, to obtain premium quality gas and to prevent clinker generation at the gasifier throat owing to the greater ash content of sludge [54]. On the other hand, gasification of biomass acquired from *Acutodesmus obliquus* algal species possessing a dry matter proportion of around 5% eventuated in gasification efficacies of around 98% at gasification temperatures of 600°C [55].

10.3.5.4 Torrefaction

Torrefaction is a thermochemical transformation method that is associated with the processing of biomass feedstock within a torrefier to yield a carbonized substance capable of being employed in soil amelioration or as a fuel. It technologically involves subjecting the feed material to an engineered reactor where heat is supplied from an exterior fuel source that is implicitly or explicitly implemented on the incoming biomass experiencing a transition to a torrefied substance. During this process, the starting material encounters several fundamental transformative reactions including vaporization, polymerization, and carbonation. Torrefied substances leaving the reactor comprise incondensable evanescent organic substances, torrefaction gases (inorganic), and a solid biocoal (Figure 10.4). The preliminary objective of torrefaction is to enhance the physical transport as well as fuel characteristics of the treated biomass. Torrefied biomass is endowed with ameliorated heating rates and grindability, reduced moisture absorption capability, and molecular compositional H/C and O/C proportions. Moreover, torrefied biomass possesses enhanced combustion and ignition characteristics owing to a consistent size distribution post-grinding [56].

Poudel *et al.* [57] carried out the torrefaction of SS, a non-lignocellulosic residue employing a horizontal tube reactor in an inert atmosphere of nitrogen, with the residence time and temperatures spanning from 0 to 50 min and 150°C to 400°C, respectively. The mass and energy yields were curtailed with an enhancement in the torrefaction temperature. The components possessing oxygen were liberated at temperatures lower than that of the gaseous hydrocarbon derivatives, and temperatures of around 300°C to 350°C were found to be optimal for sewage sludge [57].

In another study, Bach *et al.* [58] studied the wet torrefaction of a microalgal species, namely, *Chlorella vulgaris* ESP-31, accompanied by microwave heating at various time periods (5, 10, and 15 min), and temperatures (160°C, 170°C, and 180 °C). The empirical results exhibited an enhancement in fuel characteristics of microalgae post-wet torrefaction

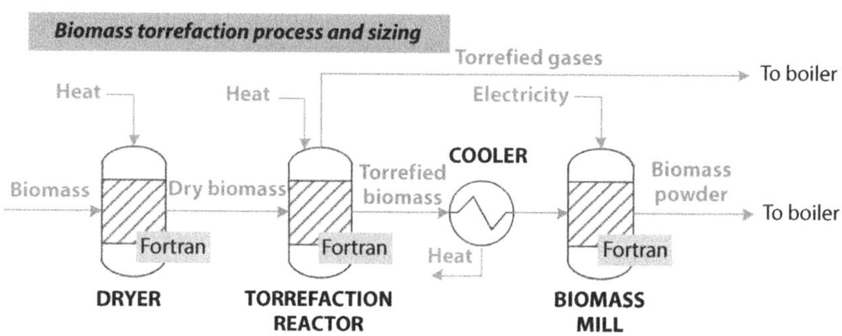

Figure 10.4 Diagrammatic illustration of the torrefaction phenomenon of waste-water-derived biomass. Reprinted with permission from [61]. Copyright 2014, Elsevier.

including diminished ash content, enhanced hydrophobicity, and calorific content. The calorific value of these microalgal species could be augmented to almost 21% and around 61.5% of the biomass energy could be preserved post-torrefaction [58].

10.4 Conclusion and Future Prospects

Wastewater-derived biomass as a substitutive energy source is remunerative and possesses tremendous opportunities to be advanced into an economically practicable solution to waste management and energy concerns. Aquatic biomass being copious worldwide permits the development of biorefineries in any geographic zone. As an energy source, biomass tends to affect the heat, electricity, and conveyance fuel necessities, albeit stringent government interventions are essential to regulate biomass feedstock accumulation, circulation, and transferral. Enormous efforts are required to ensure more cost-effective, scalable, apprehensible, facilely aggrandizable transformation techniques possessing enhanced productiveness. Significant attention is required to enhance the lignocellulosic and algal biomass, municipal sludge pre-treatment, and transformation methods. The selection of the optimal transformation methodology will be determined by the energy efficacy, production cost, facile operation, along ecological implications. This chapter is limited to the composition of different biomass feedstock along with the major biomass-to-energy conversion techniques encompassing the biochemical, thermochemical, physicochemical, physical, and chemical methodologies to yield an extensive array of products spanning from biofuel (biodiesel and bio-oil), biochar, biogas, electricity, etc., contingent with the transformation method being employed. Howbeit, an in-depth analysis of the repositioning and transference of the biomass feedstock, policies, and computer simulation predicated approximations to enhance bioenergy production are imperative.

Acknowledgments

The authors would like to express their gratitude toward Dr. C. P. Ramanarayanan, Vice-Chancellor of DIAT (DU), Pune, for their constant support and motivation. The first author would also like to acknowledge Dr. Amrita Nighojkar, Ms. Niranjana Jayaprakash, Mr. Jigar Patadiya, and Ms. Alsha Subash for providing the necessary technical support and guidance throughout the course of this book chapter writing.

References

1. World Population Prospects 2022: Summary of Results. *World Population Prospects 2022: Summary of Results*, 52, 2022.
2. Gielen, D., Boshell, F., Saygin, D., Bazilian, M.D., Wagner, N., and Gorini, R., The role of renewable energy in the global energy transformation. *Energy Strategy Rev.*, 24, 38, 2019.
3. Shafiee, S., and Topal, E., When will fossil fuel reserves be diminished? *Energy Policy*, 37, 181, 2009.
4. Ormerod, W.G., P. Freund, A. Smith, J.D. *Ocean Storage of CO2,* International Energy Agency, Greenhouse Gas R&D Programme, 2002.
5. European Commission Energy 2020 A strategy for competitive, sustainable and secure energy, European Commission: Brussels, Belgium, 2010.
6. Robert K. M. Hay, J.R.P. *The Physiology of Crop Yield*, Wiley-Blackwell, 2006.
7. Yana, S., Nizar, M., Irhamni, and Mulyati, D., Biomass waste as a renewable energy in developing bio-based economies in Indonesia: A review. *Renew. Sustain. Energy Rev.*, 160, 112268, 2022.
8. Chen, H., Xia, A., Zhu, X., Huang, Y., Zhu, X., and Liao, Q., Hydrothermal hydrolysis of algal biomass for biofuels production: A review. *Bioresour. Technol.*, 344, 126213, 2022.
9. Vassilev, S.V., Vassileva, C.G., and Vassilev, V.S., Advantages and disadvantages of composition and properties of biomass in comparison with coal: An overview. *Fuel*, 158, 330, 2015.
10. Gusiatin, Z.M., & Pawłowski, A., Biomass for fuels – classification and composition, in: *Biomass for Biofuels*, K. Bułkowska, Z.M. Gusiatin, E. Klimiuk (Eds.), CRC Press, pp. 15–28, 2016.
11. Amjith, L., and Bavanish, B., A review on biomass and wind as renewable energy for sustainable environment. *Chemosphere*, 293, 133579, 2022.
12. Raheem, A., Sikarwar, V.S., He, J., Dastyar, W., Dionysiou, D.D., Wang, W., and Zhao, M., Opportunities and challenges in sustainable treatment and resource reuse of sewage sludge: A review. *Chem. Eng. J.*, 337, 616, 2018.
13. Moraes, E.S., Reis, G.F., Cruz, J., Cozzolino, K., Neto, A.F.G., Andrade-Filho, T., Neto, A.M.J.C., Thermodynamics study of biokerosene from coconut and palm kernel oils and JP-8 aircraft fuels in the gas phase by the DFT method. *J. Mol. Model.* 26, 79, 2020.
14. Rulkens, W., Sewage sludge as a biomass resource for the production of energy: Overview and assessment of the various options. *Energy Fuels*, 22, 9, 2008.
15. Tursi, A., and Olivito, F., Biomass conversion, in: *Advances in Bioenergy and Microfluidic Applications,* M.R. Rahimpour, R. Kamali, M.A. Makarem, M.K.D. Manshadi (Eds.), Elsevier, pp. 3–39, 2021.
16. Assemany, P., Marques, I., Calijuri, M.L., and Reis, A., Complementarity of substrates in anaerobic digestion of wastewater grown algal biomass. *Waste Biomass Valorization*, 11, 5759, 2020.

17. Sawatdeenarunat, C., Surendra, K.C., Takara, D., Oechsner, H., and Khanal, S.K., Anaerobic digestion of lignocellulosic biomass: Challenges and opportunities. *Bioresour. Technol.*, 178, 178, 2015.
18. Yuan, H., Tan, L., Kida, K., Morimura, S., Sun, Z.-Y., and Tang, Y.-Q., Potential for reduced water consumption in biorefining of lignocellulosic biomass to bioethanol and biogas. *J. Biosci. Bioeng.*, 131, 461, 2021.
19. Sharma, Y.C., and Singh, V., Microalgal biodiesel: A possible solution for India's energy security. *Renew. Sustain. Energy Rev.*, 67, 72, 2017.
20. Naik, S.N., Goud, V. V., Rout, P.K., and Dalai, A.K., Production of first and second generation biofuels: A comprehensive review. *Renew. Sustain. Energy Rev.*, 14, 578, 2010.
21. Mwaurah, P.W., Kumar, S., Kumar, N., Attkan, A.K., Panghal, A., Singh, V.K., and Garg, M.K., Novel oil extraction technologies: Process conditions, quality parameters, and optimization. *Compr. Rev. Food Sci. Food Saf.*, 19, 3, 2020.
22. Duongbia, N., Chaiwongsar, S., Chaichana, C., and Chaiklangmuang, S., Acidic hydrolysis performance and hydrolyzed lipid characterizations of wet Spirulina platensis. *Biomass Convers. Biorefin.*, 9, 305, 2019.
23. Joshi, S.M., and Gogate, P.R., Intensification of dilute acid hydrolysis of spent tea powder using ultrasound for enhanced production of reducing sugars. *Ultrason. Sonochem.*, 61, 104843, 2020.
24. Martins, M.G., da Silva Arouche, T., Neto, A.F.G., da Cruz, J.N., da Costa, F.L.P., Fernandes, L.L., de Carvalho Junior, R.N., da Silva Costa, J.F., de Jesus Chaves Neto, A.M., Density functional theory for the thermodynamic gas-phase investigation of butanol biofuel and its isomers mixed with gasoline and ethanol. *J. Mol. Model.*, 27, 80, 2021.
25. Castro, Y.A., Ellis, J.T., Miller, C.D., and Sims, R.C., Optimization of wastewater microalgae saccharification using dilute acid hydrolysis for acetone, butanol, and ethanol fermentation. *Appl. Energy*, 140, 14, 2015.
26. Okolie, J.A., Nanda, S., Dalai, A.K., Berruti, F., and Kozinski, J.A., A review on subcritical and supercritical water gasification of biogenic, polymeric and petroleum wastes to hydrogen-rich synthesis gas. *Renew. Sustain. Energy Rev.*, 119, 109546, 2020.
27. Fortunati, E., Luzi, F., Puglia, D., and Torre, L., Extraction of lignocellulosic materials from waste products, in: *Multifunctional Polymeric Nanocomposites Based on Cellulosic Reinforcements*, D. Puglia, E. Fortunati and J.M. Kenny (Eds.), Elsevier, pp. 1–38, 2016.
28. Okolie, J.A., Rana, R., Nanda, S., Dalai, A.K., and Kozinski, J.A., Supercritical water gasification of biomass: A state-of-the-art review of process parameters, reaction mechanisms and catalysis. *Sustain. Energy Fuels*, 3, 578, 2019.
29. Wang, Y., Chen, G., Li, Y., Yan, B., and Pan, D., Experimental study of the bio-oil production from sewage sludge by supercritical conversion process. *Waste Manage.*, 33, 2408, 2013.

30. Vaquero, M., Rajauria, G., and Tiwari, B., Conventional extraction techniques: Solvent extraction, in: *Sustainable Seaweed Technologies*, M.D. Torres, S. Kraan, H. Dominguez (Eds.), Elsevier, pp. 171–189, 2020.
31. Kumari, N., and Singh, R.K., Biofuel and co-products from algae solvent extraction. *J. Environ. Manage.*, 247, 196, 2019.
32. Martin, G.J.O., Energy requirements for wet solvent extraction of lipids from microalgal biomass. *Bioresour. Technol.*, 205, 40, 2016.
33. Mercer, P., and Armenta, R.E., Developments in oil extraction from microalgae. *Eur. J. Lipid Sci. Technol.*, 113, 539, 2011.
34. Olsson, J., Feng, X.M., Ascue, J., Gentili, F.G., Shabiimam, M.A., Nehrenheim, E., and Thorin, E., Co-digestion of cultivated microalgae and sewage sludge from municipal waste water treatment. *Bioresour. Technol.*, 171, 203, 2014.
35. Ramachandra, T.V., and Hebbale, D., Bioethanol from macroalgae: Prospects and challenges. *Renew. Sustain. Energy Rev.*, 117, 109479, 2020.
36. Gao, K., Orr, V., and Rehmann, L., Butanol fermentation from microalgae-derived carbohydrates after ionic liquid extraction. *Bioresour. Technol.*, 206, 77, 2016.
37. Phanduang, O., Lunprom, S., Salakkam, A., and Reungsang, A., Anaerobic solid-state fermentation of bio-hydrogen from microalgal Chlorella sp. biomass. *Int. J. Hydrogen Energy*, 42, 9650, 2017.
38. Tursi, A., A review on biomass: Importance, chemistry, classification, and conversion. *Biofuel Res. J.*, 6, 962, 2019.
39. Chisti, Y., Biodiesel from microalgae. *Biotechnol. Adv.*, 25, 294, 2007.
40. Razzak, S.A., Hossain, M.M., Lucky, R.A., Bassi, A.S., and de Lasa, H., Integrated CO_2 capture, wastewater treatment and biofuel production by microalgae culturing—A review. *Renew. Sustain. Energy Rev.*, 27, 622, 2013.
41. Orr, V.C.A., Plechkova, N. V., Seddon, K.R., and Rehmann, L., Disruption and wet extraction of the microalgae Chlorella vulgaris using room-temperature ionic liquids. *ACS Sustain. Chem. Eng.*, 4, 591, 2016.
42. Ji, M.-K., Shanab, R.A.I., Kim, S.-H., Salama, E.-S., Lee, S.-H., Kabra, A.N., Lee, Y.-S., Hong, S., and Jeon, B.-H., Cultivation of microalgae species in tertiary municipal wastewater supplemented with CO_2 for nutrient removal and biomass production. *Ecol. Eng.*, 58, 142, 2013.
43. Hu, W., Zhou, X., Tan, J., Hou, J., Xie, Y., Wang, X., Wang, Y., and Zhang, Y., In situ transesterification of wet sewage sludge via hydrothermal process: Biodiesel production and residue utilization. *Biomass Bioenergy*, 141, 105715, 2020.
44. Merzari, F., Langone, M., Andreottola, G., and Fiori, L., Methane production from process water of sewage sludge hydrothermal carbonization. A review. Valorising sludge through hydrothermal carbonization. *Crit. Rev. Env. Sci. Technol.*, 49, 947, 2019.
45. Smith, A.M., Whittaker, C., Shield, I., and Ross, A.B., The potential for production of high quality bio-coal from early harvested Miscanthus by hydrothermal carbonisation. *Fuel*, 220, 546, 2018.

46. Yu, K.L., Show, P.L., Ong, H.C., Ling, T.C., Chi-Wei Lan, J., Chen, W.-H., and Chang, J.-S., Microalgae from wastewater treatment to biochar – Feedstock preparation and conversion technologies. *Energy Convers. Manage.*, 150, 1, 2017.
47. Heilmann, S.M., Davis, H.T., Jader, L.R., Lefebvre, P.A., Sadowsky, M.J., Schendel, F.J., von Keitz, M.G., and Valentas, K.J., Hydrothermal carbonization of microalgae. *Biomass Bioenergy*, 34, 875, 2010.
48. Mehrabadi, A., Craggs, R., and Farid, M.M., Pyrolysis of wastewater treatment high rate algal pond (WWT HRAP) biomass. *Algal Res.*, 24, 509, 2017.
49. Azizi, K., Keshavarz Moraveji, M., and Najafabadi, H., A review on bio-fuel production from microalgal biomass by using pyrolysis method. *Renew. Sustain. Energy Rev.*, 82, 3046, 2018.
50. Hallenbeck, P.C., Grogger, M., Mraz, M., and Veverka, D., Solar biofuels production with microalgae. *Appl. Energy*, 179, 136, 2016.
51. Maddi, B., Viamajala, S., and Varanasi, S., Comparative study of pyrolysis of algal biomass from natural lake blooms with lignocellulosic biomass. *Bioresour. Technol.*, 102, 11018, 2011.
52. Moreira, R., Bimbela, F., Gil-Lalaguna, N., Sánchez, J.L., and Portugal, A., Clean syngas production by gasification of lignocellulosic char: State of the art and future prospects. *J. Ind. Eng. Chem.*, 101, 1, 2021.
53. Sikarwar, V.S., Zhao, M., Clough, P., Yao, J., Zhong, X., Memon, M.Z., Shah, N., Anthony, E.J., and Fennell, P.S., An overview of advances in biomass gasification. *Energy Environ. Sci.*, 9, 2939, 2016.
54. Dogru, M., Midilli, A., and Howarth, C.R., Gasification of sewage sludge using a throated downdraft gasifier and uncertainty analysis. *Fuel Process. Technol.*, 75, 55, 2002.
55. Patzelt, D.J., Hindersin, S., Elsayed, S., Boukis, N., Kerner, M., and Hanelt, D., Hydrothermal gasification of Acutodesmus obliquus for renewable energy production and nutrient recycling of microalgal mass cultures. *J. Appl. Phycol.*, 27, 2239, 2015.
56. Barskov, S., Zappi, M., Buchireddy, P., Dufreche, S., Guillory, J., Gang, D., Hernandez, R., Bajpai, R., Baudier, J., Cooper, R., and Sharp, R., Torrefaction of biomass: A review of production methods for biocoal from cultured and waste lignocellulosic feedstocks. *Renew. Energy*, 142, 624, 2019.
57. Poudel, J., Ohm, T.-I., Lee, S.-H., and Oh, S.C., A study on torrefaction of sewage sludge to enhance solid fuel qualities. *Waste Manage.*, 40, 112, 2015.
58. Bach, Q.-V., Chen, W.-H., Lin, S.-C., Sheen, H.-K., and Chang, J.-S., Wet torrefaction of microalga Chlorella vulgaris ESP-31 with microwave-assisted heating. *Energy Convers. Manage.*, 141, 163, 2017.
59. Park, K.Y., Lee, K., and Kim, D., Characterized hydrochar of algal biomass for producing solid fuel through hydrothermal carbonization. *Bioresour. Technol.*, 258, 119, 2018.

60. Zhang, Y., Wan, L., Guan, J., Xiong, Q., Zhang, S., and Jin, X.A., Review on biomass gasification: Effect of main parameters on char generation and reaction. *Energy Fuels*, 34, 13438, 2020.
61. Li, J., Zhang, X., Kruczek, H., Yang, W., Kruczek, P., and Blasiak, W., Process simulation of co-firing torrefied biomass in a 220 MWe coal-fired power plant. *Energy Convers. Manage.*, 84, 503, 2014.
62. Dalena, F., Senatore, A., Tursi, A., and Basile, A., Bioenergy production from second- and third-generation feedstocks, in: *Bioenergy Systems for the Future*, Elsevier, pp. 559–599, 2017.
63. Levine, R.B., Sierra, C.O.S., Hockstad, R., Obeid, W., Hatcher, P.G., and Savage, P.E., The use of hydrothermal carbonization to recycle nutrients in algal biofuel production. *Environ. Prog. Sustain. Energy*, 32, 962, 2013.
64. Bach, Q.-V., Tran, K.-Q., Khalil, R.A., Skreiberg, Ø., and Seisenbaeva, G., Comparative assessment of wet torrefaction. *Energy Fuels*, 27, 6743, 2013.

11

Recycled Wastewater from Sewage Treatment Plants for Sustainable Agriculture

Anuska Raichoudhury[1], Radha Sankar Mal[2], Ranjay Kumar Thakur[1], Subhankar Mishra[1], Mukesh Singh[3*] and Amit Biswas[4]

[1]*Department of Food Technology, Haldia Institute of Technology, Haldia, Purba Medinipur, West Bengal, India*
[2]*Department of Biotechnology, Haldia Institute of Technology, Haldia, Purba Medinipur, West Bengal, India*
[3]*Department of Botany, Kabi Nazrul College, Murarai, Birbhum, India*
[4]*Department of Agricultural Engineering, Haldia Institute of Technology, Haldia, Purba Medinipur, West Bengal, India*

Abstract

The global population is increasing at an alarming rate, which puts immense pressure on the planet. One of the major resources under threat is freshwater. To alleviate this stress, it is necessary to adopt sustainable practices for water use. Recycling wastewater from sewage treatment plants (STPs) for sustainable agriculture is one such practice that can solve this critical issue. The present chapter explores the feasibility of using recycled wastewater from STPs for irrigation purposes in agriculture. Wastewater can be treated by a three-step clarification process to take out all the physical, chemical, and microbial contaminants present in it. The pre-treatment process includes the separation of solid large particles followed by settling, sedimentation, and separation of slug from scum. Organic wastes are removed in the secondary stage by aerobic and anaerobic bacteria. Biogas and activated sludge are considered the byproduct of this process. Tertiary purification is obtained from chlorination. The current chapter elaborates on recycled water quality and its effect on agricultural crop production, soil health, and the environment. The study also discusses the economic viability of implementing this approach in agricultural production.

**Corresponding author*: msingh006@gmail.com

Inamuddin, Tariq Altalhi, Mohammad Luqman and Joseph Kapuku (eds.) Sewage and Biomass from Wastewater to Energy, (283–326) © 2024 Scrivener Publishing LLC

Keywords: Agri-food chain, irrigation, recycled wastewater, sewage treatment plant, soil health, sustainable agriculture, water scarcity

11.1 Introduction

Freshwater shortage is a very critical issue for many countries around the world, and this is becoming increasingly important to find innovative ways to conserve and reuse water [1]. One way to address this issue is by recycling wastewater from sewage treatment plants for sustainable agriculture [2]. This approach offers several benefits, including reducing freshwater consumption, improving soil quality, and reducing pollution. As per the United Nations, the agriculture sector utilizes more than one-third (i.e., around 70%) of available freshwater, which makes it a significant contributor to water scarcity [3]. Demand for food, as well as water, will increase in the future along with the population. Therefore, it is crucial to find ways to conserve and reuse water in agriculture. Recycling wastewater from sewage treatment plants for crop production is not a new idea, and this is already practiced in several countries worldwide, including Israel, Spain, and Australia [4]. The wastewater can be treated and purified to remove contaminants, such as suspended solids, pathogens, and heavy metals before it is used for irrigation [2]. The treated water can be applied directly for crops or groundwater recharge, which can be later utilized for crop irrigation [5]. Recycling wastewater for agriculture can also improve soil quality by providing essential nutrients (i.e., nitrogen, phosphorus, and potassium) for crop intensification. Studies have shown that using treated wastewater for irrigation can increase crop yields and improve soil fertility [6, 7]. Furthermore, recycling wastewater for agriculture can reduce pollution by diverting wastewater from rivers and other bodies of water. Wastewater contains various pollutants that may contribute to eutrophication, a process that causes excessive growth of algae and other aquatic plants, which can harm aquatic ecosystems. In conclusion, recycling wastewater from sewage treatment plants for sustainable agriculture offers several benefits, including reducing water consumption, improving soil quality, and reducing pollution of water bodies. However, this is crucial to carry on research for finding innovative ways to conserve and reuse water, especially in agriculture, where water is a critical resource.

11.1.1 Worldwide Water Crisis with Its Effect on Agriculture

The scarcity of freshwater is a universal problem that affects many industries, including agriculture [8]. The availability of freshwater for agriculture is becoming increasingly limited as the world population grows and food demand rises [5]. According to the United Nations, approximately 2.2 billion persons suffer due to a lack of safe drinking water and about 50% population in the world belongs to such areas where water shortage is a severe problem. Climate change is exacerbating the crisis by causing more frequent and severe droughts in some areas and intense rainfall in others, resulting in floods and soil erosion. Agriculture is the largest freshwater user, accounting for roughly 70% of total freshwater withdrawals in the world. Crop production requires water, and as water scarcity worsens, agricultural productivity and food security are jeopardized [1]. Farmers in some areas are forced to rely on nonrenewable groundwater sources, resulting in aquifer depletion and saltwater intrusion. Furthermore, water scarcity can lead to conflicts over water resources, especially in areas where freshwater is scarce.

11.1.2 Potential of Recycled Wastewater from Sewage Treatment Plants (STPs) for Sustainable Agriculture

Nowadays, there has been a lot of interest in using recycled wastewater from sewage treatment plants for sustainable agriculture. This is because wastewater is a valuable resource that can be treated and reused to provide water and nutrients to plants, particularly in areas with limited access to freshwater sources. According to the study of Boukhris, using treated wastewater in agriculture yielded promising results in terms of enhancing crop yields and reducing dependence on freshwater sources [9]. Additionally, using recycled wastewater for agriculture reduces the volume of wastewater released into different water bodies, which may negatively affect the environment and public health. It should be noted that using recycled wastewater in agriculture necessitates proper treatment and management to ensure that it is safe for human and animal consumption. The World Health Organization (WHO) provided guidelines for safely utilizing wastewater in agriculture, including measures to reduce contamination and chemical exposure [10].

11.1.3 Aims and Scope of the Chapter

The chapter is established with some specific aims and objectives:

1. To highlight recycled wastewater's importance as a valuable resource for sustainable agriculture in regions facing water scarcity and limited access to freshwater sources.
2. To discuss the benefits and challenges associated with using recycled wastewater for agriculture, including the potential for improved crop yields and reduced freshwater demand, as well as the need for proper treatment and management to ensure the safety of the water.
3. To present various treatment technologies and management practices that can be used to ensure the safety of recycled wastewater for agricultural use, including disinfection, nutrient removal, and soil and crop management practices.
4. To assess the potential risks associated with the utilization of recycled wastewater for agriculture, together with risks to human and animal health, and discuss several measures to mitigate these risks.
5. To examine case studies and best practices that demonstrate the successful use of recycled wastewater for sustainable agriculture.

The scope of the chapter would include a review of the relevant literature, including peer-reviewed scientific studies, technical reports, and policy documents. The chapter will also include case studies and examples from around the world to illustrate the potential benefits and challenges of using recycled wastewater for sustainable agriculture. Additionally, the chapter will discuss the importance of stakeholder engagement and public education in promoting the safe and sustainable use of recycled wastewater for agriculture.

11.2 Quality of Recycled Wastewater for Irrigation

Irrigation is known as the 'supply of water' in fields and lands to grow crops conducting through a specific channel. It involves the development of a food chain, creating playgrounds, golf courses, parks, landscapes, racecourses, etc., according to the demand. Due to the deficiency of pure water sources, nowadays, recycled water is being utilized for irrigation in developing nations.

As it is correlated with the agri-food chain, the quality of recycled water is an important matter of concern. Despite all the benefits, it may consist of several environmental and health hazards. Inappropriate maintenance can deteriorate soil health by increasing the level of salt, and heavy metals and also decreases soil hydraulic conductivity which directly affects the irrigation process as well as the agri-food chain [11]. However, different types of tools are used to decrease the rate risks due to the use of recycled wastewater [11].

11.2.1 Characteristics of Recycled Wastewater

The major components of wastewater such as pathogens, salts, micro-metals, and different types of chemical compounds are contributed from sources like industrial and domestic wastewater. On the other hand, reclaimed water quality is influenced primarily, by the municipal water compositions, mercantile and households' characteristics which contribute to the wastewater. It fluctuates directly with the surrounding areas and the different work cultures [12]. Hence, the recycling process majorly focuses on the exhibition of relatively abundant salts, chemical impurities, and parasites which are probably harmful to plant growth and enhance the risk of ecological hazards as well as public health [11]. The strength of recycled water in irrigation is mainly classified into three types including minor, moderate, and severe, according to the presence of nitrogen, phosphorus, microbes, and the level of BOD5, TDS, etc. It is anticipated that the constituents of the strength of recycled water will be within the specified range of value. However, for irrigation, reclaimed water must be according to the plant type [12].

11.2.2 Methods of Treating Wastewater for Irrigation Purposes

Recycling of water proceeds in four main steps (Figure 11.1) including 1) preliminary, 2) primary, 3) secondary, or biological 4) tertiary.

11.2.2.1 Preliminary Treatment

This is the main step to omit unkempt solids such as plastic, cans, dead bodies of animals, etc., by screening, grit removal, and skimming.

Screening: In this process, the solid suspended wastes are screened out with the help of the screens present in the screener. This process is carried out by passing the wastewater through different screens containing different pore sizes to remove the different sizes of large solid waste.

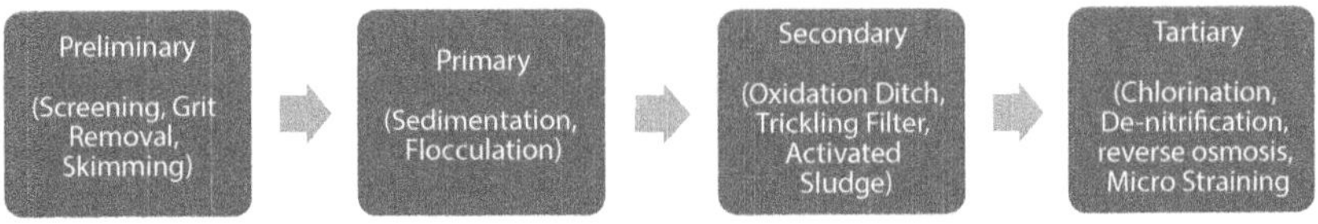

Figure 11.1 Outline of the total process [redrawn].

Grit Removal: It includes the removal of large, nonbiodegradable particles or grits by sedimentation process due to gravitational force. The grit chamber allows the waste to stay for some time and settle down.

Skimming: This includes the removal of fatty acid, oil, and greasy material by coagulating them with the help of air pressure. Thus, the coagulated waste matter is removed from the fluid.

11.2.2.2 Primary Method

This removes the fine solid suspended organic parties from wastewater by sedimentation, chemical flocculation, and mechanical flocculation.

Sedimentation: This method is considered one of the most significant processes, which helps remove the fine suspended particles from the wastewater. Once the insoluble material settles down at the bottom of the tank due to gravity, the supernatant or scum can easily be separated from the sludge. The process is influenced by the size, shape, specific gravity of the particle, and viscosity of the sewage.

Mechanical Flocculation: This process is done in the flocculation tank by rotating the sewage at the speed of 0.43 m/s with the help of rotating paddles. Due to this, the small particles bind to each other and create large size particles and settle down to the bottom.

Chemical Flocculation: Here, some chemicals are added from the outside as a coagulating agent to form the precipitate. For example, alum is added to form the precipitate of aluminum hydroxide [$Al(OH)_3$] and then it can be removed mechanically.

11.2.2.3 Secondary or Biological Method

This technique aims to remove dissolved and extremely fine biological debris, by using living microorganisms such as bacteria, algae, fungi, nematodes, and protozoa. These microorganisms use the nutrients of the wastes for their metabolism and degrade them. The techniques involved are trickling filter, activated sludge treatment, and oxidation ditch.

Trickling Filter: It has a bed, of course, hard and porous material over which sewage water is sprayed. The biomass, i.e., microorganisms attached to the media surface grows and forms a layer, referred to as biological film or microbial slime. As the sewage water trickles through the biofilm, the organic matter gets oxidized to CO, and NO, by microbial metabolism. This oxidation is carried out by the aerobic organisms that are present on the upper portion of the biological film. The biological film is rich in the bacteria *Pseudomonas*, *Flavobacterium*, *Alcaligenes*, and algae *Chlorella*, Ectothrix, and *Stigeoclonium*.

Activated Sludge Treatment: This system consists of an aeration tank, a settling tank, and a sludge recycling system (where the sludge is being activated). Activated sludge contains a large number of microorganisms that oxidize the organic wastes present in the sewage to form OC_2, H_2O, and NO_3 with the help of a continuous oxygen supply in the aeration tank.

Oxidation Ditch: The oxidation ditch consists of a circular canal with an inlet and outlet. In this method, sewage from the primary treatment plant is placed in an oxidation ditch and then it is agitated with the help of a mechanical rotator and then left for about 12 to 24 hours.

11.2.2.4 Tertiary Treatment

Removal of Suspended Solids:

A. Micro-straining: This technique involves adding sewage to a rotating drum filter before turning the drum. Clearwater exits the drum as it rotates, but suspended particulates stay inside.

B. Chemical coagulation and filtration: In this technique, sewage is mixed with precipitating agents like alum. Al(OH) surfaces absorb fine suspended solids, which then precipitate together with the adsorbed solids and are separated by filtration.

Removal of Dissolved Solids:

A. Adsorption by activated carbon: Dissolved solids like salts can be removed by filtering the water through a filter containing activated carbon particles.

B. Reverse osmosis: Reverse osmosis removes dissolved solids like NaCl and microbial cells.

Removal of Nitrate and Phosphate: Eutrophication is the process by which the water bodies become highly enriched with minerals and nutrients which induce excessive growth of algae. First, sewage is placed in a tank containing nitrifying bacteria like *Nitrosomonas*, *Nitrosococcus*, and *Nitrobacter*. These bacteria convert ammonium salt and nitrite into nitrate. Then, the sewage is placed into a second tank containing denitrifying bacteria like *Thiobacillus denitrificans* and *Micrococcus denitrificans*. These bacteria convert nitrate into nitrogen gas that leaves the sewage. Phosphate is also removed by bacteria (*Pseudomonas* and *Bacillus* etc.) by microbial absorption process.

Removal of Microorganisms: Finally, microorganisms in sewage are killed by disinfection.

A. Chemical agents: Chlorine is most commonly used. Bromine, iodine, ozone, alcohols, phenols, heavy metals, hydrogen peroxide, alkalis, and acids are sometimes employed.

B. Physical agents: Heat and light (ultraviolet rays).

C. Radiation agents: The gamma rays.

11.2.3 Types of Treatment Processes for Recycled Wastewater

For recycled wastewater, there are several treatment processes available, each with advantages and disadvantages. Here are some of the most common recycled wastewater treatment processes:

11.2.3.1 Vermi-Biofiltration

This is a sustainable wastewater treatment system in agriculture that employs earthworms to improve natural filtration and breakdown of organic contaminants in the water. This approach is very successful for treating agricultural wastewater, which can be heavy in fertilizers, pesticides, and other toxic substances. Vermi-biofiltration harnesses earthworms to encourage the natural filtration (Figure 11.2) and breakdown of organic pathogens in wastewater, resulting in enhanced water quality [13].

Vermi-biofiltration may remove up to 90% of organic matter and up to 80% of nitrogen and phosphorus from agricultural wastewater, making it a viable alternative to existing treatment methods [13]. Earthworms boost soil health and can give extra benefits such as increased crop yields. Overall, vermin-biofiltration is a promising technology for long-term wastewater treatment in agriculture, with both economic and environmental advantages.

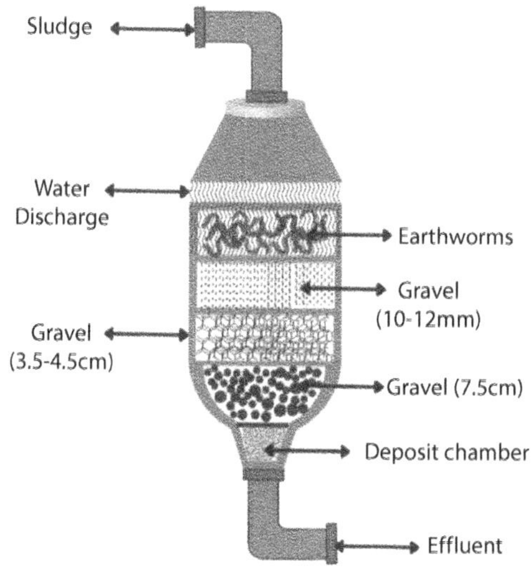

Figure 11.2 Vermi-biofiltration working principle [redrawn].

11.2.3.2 Membrane Bioreactor

Membrane filtration is a physical process that filters impurities from wastewater using semi-permeable membranes (Figure 11.3). This method effectively removes particles, bacteria, and viruses. Membrane filtering techniques include reverse osmosis, microfiltration, ultrafiltration, Analytical methods used for different parameters are listed below, guideline for irrigation water quality and nanofiltration. The membrane bioreactors were employed to treat agricultural wastewater with remarkable successful results of removal efficiencies exceeding 90% for various pollutants [14]. The study also discovered that the treated water met agricultural irrigation quality standards, with no negative effects on crop growth observed. Overall, MBRs are a promising technology for the efficient and sustainable recycling of water for agricultural purposes, with significant water conservation and environmental benefits (Figure 11.3).

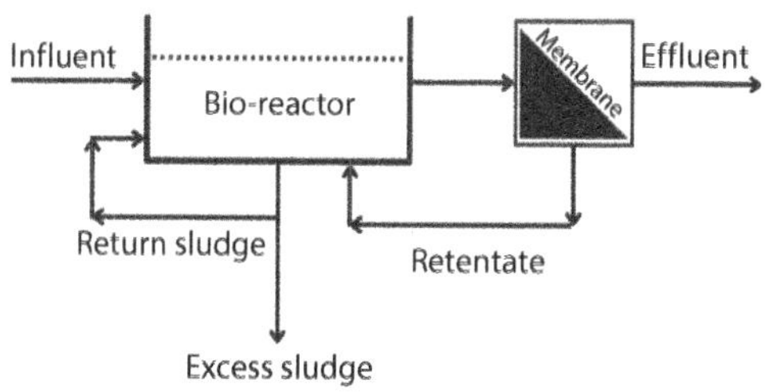

Figure 11.3 Membrane bioreactor method [redrawn].

11.2.3.3 Constructed Wetlands

Wetlands are one kind of organic wastewater treatment method that can be utilized to recycle water for agricultural use. To remove pollutants from wastewater, the method employs plants, microorganisms, and natural processes (Figure 11.4). The water is then reused for irrigation, helping to reduce the demand for freshwater resources. The interaction of plants, microorganisms, and the physical and chemical properties of the wastewater underpins the operation of constructed wetlands. The plants and microorganisms that live in the soil and root zone treat the wastewater as it flows through the wetland. Through a process known as bioremediation, these organisms break down and remove pollutants such as nitrogen and phosphorus. Reclaiming water for cultivation has been demonstrated to be an effective and long-term method of treating wastewater in wetlands. Pollutant reductions and crop yield increases have been reported in studies, as have improved water quality and crop yields. Wetlands that have been built are effective at treating wastewater and recycling water for agricultural use. According to studies, they can remove up to 99% of pollutants while also reducing water consumption in agriculture. The method benefits the environment, public health, and water resources. Significant reductions in pollutants, improved water quality, and increased crop yields have been reported in studies [15].

11.2.3.4 Waste Stabilization Pond

The waste stabilization pond is a type of wastewater treatment system that treats wastewater using natural processes before recycling it for agricultural use. The method employs a series of ponds in which microorganisms break down and digest organic matter in wastewater via a process known as microbial degradation. The effluent produced is then reused for irrigation, providing a long-term source of water for agricultural activities. The waste

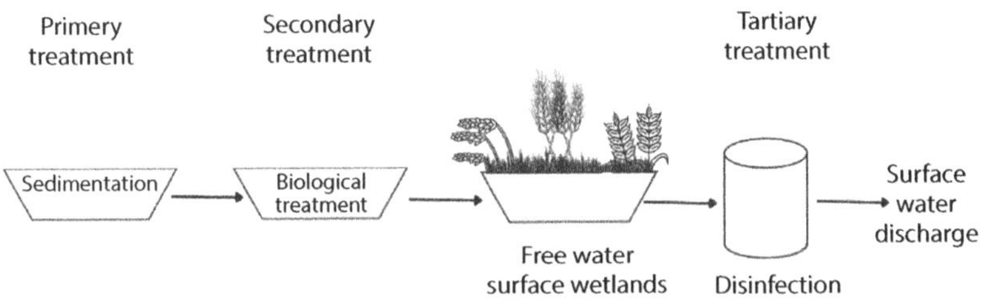

Figure 11.4 Constructed wetland method [redrawn].

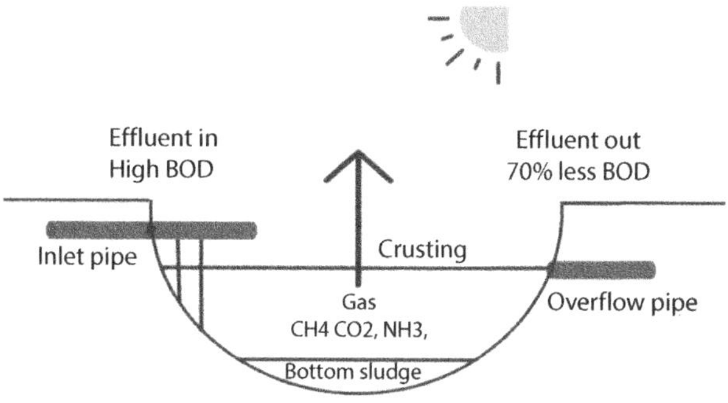

Figure 11.5 Waste stabilization pond method [redrawn].

stabilization pond method for the treatment of wastewater in agriculture is a cost-effective and environmentally sustainable option (Figure 11.5). This method also helps to relieve pressure on freshwater sources, which are becoming increasingly scarce due to climate change and other factors. This method was found to be effective in lowering pollutant levels in wastewater, with an overall removal efficiency of up to 98% [16]. The study also discovered that the effluent from the ponds met the Mexican government's irrigation water standards. Finally, the West Stabilization Pond method offers a viable option for recycling water for agricultural use, providing a cost-effective and achievable environmental response to the growing issue of scarcity of water [16].

11.2.3.5 Nanofiltration

Nanofiltration is a membrane filtration technology that separates ions and molecules from water by using a membrane with pores ranging from 1 to 10 nm in size (Figure 11.6). According to the size exclusion principle, which governs how processes operate, molecules that are larger than the size of a membrane pore are trapped, while the smaller ones flow through. Water that has been nano-filtered is free of suspended solids, bacteria, and viruses and has a lower concentration of dissolved salts and organic compounds. As a result, it is an excellent method for recycling water for agricultural purposes, as it improves water quality, reduces scaling and fouling in irrigation systems, and increases crop yields. Nano-filtration effectively removed 97% of dissolved solids from agricultural wastewater, allowing it to be reused in irrigation. The study also discovered that recycled water increased crop yield while decreasing the need for fertilizers and pesticides. To summarize, nano-filtration is an effective and efficient method

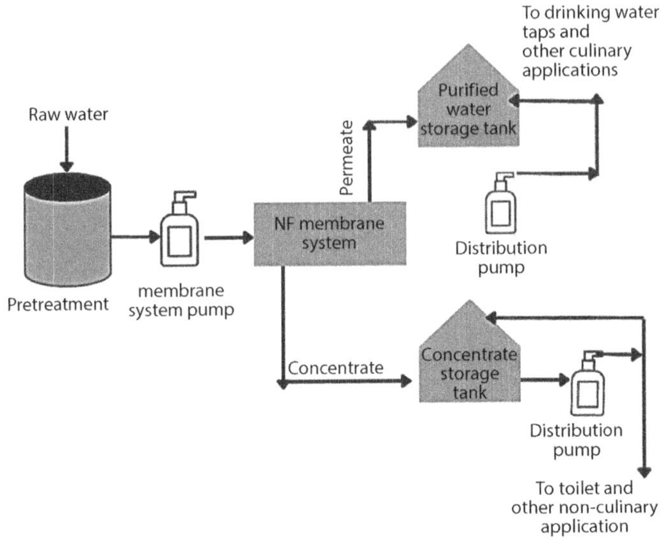

Figure 11.6 Nano-filtration method [redrawn].

of recycling water for agriculture, with numerous environmental and crop production benefits.

11.2.3.6 Advanced Oxidation Processes (AOPs)

Chemical reactions are used in advanced oxidation processes (AOPs) to remove contaminants from wastewater. Organic matter, pharmaceuticals, and personal care products can all be removed using these methods. Ozone, hydrogen peroxide, and ultraviolet radiation are the most commonly used AOPs.

11.2.3.7 Adsorption

Adsorption is the removal of impurities from wastewater by adhering to a surface. In this process, adsorbents such as activated carbon, zeolites, and clays are used.

11.2.3.8 Electrochemical Processes

Electrochemical processes remove contaminants from wastewater by using an electric current. Electro-coagulation, electro-oxidation, and electrochemical disinfection are the three most common electrochemical processes.

Figure 11.7 Fresh water and recycled water figure read on from the work of Hou and coworkers, 2021 [54].

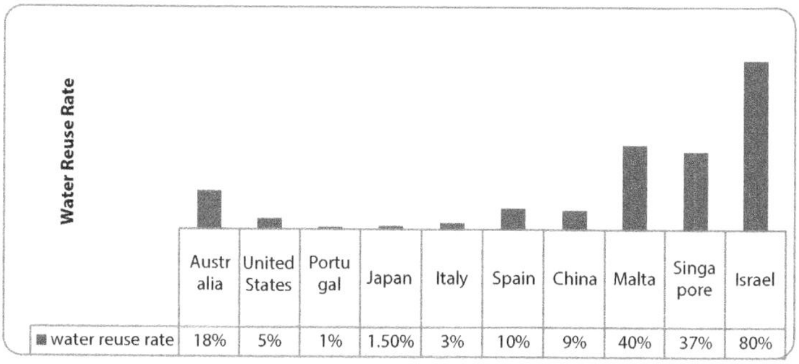

Figure 11.8 Water use in agriculture. The data is tabulated from, Nations, F. and A.O. of the U, 1993 [66].

11.2.4 Suitability of Recycled Wastewater for Irrigation

A sufficient quantity of usable water is essential for agricultural irrigation purposes. Because of the accessibility of good quality water supply, quality standardization has always been neglected. However, the situation changes with the development of the territory. Specific chemical, physical, and biological parameters are used to characterize the quality of recycled water [17, 18]. The total quantity and type of salt, the amount of phosphate and nitrogen, the electrical conductivity (EC), pH, residual sodium carbonate (RSC), soluble sodium percentage (SSP), sodium adsorption ratio (SAR), and exchangeable sodium percentage (ESP) are some crucial factors that determine the suitability of water for agricultural irrigation purpose [18, 19].

11.2.5 Problems Related to Recycled Water Quality

11.2.5.1 Salinity

The plant roots typically absorb moisture by the osmosis process through root cell membranes. Osmosis is the process of a solution moving across a semi-permeable membrane from a solution with a relatively low concentration of salt (irrigation water) to a solution with a relatively high concentration of salt (in root cells) to achieve equilibrium in the two solutions [20]. The plant needs to work harder to draw water from the soil when irrigation water is reasonably salty. The plant quickly starts to wilt due to a lack of water, and its development slows down, resulting in lower yields. Reverse osmosis can be brought on by very salty water. As a consequence, the plant experiences severe stress by losing moisture. Because of this, the signs of high salt damage are comparable to those of high moisture stress, such as leaf tip dieback, burnt, yellowed, and turned brown or black edges, followed by leaf fall of dead leaves. The maximum salinity is said to be 2.4 ds/m (NSW Department of Primary Industries) according to water quality guidelines [21, 22].

11.2.5.2 Rate of Infiltration

Water penetrates the soil with a certain velocity or speed, which is known as the infiltration rate. The average steady infiltration rates for various soil types are >0.8 inches per hour in gravel and coarse sands, and 0.4 to 0.8 inches per hour in sandy loams. In loams, 0.2 to 0.4 inches per hour is expected [23].

11.2.5.3 Toxicity

It occurs when certain ions (sodium, chloride, or boron) from the soil or water build up in a sensitive crop to quantities that result in crop damage and decreased yields [19].

11.2.5.4 Miscellaneous

Excessive nutrients lower yield or quality; unattractive coatings on fruit or foliage decrease marketability; excessive equipment corrosion increases maintenance and repairs [20, 21].

11.3 Recycled Wastewater Effect on Soil Health

Recycled water, often referred to as reclaimed or reused water, is becoming a more common source of irrigation water for gardens and crops. However, its impact on soil health has been a source of contention among scientists and practitioners. We will discuss the influence of reclaimed water on soil health in this essay, drawing on relevant literature and research. According to research, the consequences of utilizing recycled water for irrigation on soil health can be both beneficial and detrimental. On the other hand, recycled water includes vital elements like phosphorus, nitrogen, and potassium that can increase the fertility of the soil as well as the growth of plants [24, 25]. Additionally, using recycled water can lessen the amount of treated wastewater released into the environment and help conserve freshwater resources, reducing water pollution. On the other side, using recycled water for irrigation may be detrimental to the health of the soil. Recycled water frequently contains high levels of salts and trace elements like boron, copper, and zinc, which can accumulate in the soil and cause salinization and toxicity [24, 25]. Additionally, recycled water may contain pathogens and organic contaminants, which can endanger human and animal health and disrupt soil microbial activity and diversity [26].

11.3.1 Recycled Wastewater Impact on Soil Properties and Microbial Activity

The texture is a key soil property that influences water and nutrient retention as well as plant growth. According to research, using recycled water for irrigation can change soil texture by increasing salinity and changing clay content. For instance, using reclaimed water for irrigation in sandy loam soil increased soil salinity while decreasing clay content [27]. The use of reused water for watering in loamy sand soil was demonstrated to enhance soil salinity while lowering silt and clay concentration [28]. Another important soil property that influences plant growth and nutrient availability is pH. By increasing the concentration of dissolved ions such as sodium, bicarbonate, and chloride, the use of recycled water for irrigation can affect soil pH. For example, using recycled water for irrigation in sandy loam soil increased soil pH because of the aggregation of sodium and bicarbonate ions in the soil [29]. Organic matter content is an essential soil property, which influences soil structure, nutrient cycling, and water retention. By altering soil microbial activity and nutrient availability, the use of recycled water for irrigation can affect soil organic matter content. For example, the utilization of reclaimed water for irrigation in sandy loam soil increased soil organic matter content by promoting microbial activity and nutrient availability [30]. Irrigation with recovered water may have a significant influence on soil properties such as texture, pH, and organic matter content. To certify the long-term viability of recycled water for irrigation, soil properties must be monitored and managed, and appropriate management practices must be implemented to minimize negative impacts on soil health.

11.3.2 Changes in Soil Fertility and Nutrient Availability

Treated water mainly consists of nutrients such as nitrogen, phosphorus, and potassium, which can have an impact on soil fertility and nutrient availability. Several research studies were carried out to look at how recycled water affects soil fertility and nutrient availability, with varying degrees of success. For example, recycled water irrigation promoted soil microbial activity and increased soil biological content, which increased the accessibility of essential nutrients in sandy loam soil [30]. However, it was shown that utilizing recycled water for seeding diminished the quantity of organic matter in the soil and the availability of several minerals, including phosphorus, in loamy sand soil [28]. It was discovered that using recycled water for irrigation in sandy loam soil reduced soil quality and fertility, including a decrease in soil organic matter and available phosphorus [27]. The impact of treated water on soil fertility and nutrient availability varies depending on the type of soil, frequency of irrigation, and quality of water. Proper management practices, such as monitoring soil fertility and nutrient levels, may aid in ensuring that the use of recycled water has no negative impact on soil health. Even though utilizing recycled water for irrigation might give plants nutrients, the effects on soil fertility and nutrient availability can differ and should be carefully monitored.

11.3.3 Impact on Crop Quality and Food Safety

Pathogens and pollutants in recycled water may have a potential impact on the quality of crops and livestock. Therefore, it is crucial to be aware of the dangers of using recycled water for irrigation. Numerous investigations into the impact of recycled water on crop quality as well as food security have been carried out. The quality of lettuce and spinach chops was unaffected by the use of recycled water for irrigation [32]. However, the study discovered that the presence of pathogenic bacteria in recycled water may endanger food safety. It came that recovered water for agricultural usage did not influence the chemical composition or sensory properties of tomato fruits [33]. However, the study discovered that using recycled water raised some specific heavy metals in soil and plant tissues, which could endanger food safety. In contrast, using reclaimed water for agricultural irrigation increased the accumulation of trace elements in vegetable crops, which could impact food safety [34]. Overall, the consequences of recycled water on crop standards and food safety vary depending on factors such as water quality, crop type, and management practices. Proper monitoring and management practices, such as regular testing of recycled water and soil, can help ensure that the addition of reclaimed water does not hurt crop quality or pose a food safety risk.

11.4 Impact of Recycled Wastewater on Crop Parameters

Recycled water usage can affect crop output in both good and adverse ways, depending on variables including water quality, soil type, and crop cultivation techniques. The impact of recycled water on agricultural output has been the subject of several researches. In some crops, like lettuce, using recycled water for irrigation can boost crop production and water productivity; however, the degree of gain depends on the crop and environmental variables [35]. Another study, by Ahmadi *et al.* [36], discovered that using recycled water increased crop yield in some cases, particularly for drought-tolerant crops like corn, but the increase in yield was not statistically significant. However, using recycled water can harm crop yield if it contains high levels of contaminants or pathogens that can interfere with plant growth and development. Hsiao discovered that using recycled water for irrigation reduced crop yield and plant growth in some cases, possibly due to high salinity and heavy metal levels in the water [37, 111]. The impact of recycled water on crop yield varies depending on a variety of factors, and proper management practices, such as regular testing of treated water and soil, are critical to ensuring that the utilization of reclaimed water does not negatively impact crop yield.

11.4.1 Comparison of Crop Yield with Recycled Water Irrigation Versus Freshwater Irrigation and the Effect of Recycled Water on Crop Growth and Development

Several studies have been conducted to compare crop yield with recycled water irrigation versus freshwater irrigation. According to one study, crop yields with recycled water were generally comparable to those with freshwater, with no significant difference in yield observed for several crops such as wheat, cotton, and maize. The study did note, however, that crop yield varied depending on the quality of recycled water used. Scientist Ben-Gal conducted another study on drip-irrigated tomato plants that compared crop yield with recycled water irrigation versus freshwater irrigation [39]. The study found no significant difference between the two irrigation techniques in terms of production; however, recycled water did cause a little reduction in the growth and development of plants, most likely because of the water's high saline levels. Recycled water can affect the progress of crops in both good and adverse ways, depending on several variables, according to many studies. The nutrients in the water, which are most likely to blame, helped pepper plants grow and develop better when they were watered with recycled water [40].

11.4.2 Effect of Different Irrigation Methods on Crop Production

The use of recycled water has received attention recently as a sustainable option to lessen the huge pressure on freshwater supplies. Water shortage is a major problem in agriculture. However, the effectiveness of different irrigation methods on crop yield with recycled water is still a topic of research and debate. Studies have shown that the choice of irrigation method can significantly affect crop yield with recycled water. For example, a study found that drip irrigation with recycled water and the yield of tomato and cucumber crops by 20 and 30%, respectively, compared to sprinkler irrigation [41]. Similarly, a study by Ismail and Almarshadi reported that drip irrigation with recycled water improved the yield of lettuce by 25% compared to furrow irrigation [42]. The improved crop yield with drip irrigation can be attributed to the efficient use of water, which reduces water loss through evaporation and runoff and ensures that water is delivered directly to the roots of the plants. In contrast, sprinkler irrigation can result in water loss through wind drift and evaporation, which can lead to uneven distribution of water and reduced crop yield.

11.4.3 Challenges Associated with Crop Yield Improvement with Recycled Wastewater

Optimizing crop yield with recycled water is a difficult task that requires addressing a variety of issues that arise when using recycled water. These difficulties can include issues with water quality, soil salinity, soil structure, and nutrients, and irrigation system clogging. Water quality is a critical factor influencing crop yield when using recycled water. Contaminants such as salts, nutrients, and pathogens can be found in recycled water, affecting plant growth and yield. High salt levels in recycled water can increase soil salinity, resulting in decreased crop growth and yield. Recycled water can also change the availability of nutrients and the structure of the soil, which will impact plant output and growth. Finally, recovered water may contain particles that reduce water efficiency by clogging irrigation systems [43]. Various strategies, such as treating recycled water to remove contaminants and reduce salinity levels, managing soil salinity through proper irrigation and fertilization practices, and implementing irrigation systems that can handle particles in recycled water, can be used to overcome these challenges. Additionally, selecting crops that can tolerate salt and other impurities can assist in increasing agricultural productivity with reclaimed water [44, 45].

11.5 Environmental Impacts of Recycled Wastewater Use in Agriculture

Wastewater that has undergone treatment to make it acceptable for a variety of non-potable purposes, such as irrigation, industrial activities, and toilet flushing, is referred to as recycled water, sometimes known as reclaimed water. Using recycled water in agriculture can provide a sustainable source of water for crop production and reduce the demand for freshwater resources. Recycled water use in agriculture, however, may have adverse effects on the environment.

One potential environmental impact of utilizing recycled water in agriculture is the buildup of salts in the soil. Recycled water typically contains high levels of dissolved salts, which can accumulate in the soil over time, leading to soil degradation and reduced crop yields [46]. To mitigate this impact, it is important to monitor soil salinity levels and implement management practices, such as soil amendments and crop rotation, to reduce salt buildup [47]. Contaminants like infections and pesticides may also harm the environment when reused water is used in cultivation. While recycled water treatment processes are designed to remove contaminants, there is still a risk of pathogens and chemicals remaining in the water [48]. Utilizing the proper treatment procedures and monitoring procedures can help to reduce this risk and guarantee that the recycled water is suitable for agricultural use [49].

In addition to having a favorable effect on the environment, using recycled water in agriculture may also reduce the need for freshwater resources and the quantity of wastewater that is released into the environment [50]. Additionally, using recycled water for irrigation can increase soil moisture levels and promote plant growth, leading to increased carbon sequestration and improved soil health [51]. Overall, depending on the individual situation and management techniques, employing recycled water in agriculture may have negative environmental effects. While there are potential environmental risks, the use of recycled water can also provide significant environmental benefits. Proper management practices, monitoring protocols, and treatment processes can help ensure that recycled water use in agriculture is safe and sustainable [52].

11.5.1 Potential Effect of Wastewater Recycling on the Environment

After properly processed to ensure useable water quality, recycled water can meet the majority of water demands. The most frequent uses of recycled water are for irrigation on golf courses, public parks, and other non-potable industries. Other non-potable applications include cooling water for oil refineries and power plants, industrial process water for businesses like paper mills and carpet dyers, toilet flushing, dust control, building projects, concrete mixing, and artificial lakes. The amount of potable water provided to decentralized locations, the amount of fertilizer required, and the amount of wastewater created, transported, and processed at wastewater treatment facilities are all reduced by the use of gray water at these sites for landscape irrigation and toilet flushing. Therefore, water reuse conserves energy, water, and money [53].

11.5.1.1 Several Benefits of Water Recycling in the Environment

Accelerated Irrigation: The recycled water can be used effectively for the fertigation of crops, while raw untreated wastewater can have a detrimental effect on crop health. Treated sewage contains huge amounts of nitrogen and phosphorus which are the most necessary fertilizers for plants and deadly for aquatic life [54].

Magnified Wetlands: Wetlands give a boost to the ecosystem in many ways, including housing species, flood reduction, intensifying water quality, and providing suitable habitats to fish populations. Dried-out wetlands may be frequently revived into lush ecosystems by adding recycled water [52].

Sustain Aquatic Habitats: There is a high requirement for freshwater to make a sustainable Aquarius inheritance. Due to the development of the territory, the amount of freshwater has declined, making the aquatic species endangered. Companies and irrigation sectors may solve this problem by using recycled water, which might assist in retaining freshwater for aquatic creatures, biodiversity, and horticulture [55].

Pollution Reduction: Only 3% of the water on our planet is recognized as fresh. Of which 1.2% are drinkable, which is also contaminated day after day. Aquatic life is being destroyed by the high load of nitrogen and metal ions. Recycled water can not only detox the environment but also can save the life of the world [51].

Storage of Energy: The total treatment process of recycling water is energy-consuming because it includes lots of series and parallel processes. To reduce the thrust of energy, we imply different methodologies such as anaerobic treatment which produces methane gas as the source of energy as well as the byproducts in the form of fertilizers which makes it more cost-effective and energy-efficient [53].

Additionally, water must be harvested, processed, and transported, often over great distances, to meet the growing demand for it. This process consumes a lot of energy. If the regional supply of water is groundwater, the groundwater level drops as more is withdrawn, which raises the energy needed to pump the water to the surface. Reclaiming water on-site or nearby conserves energy by lowering the amount of energy needed to pump water from an aquifer or transport it over great distances. When water quality is adjusted for a particular water consumption, less energy is needed to treat the water. When compared to drinking water, the quality of the water needed to flush a toilet is less rigorous and requires less effort to get. Redistributing recycled water that has been restored to inferior standards for uses that do not require it cuts the amount of energy and money required to sterilize it [52, 53].

11.5.2 Effects on Aquatic Ecosystems

Freshwater is necessary for fish and aquatic life. They are unable to survive when the water is over-polluted. Excessive levels of chemicals, such as nitrogen and phosphates, can induce excessive plant growth and the discharge of toxins into the water in streams, rivers, and other large bodies of water. Dead zones, where fish and other aquatic life can no longer live, are the result of oxygen deprivation. Wastewater treatment converts the toxins and pathogens containing water into effluent and returns it to the environment such as water bodies, which helps stabilize aquatic life by maintaining the dissolved oxygen level, salt level, etc. [54]. The good nutrients present in it help grow plants underwater for the consultation of fish. That is how the aquatic ecosystem is maintained. The use of recycled water in aquatic ecosystems can have both positive and negative effects on the environment. On the one hand, the use of recycled water can provide a sustainable source of water for aquatic ecosystems and help to reduce the demand for freshwater resources. On the other hand, recycled water may contain contaminants, such as pathogens, nutrients, and chemicals that can potentially harm aquatic organisms and the ecosystem as a whole [56]. Utilizing recycled water in aquatic environments may have unfavorable effects on them, including the introduction of nutrients like nitrogen and phosphorus that can cause eutrophication and destructive algal blooms [57]. Increased fertilizer inputs encourage the growth of algae and other aquatic plants, which can reduce oxygen levels and kill fish and other aquatic species. This process is known as eutrophication [58]. To mitigate this impact, it is important to manage nutrient inputs and use appropriate treatment processes to remove excess nutrients from recycled water [59]. The presence of pathogens and pollutants in the water is another possible drawback of using recycled water in aquatic environments. Recycled water may contain bacteria, viruses, and other pathogens that can infect aquatic organisms and humans [48]. Recycled water may also contain substances like medications and personal care items that might damage aquatic life and build up in the chain of food [60]. To minimize these risks, it is important to use appropriate treatment processes and monitoring protocols to certify that the reclaimed water is reliable for aquatic use [17, 61, 62].

11.6 Economic Viability of Recycled Wastewater Use in Agriculture

Throughout the history of mankind, agriculture has been claiming the highest area in the pie of freshwater consumption, thus clearly indicating the need for research for an alternative water resource. Wastewater treatment can resolve the problem in many ways. However, it is known that procedures like wastewater treatments consist of "Green Premium" and need large-scale funding. Alternatives to freshwater, groundwater, and desalinated water have been compared in several case studies to assess the effect caused by recycled water from an economic and monetary viewpoint [63]. The study demonstrated the financial benefit of using treated wastewater in agriculture. The benefit-to-cost ratio is larger than one even in the worst-case scenario, indicating that using recycled water is one of the most financially and economically sustainable alternative water supplies accessible. It is also one of the most financially and economically feasible [64, 65].

11.6.1 Cost Analysis and Profits of Using Recycled Wastewater for Irrigation

The use of recycled water in agriculture can bring both costs and benefits to agricultural operations. Below are some of the key costs and benefits to consider when analyzing the achievability of using recycled water:

11.6.1.1 Costs

Infrastructure Costs: The infrastructure needed to transport, store, and distribute recycled water involves a significant cost, especially for large agricultural operations.

Treatment Costs: Recycled water must be treated to guarantee that it complies with legal standards and is secure for use in irrigation. Depending on the amount of treatment needed and the kinds of pollutants present in the water, the cost of treatment might change [67].

Maintenance Costs: The infrastructure and treatment systems used to supply recycled water for irrigation require ongoing maintenance to ensure their continued operation and effectiveness [63].

11.6.1.2 Benefits

Water for irrigation can lower the cost of water for agricultural activities.

Improved Water Supply Reliability: Recycled water offers a continuous source of water that is not impacted by fluctuations in natural water sources, which can assist in boosting the reliability of water supply for agricultural activities [65, 68].

Reduced Environmental Impact: Using recycled water for irrigation can help to reduce the impact of agricultural operations on natural water sources by reducing the amount of freshwater that needs to be withdrawn for irrigation [65].

Improved Soil Health: Recycled water can contain nutrients that are beneficial for plant growth, which can improve soil health and reduce the need for synthetic fertilizers.

It is crucial to take into account both the immediate and long-term effects of the practice when weighing the costs and advantages of using recycled water for irrigation. While the initial costs of infrastructure and treatment may be significant, the long-term benefits of reduced water costs and improved water supply reliability can make the investment in recycled water infrastructure worthwhile. Additionally, the potential benefits of reduced environmental impact and improved soil health should be considered as part of the overall cost-benefit analysis [63, 65].

11.6.2 Comparison of Recycled Wastewater Cost with Freshwater

Recycled water prices might vary based on the source and quality of the water, the methods of treatment, and the distance that the water needs to travel. However, in general, recycled water is often less expensive than freshwater sources for irrigation purposes. The cost of freshwater can vary greatly depending on factors such as location, availability, and infrastructure. For example, in some areas with ample water resources, the cost of freshwater may be relatively low. In other areas with limited water resources, the cost of freshwater can be much higher, especially during periods of drought or high demand [68].

Recycled water, on the other hand, is often cheaper than fresh water because of the lower treatment cost compared to the cost of acquiring, treating, and distributing freshwater. In addition, in areas where there is limited water availability or high water costs, the utilization of reclaimed water can decrease the overall cost of water in agriculture. It is also important to keep in mind that the price of recycled water might change based on the degree of treatment needed. For instance, water treated with advanced techniques like reverse osmosis or advanced oxidation may cost more than water treated with less sophisticated techniques [65]. However, the additional cost of advanced treatment may be justified in areas where the water quality requirements are high, or when there are worries about how toxins can affect our surroundings or the well-being of humans. Overall, while the cost of recycled water may vary depending on a range of factors, it is often a cost-effective alternative to freshwater for agricultural irrigation purposes [68].

11.6.3 Economic Feasibility of Implementing Recycled Water Use in Agriculture at Different Scales

Several variables might affect whether using recycled water in agriculture is economically viable, like the size and type of the farm, the cost of water, the cost of treatment, and the availability of funding and incentives. Here are some general considerations for different scales of implementations:

11.6.3.1 Small-Scale Farms

For small-scale farms, the cost of implementing recycled water use may be a significant barrier. Small farms may not have the resources to invest in the necessary infrastructure for water treatment and distribution. However, in areas with high water costs or limited water availability, recycled water may provide a cost-effective alternative to other water sources [69].

11.6.3.2 Medium-Scale Farms

Medium-scale farms may have the resources to invest in recycled water infrastructure, but the cost-effectiveness of the investment will depend on the specific circumstances of the farm. In some cases, recycled water may provide a more cost-effective alternative to other water sources, while in other cases, the cost of treatment and distribution may be prohibitively high [42, 70].

11.6.3.3 Large-Scale Farms

Large-scale farms may be better positioned to implement recycled water use due to their greater financial resources and economies of scale. Recycled water may be more cost-effective for large farms, particularly in areas with high water costs or limited water availability [69, 71].

In general, the economic feasibility of implementing recycled water use in agriculture can be improved by the availability of funding and incentives. Government grants and subsidies can help to offset the costs of infrastructure investment, while incentives such as tax credits or reduced water rates can help to encourage farmers to adopt recycled water utilization. Finally, It is essential to note that, the economic feasibility of recycled water use in agriculture is not the only consideration. There are also important environmental and social benefits to using recycled water, such as reducing water stress on natural ecosystems, reducing energy consumption for water treatment, and ensuring the long-term sustainability of agricultural production [69].

11.7 Challenges Associated with Recycled Wastewater Use in Agriculture

Challenges can be associated with both the environmental and health perspective. If the satisfaction process is not done properly, there will be a strong chance of finding living microorganisms including pathogens, bacteria (e.g., *Escherichia coli* and *Salmonella* spp), viruses (e.g., *Enteroviruses*, *Rotavirus*, and *Hepatitis A*), protozoa (e.g., *Giardia Lamblia* and *Cryptosporidium parvum*), and helminths (e.g., *Taenia* spp (tapeworm), *Ancylostoma* spp (hookworm)). Some potential environmental risks are presented in Table 11.6.

Analytical methods used for different parameters are listed below, guideline for irrigation water quality (Tables 11.1 to 11.7).

Table 11.1 Strength-based classification of reclaimed water [12].

	Reclaimed water quality		
Constituents	**Minor (mg/l)**	**Moderate (mg/l)**	**Severe (mg/l)**
Total dissolved solids	Up to 600	600 to 1000	1000 to 2500
Overall nitrogen	Up to 50	50 to 100	More than 100
Overall phosphorus	Up to 10	10 to 20	More than 20
Biological oxygen demand	Up to 40	40 to 1500	More than 1500
Lubricants and oil	More than 1500 mg/L is regarded as high strength.		

Table 11.2 Analytical methods used for different parameters [21].

Parameter	**Examination processes**
Acidity, basicity (pH, pOH)	Electrometric method
Electrical conductivity (EC)	Conductivity
Total hardness (T. H)	Titrimetry using EDTA
Calcium (Ca^{2+})	Titrimetry using EDTA
Magnesium (Mg^{2+})	Titrimetry using EDTA
Sodium (Na^{+})	Flame emission spectrometry
Potassium (K^{+})	Flame emission spectrometry
Bicarbonate ($HC0_3^{-}$)	Titrimetry
Carbonate (CO_3^{2-})	Titrimetry
Chloride (Cl^{-})	Argentometry
Nitrate ($N0_3^{-}$)	Spectrophotometry (UV screening method)
Phosphate (PO_4^{3-})	Stannous chloride method
TSS	Suspended solid dry method (103°C–105°C)
Biochemical oxygen demand (BOD)	BOD5 test
Chemical oxygen demand (COD)	Closed reflux, titrimetry

Table 11.3 Guideline for irrigation water quality [21].

Parameters	Unit	Suitability range		
		Low	Moderate	High
pH		Normal range (6–9)		
EC	dS/m	Up to 1.7	1.7 to 3.0	More than 3
TSS	mg/L	Up to 50	50 to 100	More than 100
TDS	mg/L	Up to 1088	2.81088 to 2000	More than 2000
DOB	mg/L	Up to 60	110 to 400	More than 400
COD	mg/L	Up to 120	250 to 1000	More than 1000
HCO^{3-}	mg/L	Up to 90	90 to 520	More than 520
Ca^{2+}	mg/L	230	230	230
Mg^{2+}	mg/L	100	100	100
Na^{+}	mg/L	Up to 69	69 to 207	More than 207
NO_3	mg/L	Up to 5	5 to 30	More than 30
T-P	mg/L	Up to 6	6 to 20	More than 20
Cl^{-}	mg/L	Up to 142	142 to 355	More than 355
B	mg/L	Up to 0.7	0.7 to 3	More than 3
Fe^{2+}	mg/L	Up to 0.1	0.1 to 1.5	More than 1.5
Mn^{2+}	mg/L	Up to 0.1	0.1 to 1.5	More than 1.5

Table 11.4 Recycled water on soil properties [31].

Soil property	Effect of recycled water
Texture	Increase in clay content Decrease in sand content
pH	Increase in alkalinity
Organic matter content	Intensification of soil organic matter
Nutrient availability	Increase in soil nutrient availability

Table 11.5 Recycled water impact on crop yield [38].

Crop	Recycled water impact on yield
Tomato	Increase in yield
Wheat	Increase in yield
Lettuce	No significant impact on yield
Corn	Decrease in yield
Rice	Increase in yield

Table 11.6 Challenges associated with recycled water use in agriculture [107–110].

Components	Effectiveness
Salinity	It decreases plant development, harms and potentially impacts freshwater plants, and, if dumped directly with little dilution, impacts natural ecosystems such as sodium chloride.
Sodicity	This term refers to the level of sodium present in the treated wastewater which affects the information of water on heavier texture soils. Excessive sodium can be toxic to some plants during irrigation. This can be difficult to cure.
Chloride	It is a crucial element in salinity that can be hazardous to plants if sprayed directly on leaves.
Nitrogen and phosphorus	In both terrestrial and aquatic environments, an excessive nitrogen content can lead to eutrophication.
Chlorine residual	If dumped directly with little dilution, disinfection byproducts may be damaging to aquatic or marine habitats.
Hydraulic loading	Applying too much water to land can lead to secondary salinity, water logging, and excessive groundwater recharge.
Boron	If water from continued irrigation builds up in particular soils, it may cause plant toxicity in some plants.
Surfactants	Detergents can leave behind certain hazardous organic and inorganic surface-active chemicals that can persist in recycled water.

11.7.1 Technical, Regulatory, and Social Challenges Associated with Recycled Water Use in Agriculture

Reclaimed water, which can be referred to as recovered water, is cleaned wastewater that may be implemented for a variety of tasks, including agricultural irrigation. However, using recycled water in agriculture comes with several technical, legislative, and societal problems [72].

11.7.1.1 Technical Challenges

Treatment Processes: The type of treatment utilized affects the quality of water that is reclaimed and used in agriculture. Treatment processes must be capable of removing harmful pathogens, chemicals, and nutrients from the water to ensure that it is safe for crop irrigation.

Water Quality Monitoring: Recycled water must be continuously monitored to ensure that it meets water quality standards. Monitoring must be done at various stages of the treatment process, storage, and distribution to detect any issues that may impact crop growth [73, 74].

11.7.1.2 Regulatory Challenges

Legal Frameworks: A regulatory framework must be established to govern the application of reclaimed water in the agricultural sector. The regulatory framework must cover aspects such as water quality standards, treatment processes, storage, and distribution.

Public Perception: It can be difficult to change how the public views the use of recycled water in agriculture. The public can be worried about the health implications of eating crops cultivated with recycled water as well as the usage of recovered water's possible environmental effects [74].

11.7.1.3 Social Challenges

Acceptance: Recognizing the use of recycled water in agriculture is a need for farmers and other stakeholders. Changing opinions regarding the utilization of recycled water may necessitate public education and awareness initiatives.

Water Allocation: There may be competition for water resources between various stakeholders, including agriculture, domestic, and industrial users. The allocation of recycled water for agricultural use must be balanced with the needs of other stakeholders.

Research by Jayasooriya *et al.* [21] claimed that the use of recycled water in agriculture was affected by several variables, notably water availability, the quality of the water, legal frameworks, and public acceptability. The study emphasized the requirement for a comprehensive strategy that takes into account technical, governmental, and societal issues when implementing recycled water usage in agriculture [21, 72].

11.7.2 Solutions and Strategies for Addressing Challenges

It is a complicated matter that necessitates a careful evaluation of several aspects, including the chemical composition of the water, possible threats to soil and groundwater, and human health [89]. Setting strict quality standards for irrigation water and making sure that wastewater treatment methods are adjusted to eliminate organic microcontaminants, notably PBT and PMT chemicals, are critical steps in reducing the danger of contamination.

To guarantee safe water reuse for irrigation, it is also required to put in place a mix of proactive measures at the source, excellent agricultural behaviors, and extra preventative measures. The sharing of information and educating those who are intrigued by the reuse of reclaimed water for agricultural purposes may fall under this [21, 74]. Water demand management should also be taken into account in light of the growing conflicts between various water usage and the changing climate. All sectors should place a high priority on initiatives to cut back on water use, increase effectiveness, and reduce water loss. Overall, a complete and multidisciplinary strategy that takes into account all facets of water quality, treatment, and management is necessary for safe agricultural irrigation utilizing recovered water [72].

11.7.3 Case Studies of Successful Recycled Wastewater Use in Agriculture

Several case studies of affluent recycled water operating in agriculture offer valuable lessons for the harmless and effective use of regained water in irrigation. The key lessons learned from these case studies are the following.

11.7.3.1 Collaboration and Stakeholder Engagement are Critical

Successful recycled water projects require collaboration between all stakeholders, including water agencies, farmers, regulators, and the local community. Engaging stakeholders early and often can help to build trust, address concerns, and ensure the success of the project.

11.7.3.2 Effective Treatment and Monitoring are Essential

Reclaimed water must be treated to remove contaminants and ensure that it meets the necessary quality standards for agricultural use. To guarantee that the water is safe for irrigation, frequent inspections and assessments are also essential [73].

11.7.3.3 Key Agricultural Practices

To lessen the danger of contamination and guarantee the security of the food supply, farmers must adhere to excellent agricultural behaviors, such as appropriate crop selection, irrigation, and soil management [21].

11.7.3.4 Public Education and Outreach are Important

Public education and outreach programs can help raise awareness of the goods of recycled water implementation in agriculture and address concerns about safety and quality. These programs can also help to promote best practices and build community support for recycled water projects [72].

11.7.3.5 Adequate Funding and Incentives are Necessary

Adequate funding and incentives can help to overcome the initial cost barriers of recycled water projects and encourage farmers to adopt water-efficient practices. Regulatory frameworks must be in place: Effective regulatory frameworks are unavoidable to ensure that recycled water is handled safely and responsibly. Regulations should establish clear standards for water quality and agricultural practices and provide a framework for monitoring and enforcement. By applying these lessons learned, recycled water can be a safe and sustainable resource for agricultural irrigation that can help to recognize water scarcity as well as assure the long-term viability of agricultural production [21, 73].

11.8 Emerging Technologies and Innovations in Recycled Wastewater Use in Agriculture

Recycled water, also known as reclaimed water, is becoming an increasingly important source of agricultural irrigation water. As global water scarcity persists, recycled water can be utilized to conserve freshwater resources while supporting agricultural production. Several emerging technologies and innovations have been developed in the past few years to optimize the usage of recycled water in agriculture [82, 83]. The use of membrane filtration technologies to remove contaminants and reduce salinity levels in swage is one of the emerging technologies in recycled water use. Another breakthrough is the creation of smart irrigation systems that use real-time data and weather information to optimize irrigation scheduling and reduce water consumption. Furthermore, the use of bio-filtration systems, such as constructed wetlands, can aid in the further treatment of recycled water and its quality for agricultural use. Several developing nations have made remarkable evolutions in the agricultural reuse of recycled water. More than 50% of the irrigation water used in Israel comes from recovered wastewater [75, 84]. In recent years, the use of recycled water in agriculture in Australia has increased significantly, particularly in urban areas with limited access to freshwater resources. California has been a leader in employing recycled water for agriculture in the United States, having several large-scale projects in place to support the usage of recycled water in crop production [76, 77].

11.8.1 Overview of New Technologies and Innovations in the Treatment of Wastewater for Irrigation

Providing access to clean, inexpensive drinking water is one of today's concerns. The world's ever-growing population is causing a water shortage, and pollutants are poisoning all of the available water sources. The agricultural industry has a serious problem with water shortage, which calls for the adoption of cutting-edge water management techniques and products [85, 86]. Recycled water has become a reliable supply of irrigation water for crops. However, using recycled water for crop irrigation presents several challenges, including contaminants, salinity, and microbiological risks [87, 88]. To address these challenges, new technologies and innovations for recycled water treatment and irrigation are required. Several new technologies and innovations for treating recycled water in agriculture have been developed. AOPs, membrane filtration, and reverse osmosis (RO) systems are among these technologies. AOPs employ strong oxidizing agents to degrade organic compounds, whereas membrane filtration employs membranes with varying pore sizes to remove particles, microorganisms, and other contaminants [90, 91]. To remove dissolved salts and other impurities, water is forced through a semi-permeable membrane by pressure in RO systems. Nanotechnology has provided innovative water treatment solutions. Water contaminants include micro-organisms, organic suspensions, and inorganic heavy metal ions, among others [92–94]. To address this issue, various water cleaning methods have been used since antiquity. Over the past 40 years, advances in water purification techniques have been made, with the utilization of nano-materials and nano-membranes being the most important [95, 96].

Engineered materials have utilized nanoparticles and polymeric nano-membranes. In India and many other developing nations, agriculture is a significant industry that offers rural inhabitants several job possibilities and helps them achieve their goals for food and nutritional security [97]. The issues of raising food production and enhancing rural lives while safeguarding essential supplies of water for long-term usage are attempted to be addressed, especially in places that are prone to drought. Through the implementation of technologies like resource-conserving farming, micro-irrigation, and water harvesting, an integrated strategy to manage water for farming must be put into place to maximize the efficiency of water usage in agriculture and other vital services for humans and animals. The purpose of this study is to enhance understanding of the possible effects of climate change.

11.8.2 Potential Impact of New Technologies on Water, Soil, and Crop Quality in Agriculture

Precision irrigation technologies, such as drip irrigation, allow for precise water application to crops, reducing water waste due to runoff or evaporation. This can improve water quality by lowering the number of nutrients and pesticides washed into streams [98–100]. Furthermore, fertigation technologies, which involve the precise application of fertilizers via irrigation systems, can reduce the amount of fertilizer lost through leaching or runoff, improving water quality even further [78].

Another technology that has the potential to enhance water quality and efficiency in agriculture is the use of sensors and systems for remote surveillance. These technologies provide more precise irrigation scheduling and requirements for crop water by giving farmers access to real-time data on soil moisture levels, weather, and crop water needs. Additionally, irrigation with recycled sewage from sewage treatment facilities can boost the effectiveness and quality of the water. Recycled wastewater may be processed to eliminate dangerous pollutants like viruses and metals that are hazardous before being utilized for irrigation. Minimizing the quantity of wastewater released into rivers and reducing the demand for freshwater resources can assist in improving the quality of the water.

11.8.3 Prospects for Recycled Wastewater Use in Agriculture

The use of treated water for agriculture has bright prospects. As freshwater resources become increasingly scarce and food demand rises, recycled water has the potential to be a cost-effective and sustainable source of irrigation water for crop production [102, 103].

According to a study by Pham Do [79], the utility of recycled water for irrigation has the potential to reduce the demand for freshwater resources in agriculture by up to 20% while also reducing the amount of wastewater discharged into waterways. Furthermore, recycled water can provide nutrients and organic matter to improve soil health and fertility [105, 106]. Many regions around the world are already implementing the use of recycled water for irrigation. In California, for example, the use of recycled water for irrigation has increased by more than 50% in the past decade [80]. More than half of the irrigation water used for agriculture in Israel is recycled water. In addition to using surplus water for irrigation, purified wastewater may also be used for aquaculture and hydroponics. Aquaponics, a hybrid of aquaculture and hydroponics, uses recycled water from fish tanks to irrigate crops, resulting in a closed-loop system that saves water and reduces nutrient waste [81]. However, there are still issues that must be addressed before recycled water can be widely used in agriculture, such as public acceptance, regulatory barriers, and the cost of treatment and distribution infrastructure [101, 104]. To develop sustainable solutions to these challenges, policymakers, farmers, and researchers must work together.

11.9 Conclusion

Finally, using recycled wastewater from sewage treatment plants for sustainable agriculture has the potential to address the global water crisis while also promoting sustainable agricultural practices. This approach has the potential to reduce freshwater demand, improve water efficiency, and reduce the wastewater quantity discharged into waterways. Furthermore, recycled wastewater can provide valuable nutrients and organic matter, improving soil health and crop yields. However, there are still issues that must be addressed before recycled wastewater can be widely used in agriculture. These include public acceptance, regulatory barriers, and the cost of treatment and distribution infrastructure. Further research and collaboration between policymakers, farmers, and researchers are needed to overcome these challenges and ensure the successful adoption of this technology. The use of recycled wastewater for sustainable agriculture is a promising solution that can help agriculture and the environment have a more sustainable and resilient future.

References

1. Howlader, M.I., Morsada, Z., Hossain, M.M., Pollutants removal from textile wastewater by biofilter. In: Shah, M., Rodriguez-Couto, S., and Biswas, J.B.T.-A.I.R. of B. in W.T.P. (WWTPs) (eds.) *An Innovative Role of Biofiltration in Wastewater Treatment Plants (WWTPs)*, pp. 309–330. Elsevier, 2021.
2. Mudulia, M., Ray, S., Constructed Wetlands for Municipal Wastewater Treatment: a Sustainable Approach. *Int. J. Environ. Sci.*, 12, 2021.
3. Gooijer, G. De, Rast, W., Tropp, H., Aimard, V., Allaerts, G., Black, M., Boelens, R., Burchi, S., Engel, H., Harlin, J., Hellmuth, M., Hendry, S., Hudson, A., Jägerskog, A., Larsen, H., Moriarty, P., van Beek, E., van der Zaag, P., Winpenny, J., Tropp, H., de Gooijer, G., Options inside the water box. United Nations, 2012.
4. Chen, C., Li, J., Chen, P., Ding, R., Zhang, P., Li, X., Occurrence of antibiotics and antibiotic resistances in soils from wastewater irrigation areas in Beijing and Tianjin, China. *Environ. Pollut.*, 193, 94, 2014.
5. Brewster, M.R., Buros, O.K., Non-conventional Water Resources: I. Economics and Experiences in Developing Countries. *Nat. Resour. Forum*, 9, 65, 1985.
6. Leverenz, H.L., Tchobanoglous, G., Asano, T., Direct potable reuse: A future imperative. *J. Water Reuse Desalin.*, 1, 2, 2011.
7. Biswas, A., Mailapalli, D.R., Raghuwanshi, N.S., Treated municipal wastewater to fulfil crop water footprints and irrigation demand – A review. *Water Sci. Technol. Water Supply*, 21, 1398, 2021.
8. Biswas, A., Mailapalli, D.R., Raghuwanshi, N.S., APSIM-Oryza model for simulating paddy consumptive water footprints under alternate wetting and drying practice for Kharagpur, West Bengal, India. *Paddy Water Environ.*, 19, 481, 2021.
9. Bedbabis, S., Ben Rouina, B., Boukhris, M., Ferrara, G., Effect of irrigation with treated wastewater on soil chemical properties and infiltration rate. *J. Environ. Manage.*, 133, 45, 2014.
10. Prasad, V., Schwerdtfeger, U., El-Awa, F., Bettcher, D., da Costa e Silva, V., Closing the door on illicit tobacco trade, opens the way to better tobacco control. *East. Mediterr. Heal. J.*, 21, 379, 2015.

11. Rahman, M.M., Hagare, D., Maheshwari, B., Use of Recycled Water for Irrigation of Open Spaces: Benefits and Risks. In: Maheshwari, B., Thoradeniya, B., and Singh, V.P. (eds.) *Balanced Urban Development: Options and Strategies for Liveable Cities*, pp. 261–288. Springer International Publishing, Cham, 2016.
12. Aiello, R., Cirelli, G.L., Consoli, S., Effects of reclaimed wastewater irrigation on soil and tomato fruits: A case study in Sicily (Italy). *Agric. Water Manage.*, 93, 65, 2007.
13. Kumar, A., Verma, H., Singh, V.K., Singh, P.P., Singh, S.K., Ansari, W.A., Yadav, A., Singh, P.K., Pandey, K.D., Role of Pseudomonas sp. in sustainable agriculture and disease management. In: Meena, V.S., Mishra, P.K., Bisht, J.K., and Pattanayak, A. (eds.) *Agriculturally Important Microbes for Sustainable Agriculture*, pp. 195–215. Springer Singapore, Singapore, 2017.
14. Abegglen, C., Siegrist, H., Domestic wastewater treatment with a small-scale membrane bioreactor. *Water Sci. Technol.*, 53, 69, 2006.
15. Kadlec, R.H., Wallace, S., *Treatment Wetlands*, CRC Press, 2008.
16. Mara, D.D., Pearson, H.W., A hybrid waste stabilization pond and wastewater storage and treatment reservoir system for wastewater reuse for both restricted and unrestricted crop irrigation. *Water Res.*, 33, 591, 1999.
17. Howe, J., Wagner, M.R., Effects of pulpmill effluent irrigation on the distribution of elements in the profile of an arid region soil. *Environ. Pollut.*, 105, 129, 1999.
18. Tunc, T., Sahin, U., Red cabbage yield, heavy metal content, water use and soil chemical characteristics under wastewater irrigation. *Environ. Sci. Pollut. Res.*, 23, 6264, 2016.
19. Zhang, H., Li, W., Miao, P., Sun, B., Kong, F., Risk grade assessment of sudden water pollution based on analytic hierarchy process and fuzzy comprehensive evaluation. *Environ. Sci. Pollut. Res.*, 27, 469, 2020.
20. Tabatabaei, S.H., Najafi, P., Effets de l'irrigation avec des eaux usées municipales traitées sur les propriétés des sols dans les régions arides et semi-arides. *Irrig. Drain.*, 58, 551, 2009.
21. Jayasooriya, V.M., Ng, A.W.M., Muthukumaran, S., Perera, B.J.C., Optimal Sizing of Green Infrastructure Treatment Trains for Stormwater Management. *Water Resour. Manage.*, 30, 5407, 2016.
22. Güneş, G., Evaluation of Bartın river water quality index and suitability as irrigation water with physicochemical parameters. *Environ. Res. Technol.*, 5, 357, 2022.
23. Xu, Z., Zhou, S., Zhang, C., Yang, M., Jiang, M., A Bayesian network model for suitability evaluation of underground space development in urban areas: The case of Changsha, China. *J. Clean. Prod.*, 418, 138135, 2023.
24. Page, A.L., Chang, A.C., Fate of Wastewater Constituents in Soil and Groundwater: Trace Elements. In: A. L. Page, A.C.C. (ed.) *Irrigation With Reclaimed Municipal Wastewater–A Guidance Manual*, pp. 13-1-13–16. CRC Press, 2019.

25. Tunc, T., Sahin, U., The changes in the physical and hydraulic properties of a loamy soil under irrigation with simpler-reclaimed wastewaters. *Agric. Water Manage.*, 158, 213, 2015.
26. Liang, P., Jingan, X., Liying, S., The effects of reclaimed water irrigation on the soil characteristics and microbial populations of plant rhizosphere. *Environ. Sci. Pollut. Res.*, 29, 17570, 2022.
27. Kapoor, A.K., Pal, R., Predicting salinization and sodification of a bare sandy loam soil after irrigation with poor-quality water interspersed with rain. *Soil Sci.*, 141, 281, 1986.
28. Yuan, T., Fengmin, L., Puhai, L., Economic analysis of rainwater harvesting and irrigation methods, with an example from China. *Agric. Water Manage.*, 60, 217, 2003.
29. Al-Hamaiedeh, H., Bino, M., Effect of treated grey water reuse in irrigation on soil and plants. *Desalination*, 256, 115, 2010.
30. Zhang, P., Wei, T., Jia, Z., Han, Q., Ren, X., Li, Y., Effects of straw incorporation on soil organic matter and soil water-stable aggregates content in semi-arid regions of Northwest China. *PLoS One*, 9, e92839, 2014.
31. Yerli, C., Cakmakci, T., Sahin, U., Soil CO2 Emission Linearly Increases with Organic Matter Added Using Stabilized Sewage Sludge under Recycled Wastewater Irrigation Conditions. *Water. Air. Soil Pollut.*, 234, 2023.
32. Rahman, M.S., Hoque, M.N., Puspo, J.A., Islam, M.R., Das, N., Siddique, M.A., Hossain, M.A., Sultana, M., Microbiome signature and diversity regulates the level of energy production under anaerobic condition. *Sci. Rep.*, 11, 2021.05.11.443548, 2021.
33. Li, Q., Harris, B., Aydogan, C., Ang, M., Tade, M., Feasibility of recharging reclaimed wastewater to the coastal aquifers of Perth, Western Australia. *Process Saf. Environ. Prot.*, 84, 237, 2006.
34. Imtiyaz, M., Mgadla, N.P., Chepete, B., Manase, S.K., Response of six vegetable crops to irrigation schedules. *Agric. Water Manage.*, 45, 331, 2000.
35. Maheshwari, B.L., Grewal, H.S., Magnetic treatment of irrigation water: Its effects on vegetable crop yield and water productivity. *Agric. Water Manage.*, 96, 1229, 2009.
36. Ahmadi, S.H., Agharezaee, M., Kamgar-Haghighi, A.A., Sepaskhah, A.R., Compatibility of root growth and tuber production of potato cultivars with dynamic and static water-saving irrigation managements. *Soil Use Manage.*, 33, 106, 2017.
37. Hsiao, Y. C., Huang, T. L., Chung, R. S., Effect of treated wastewater irrigation on soil properties, crop yield, and environment. *Agric. Ecosyst. Environ.*, 239, 80–89, 2017.
38. Serikbayeva, A., Taizhanova, L., Suleimenova, B., Altybayeva, Z., Seidalieva, L., Intensification of the Wastewater Treatment Process of a Bitumen Plant with the Production of Recycled Water. *J. Ecol. Eng.*, 24, 295, 2023.

39. Ityel, E., Lazarovitch, N., Silberbush, M., Ben-Gal, A., An artificial capillary barrier to improve root-zone conditions for horticultural crops: Response of pepper, lettuce, melon, and tomato. *Irrig. Sci.*, 30, 293, 2012.
40. Moraes, E.S., Reis, G.F., Cruz, J., Cozzolino, K., Neto, A.F.G., Andrade-Filho, T., Neto, A.M.J.C., Thermodynamics study of biokerosene from coconut and palm kernel oils and JP-8 aircraft fuels in the gas phase by the DFT method. *J. Mol. Model.*, 26, 79, 2020.
41. Rahil, M.H., Qanadillo, A., Effects of different irrigation regimes on yield and water use efficiency of cucumber crop. *Agric. Water Manage.*, 148, 10, 2015.
42. Ismail, S.M., Almarshadi, M.H., Maximizing Productivity And Water Use Efficiency Of Alfalfa Under Precise Subsurface Drip Irrigation In Arid Regions. *Irrig. Drain.*, 62, 57, 2013.
43. Bertram, G.C., *The state of food and agriculture, 1966*, United Nations 1967.
44. Chauhan, J.S., Kumar, S., Wastewater ferti-irrigation: an eco-technology for sustainable agriculture. *Sustain. Water Resour. Manage.*, 6, 2020.
45. Fu, L., Wang, J., Wang, S., Peng, H., Gui, Z., Study of Water Resource Allocation and Optimization Considering Reclaimed Water in a Typical Chinese City. *Sustainability*, 15, 819, 2023.
46. Chen, W., Lu, S., Pan, N., Jiao, W., Impacts of long-term reclaimed water irrigation on soil salinity accumulation in urban green land in Beijing. *Water Resour. Res.*, 49, 7401, 2013.
47. Liu, Z., Zhou, W., Shen, J., Li, S., He, P., Liang, G., Soil quality assessment of Albic soils with different productivities for eastern China. *Soil Tillage Res.*, 140, 74, 2014.
48. 1Barrow, G., Bacterial Indicators and Standards of Water Quality in Britain. *Bact. Indic. Hazards Assoc. with Water*, 289, 2009.
49. Zhong, Z., Micropollutant abatement by UV photolysis of chloramines towards potable water reuse, http://lbezone.ust.hk/bib/991012752656003412, (2019).
50. Halalsheh, M., Kassab, G., Shatanawi, K., Al-Shareef, M., Development of sanitation safety plans to implement World Health Organization guidelines: Jordanian experience. In: Hettiarachchi, H. and Ardakanian, R. (eds.) *Safe Use of Wastewater in Agriculture: From Concept to Implementation*, pp. 101–130. Springer International Publishing, Cham, 2018.
51. Singh, P.K., Deshbhratar, P.B., Ramteke, D.S., Effects of sewage wastewater irrigation on soil properties, crop yield and environment. *Agric. Water Manag.*, 103, 100, 2012.
52. Khoury, A., Post-Conflict Reconciliation in the Middle East and North Africa (MENA) Region. In: International Relations. Oxford University Press (2018).
53. Agunwamba, J.C., Analysis of socioeconomic and environmental impacts of waste stabilization pond and unrestricted wastewater irrigation: Interface with maintenance. *Environ. Manage.*, 27, 463, 2001.

54. Hou, T., Zhao, J., Lei, Z., Shimizu, K., Zhang, Z., Addition of air-nanobubble water to mitigate the inhibition of high salinity on co-production of hydrogen and methane from two-stage anaerobic digestion of food waste. *J. Clean. Prod.*, 314, 127942, 2021.
55. Wang, Z., Li, J., Li, Y., Using Reclaimed Water for Agricultural and Landscape Irrigation in China: a Review. *Irrig. Drain.*, 66, 672, 2017.
56. Aiello, R., Cirelli, G.L., Consoli, S., Effects of reclaimed wastewater irrigation on soil and tomato fruits: A case study in Sicily (Italy). *Agric. Water Manage.*, 93, 65, 2007.
57. Schindler, D.W., Recent advances in the understanding and management of eutrophication. *Limnol. Oceanogr.*, 51, 356, 2006.
58. Cloern, J.E., Our evolving conceptual model of the coastal eutrophication problem. *Mar. Ecol. Prog. Ser.*, 210, 223, 2001.
59. Agency, E.P., Water Recycling and Reuse: The Environmental Benefits. In: *Water Encyclopedia*, pp. 610–613, 2004.
60. Du, C., Guo, W., Li, G., Bai, M., Zhu, Q., Tian, Z., Li, M., Zhao, C., Zhang, L., Biomanipulation as a strategy for minimizing ecological risks in river supplied with reclaimed water. *Environ. Res.*, 228, 115801, 2023.
61. Domini, M., Langergraber, G., Rondi, L., Sorlini, S., Maswaga, S., Development of a sanitation safety plan for improving the sanitation system in peri-urban areas of Iringa, Tanzania. *J. Water Sanit. Hyg. Dev.*, 7, 340, 2017.
62. Haj-Amor, Z., Araya, T., Kim, D.G., Bouri, S., Lee, J., Ghiloufi, W., Yang, Y., Kang, H., Jhariya, M.K., Banerjee, A., Lal, R., Soil salinity and its associated effects on soil microorganisms, greenhouse gas emissions, crop yield, biodiversity and desertification: A review. *Sci. Total Environ.*, 843, 156946, 2022.
63. Arena, C., Genco, M., Mazzola, M.R., Environmental benefits and economical sustainability of urban wastewater reuse for irrigation—a cost-benefit analysis of an existing reuse project in Puglia, Italy. *Water (Switzerland)*, 12, 1, 2020.
64. López-Serrano, M.J., Velasco-Muñoz, J.F., Aznar-Sánchez, J.A., Román-Sánchez, I.M., Financial evaluation of the use of reclaimed water in agriculture in Southeastern Spain, a mediterranean region. *Agronomy*, 11, 2218, 2021.
65. Huang, S., Zhang, N., Li, Y., Song, P., Liu, Z., Pilot electrochemical prevention of reclaimed water irrigation clogging: Function interactions and microbial metabolism. *J. Clean. Prod.*, 336, 130436, 2022.
66. Nations, F. and A.O. of the U., Water resources: Economics and policy. United Nations Publications 1993.
67. Jimmy Akinfemi OSUNBITAN, Short term effects of cassava processing waste water on some chemical properties of loamy sand soil in Nigeria. *J. Soil Sci. Environ. Manage.*, 3, 2012.

68. Paradelo, R., Villada, A., Barral, M.T., Heavy metal uptake of lettuce and rye-grass from urban waste composts. *Int. J. Environ. Res. Public Health*, 17, 2887, 2020.
69. Lamei, A., Van Der Zaag, P., Imam, E., Integrating wastewater reuse in water resources management for hotels in arid coastal regions - Case study of Sharm El Sheikh, Egypt. *Water Sci. Technol.*, 60, 2235, 2009.
70. Qadir, M., Boers, T.M., Schubert, S., Ghafoor, A., Murtaza, G., Agricultural water management in water-starved countries: Challenges and opportunities. *Agric. Water Manage.*, 62, 165, 2003.
71. Feder, F., Irrigation with treated wastewater in humid regions: Effects on Nitisols, sugarcane yield and quality. *Agric. Water Manage.*, 247, 106733, 2021.
72. Adeyemo, A.A., Adeoye, I.O., Bello, O.S., Adsorption of dyes using different types of clay: a review. *Appl. Water Sci.*, 7, 543, 2017.
73. Devatha, C.P., Pavithra, N., Isolation and identification of Pseudomonas from wastewater, its immobilization in cellulose biopolymer and performance in degrading Triclosan. *J. Environ. Manage.*, 232, 584, 2019.
74. Menyuka, N.N., Sibanda, M., Bob, U., Perceptions of the challenges and opportunities of utilising organic waste through urban agriculture in the durban south basin. *Int. J. Environ. Res. Public Health.*, 17, 1158, 2020.
75. Dupont, D.P., Water use restrictions or wastewater recycling? A Canadian willingness to pay study for reclaimed wastewater. *Water Resour. Econ.*, 1, 61, 2013.
76. Narayanamoorthy, A., Financial Performance of India's Irrigation Sector: Past, Present and Future. In: Narayanamoorthy, A. (ed.) *Global Issues in Water Policy*, pp. 139–158. Springer International Publishing, Cham, 2022.
77. Ahmed, F., Johnson, D., Hashaikeh, R., Hilal, N., Barriers to Innovation in Water Treatment. *Water (Switzerland)*, 15, 773, 2023.
78. Ezzati, G., Kyllmar, K., Barron, J., Long-term water quality monitoring in agricultural catchments in Sweden: Impact of climatic drivers on diffuse nutrient loads. *Sci. Total Environ.*, 864, 160978, 2023.
79. Pham Do, K.H., Water Resource Management: Challenges and Opportunities with Game Theory Approaches. In: Oxford Research Encyclopedia of Environmental Science. Oxford University Press (2022).
80. Laudon, L., State Water Resources Control Board, California Agreement in Principle 1995 summary report. Office of Scientific and Technical Information (OSTI) 1996.
81. Rakocy, J.E., Aquaponics-Integrating Fish and Plant Culture. In: Tidwell, J.H. (ed.) *Aquaculture Production Systems*, pp. 344–386. Wiley, 2012.
82. Erratum. *J. Environ. Sci. Heal. Part C Environ. Carcinog. Rev*, 1, 263, 1983.
83. Shtaya, M.J.Y., Qubbaj, T., Effect of different soilless agriculture methods on irrigation water saving and growth of lettuce (Lactuca sativa). *Res. Crop*, 23, 156, 2022.

84. Alramthi, S.M., Ali, G.H., Elthagafi, A.M., Eldosari, S.H., Zhu, B.K., Safaa, H.M., Oxidation Ditches for Recycling and Reusing Wastewater Are Critical for Long-Term Sustainability—A Case Study. *Sustainability*, 14, 2022.
85. Al-Jayyousi, O., Bergkamp, G., Water Management in the Jordan River Basin: Towards an Ecosystem Approach. In: Varis, O., Biswas, A.K., and Tortajada, C. (eds.) *Water Resources Development and Management*, pp. 105–121. Springer Berlin Heidelberg, Berlin, Heidelberg, 2008.
86. Anderson, J., The environmental benefits of water recycling and reuse. *Water Sci. Technol. Water Supply*, 3, 1, 2003.
87. Vincent-Lancrin, S., González-Sancho, C., Bouckaert, M., de Luca, F., Fernández-Barrerra, M., Jacotin, G., Urgel, J., Vidal, Q., Technical Annex. In: *The State of Food and Agriculture 2019*, pp. 339–356. United Nations, 2019.
88. Barnes, D., Merz, S.K., Mollenkopf, T., The Australian Urban Water Sector A 'World' of Approaches. *Proc. Water Environ. Fed.*, 2009, 1734, 2012.
89. Rochelle-Newall, E., Nguyen, T.M.H., Le, T.P.Q., Sengtaheuanghoung, O., Ribolzi, O., A short review of fecal indicator bacteria in tropical aquatic ecosystems: Knowledge gaps and future directions. *Front. Microbiol.*, 6, 1, 2015.
90. Kitanou, S., Qabli, H., Zdeg, A., Taky, M., Elmidaoui, A., Performance of external membrane bioreactor for wastewater treatment and irrigation reuse. *Desalin. Water Treat.*, 78, 19, 2017.
91. Tomei, M.C., Del Moro, G., Di Iaconi, C., Mascolo, G., Enhanced Biological Wastewater Treatment to Produce Effluents Suitable for Reuse. In: Fatta-Kassinos, D., Dionysiou, D.D., and Kümmerer, K. (eds.) *Handbook of Environmental Chemistry*, pp. 79–106. Springer International Publishing, Cham, 2015.
92. Nagtode, V.S., Cardoza, C., Yasin, H.K.A., Mali, S.N., Tambe, S.M., Roy, P., Singh, K., Goel, A., Amin, P.D., Thorat, B.R., Cruz, J.N., Pratap, A.P., Green Surfactants (Biosurfactants): A Petroleum-Free Substitute for Sustainability Comparison, Applications, Market, and Future Prospects. *ACS Omega*, 8, 11674, 2023.
93. Iyer, A., Mody, K., Jha, B., Biosorption of heavy metals by a marine bacterium. *Mar. Pollut. Bull.*, 50, 340, 2005.
94. Cui, Y., Lin, H.H., Xie, Y., Liu, S.H., Application study of crop yield prediction based on AquaCrop model in black soil region of Northeast China. *Acta Agron. Sin.*, 47, 159, 2021.
95. Fishel, F.M., Sanchez, T., Worker Protection Standard - Certified Crop Advisor Exemption. *Edis.*, 2016, 2016.
96. Challinor, A.J., Watson, J., Lobell, D.B., Howden, S.M., Smith, D.R., Chhetri, N., A meta-analysis of crop yield under climate change and adaptation. *Nat. Clim. Change*, 4, 287, 2014.
97. Rajivgandhi, G., Vimala, R.T.V., Maruthupandy, M., Alharbi, N.S., Kadaikunnan, S., Khaled, J.M., Manoharan, N., Li, W.J., Enlightening the characteristics of bioflocculant of endophytic actinomycetes from marine

algae and its biosorption of heavy metal removal. *Environ. Res.*, 200, 111708, 2021.
98. Martins, M.G., da Silva Arouche, T., Neto, A.F.G., da Cruz, J.N., da Costa, F.L.P., Fernandes, L.L., de Carvalho Junior, R.N., da Silva Costa, J.F., de Jesus Chaves Neto, A.M., Density functional theory for the thermodynamic gas-phase investigation of butanol biofuel and its isomers mixed with gasoline and ethanol. *J. Mol. Model.*, 27, 80, 2021.
99. 1. Dacewicz, E., Chmielowski, K., The Importance of Media in Wastewater Treatment. In: Zhu, I.X. (ed.) *Sewage*, InTech, 2018.
100. Avnery, S., Mauzerall, D.L., Liu, J., Horowitz, L.W., Response to "Comments on 'Global crop yield reductions due to surface ozone exposure: 1. Year 2000 crop production losses and economic damage' and 'Global crop yield reductions due to surface ozone exposure: 2. Year 2030 potential crop production losses and economic damage under two scenarios of O3 pollution." *Atmos. Environ.*, 71, 410, 2013.
101. Wolpert-Barraza, E., Coediting the Revista De Investigación Clínica With Ruben Lisker, 1974-1987. *Rev. Invest. Clin.*, 73, 335, 2021.
102. Komolafe, O., Mrozik, W., Dolfing, J., Acharya, K., Vassalle, L., Mota, C.R., Davenport, R., Fate of four Different Classes of Chemicals Under Aerobic and Anaerobic Conditions in Biological Wastewater Treatment. *Front. Environ. Sci.*, 9, 1, 2021.
103. Ali, D. al deen A., Palaniandy, P., Feroz, S., Advanced Oxidation Processes (AOPs) to Treat the Petroleum Wastewater. In: Aziz, H.A. and Abu Amr, S.S. (eds.) *Advanced Oxidation Processes (AOPs) in Water and Wastewater Treatment*, pp. 99–122. IGI Global, Hershey, PA, USA, 2018.
104. Fader, M., Gerten, D., Thammer, M., Heinke, J., Lotze-Campen, H., Lucht, W., Cramer, W., Internal and external green-blue agricultural water footprints of nations, and related water and land savings through trade. *Hydrol. Earth Syst. Sci.*, 15, 1641, 2011.
105. Sachs, J., *OECD-FAO Agricultural Outlook 2009*, 2009.
106. Kerstin, M., Food and Agriculture Organization of the United Nations (FAO). In: Max Planck Encyclopedia of Public International Law. Oxford University Press (2006).
107. Amador, A.T., Neto, A.F.G., Cruz, J.N., Magno, F.N.B., Marques, F.C., Da Silva, C.A.B., Nero, J. Del, Neto, A.M.J.C., Molecular Electronics Including Temperature Effects Based on Dyes Pigments. *J. Nanosci. Nanotechnol.*, 19, 3631, 2019.
108. Müller, L.J., Kätelhön, A., Bachmann, M., Zimmermann, A., Sternberg, A., Bardow, A., A Guideline for Life Cycle Assessment of Carbon Capture and Utilization. *Front. Energy Res.*, 8, 1, 2020.
109. Helmecke, M., Fries, E., Schulte, C., Regulating water reuse for agricultural irrigation: risks related to organic micro-contaminants. *Environ. Sci. Eur.*, 32, 2020.

110. Lee, K., Jepson, W., Drivers and barriers to urban water reuse: A systematic review. *Water Secur.*, 11, 100073, 2020.
111. Järup, L., Hazards of heavy metal contamination. *Br. Med. Bull.*, 68, 167, 2003.
112. Bhandari, P.B., Rural livelihood change? Household capital, community resources and livelihood transition. *J. Rural Stud.*, 32, 126, 2013.

12

Recovery of Value-Added Products from Sewage Sludge Using Biological Processes

Nadia Akram*, Khalid Mahmood Zia, Muhammad Usman, Fozia Anjum and Sana Pervaiz

Department of Chemistry, Government College University Faisalabad, Faisalabad, Pakistan

Abstract

Sustainable WAS (waste-activated sludge) management is successful in drawing the attention of scientists to transforming wastewater treatment facilities into biorefineries. The bio-product quality is crucial to the success of innovative WAS-based manufacturing methods. This chapter involves a comprehensive evaluation of the production pathways of several products that can be extracted from WAS. The study investigates the current limitations associated with these production routes and proposes future research directions to facilitate sustainable management of such expanding waste streams. The practice of collecting food waste separately is becoming increasingly popular. The chapter has been organized with the aim of developing a comprehensive layout of the several methods being developed or employed for resource recovery from food waste (FW). These methods include the recovery of energy and other components.

Keywords: Bioplastics, biorefinery, waste-activated sludge, biogas, bio-hydrogen

12.1 Waste Activated Sludge

Modern activated sludge wastewater treatment generates WAS (waste-activated sludge), which poses a serious disposal challenge for water management authorities everywhere. As an illustration, the rates of biosolids production per capita in the regions of the United States, Europe, and Asia,

**Corresponding author*: nadiaakram@gcuf.edu.pk

Inamuddin, Tariq Altalhi, Mohammad Luqman and Joseph Kapuku (eds.) Sewage and Biomass from Wastewater to Energy, (327–346) © 2024 Scrivener Publishing LLC

along with 2015 United Nations data on global population, sewerage, and wastewater treatment, suggest that present rates of biosolids production on a global scale are somewhere in the range of 25 to 60 Mt dry solids per year [1]. The trash accumulation associated with WAS is expected to grow, calling for more long-term, viable alternatives to current waste management practices. Consistent with international cleanliness goals, the world's wastewater collection and treatment infrastructure are expected to significantly expand in the next few decades [2, 3]. Traditional uses of WAS in agriculture include soil conditioning and application; however, concerns about the spread of disease from pathogens and toxic chemicals, as well as complaints about unpleasant odors, have limited the extent as well as the acceptability of this practice [4–6]. When combined with primary sludge, the management of WAS with a focus on sustainability has conventionally been implemented in the context of agricultural land or as a means of generating methane via anaerobic digestion (AD). This phenomenon involves higher concentrations of nutrients that are currently available. WAS is often used in low-value end applications like the restoration of mining sites. Detection of impurities in processed sewage sludge at lower concentrations ranging from grams per kilogram to milligrams per kilogram raised concerns about the possibility of toxic substances accumulating and spreading via the food chain as impurities. WAS results from the purification of sewage in populated areas, the state of health among the general public. Incineration and thermal energy recovery for electricity generation is another popular sustainable approach to WAS management; since biosolids have a low heating value and a high moisture content, they consume a lot of energy, necessitating the need for additional fuel to keep incineration facilities running. Carbon byproducts from incineration can serve as environmental adsorbents or be used in alternative construction materials (for example, bricks and artificial grass) [7].

12.2 Amino Acids and Proteins

The organic matter present in WAS primarily comprises intricate compounds such as lipids, polysaccharides, and proteins, which collectively account for approximately 70 to 80% of its composition. Notably, the protein fraction constitutes nearly 50% of the bacterial cells' dry weight. This implies that WAS can be potentially utilized as a protein source or amino acids. In comparison to conventional protein sources, protein derived from WAS (waste-activated sludge) exhibits a higher potential for serving as a principal constituent in the manufacture of animal feeds. Nonetheless, it is imperative to conduct efficient detoxification procedures on sludge, which involve sterilization and elimination of heavy metals and other harmful substances. The initial and crucial stage in obtaining protein from WAS involves the successful solubilization of intracellular material, which serves as the basis for numerous extraction methods documented in scholarly works [8]. Thermal digestion, coupled with centrifugal separation, represents a straightforward methodology. Thermal digestion has the potential to improve the dewatering properties of WAS (waste-activated sludge) by disrupting flocs. The subsequent benefits are particularly noteworthy and surpass all others: The advantages of this approach include a low treatment cost, the absence of the need for new reagents, and the absence of any additional waste stream. The thermal extraction method possesses an inherent limitation in its capacity to eliminate numerous detrimental organic compounds and heavy metals that are present in WAS (waste-activated sludge). To comply with the progressively stringent regulations on feed health and safety, a greater number of chemically intricate methods have been developed. The alkali solubilization method employing NaOH and NaCl, along with precipitation, was initially explored by Chishti *et al.* for the purpose of protein extraction from WAS. The results of the study indicate that ammonium sulfate (40%) exhibited the highest efficacy as a precipitant, resulting in a protein recovery rate of approximately 91%. This finding was in contrast to the other precipitants tested, which included HCl, sodium lignosulfonate, H_2SO_4, CH_3COOH, and NH_4SO_3. The technique involved the use of papain enzyme for hydrolysis. A 6% enzyme-to-substrate ratio, a pH of 6.5, a 4:1 liquid-to-solid ratio, 5.5 hours of hydrolysis time, and 55°C of hydrolysis temperature were described as the optimal conditions for papain-assisted hydrolysis. The results of the investigations show that all necessary amino acids exist in the products required for animal nutrition, thereby qualifying them as a viable supplement for animal feed, in comparison to protein feeds available in the market. The concentration of heavy metal in the original WAS was significantly reduced by all of the methods investigated, notably through alkaline precipitation. Scientists conducted research to evaluate the effectiveness of mechanical techniques, such as cationic exchange resin and triton, in extracting proteins from two distinct activated sludges [9–13]. Hwang *et al.* utilized the ultrasonic-alkaline method to achieve minimal concentrations of harmful pollutants and microorganisms. Additionally, the study evaluated the appropriateness of these techniques for conventional quantification methods. This was demonstrated by means of comparing various extraction methodologies. Improved competence in methodology may involve an existing mechanical process. The beneficial isolation of proteins was observed through the effective utilization of surfactants, emphasizing its significance. Triton/Triton and Ultraturax/Triton extraction methods produced protein extracts. The ratios of protease activity to extracted protein may exhibit significant variations depending on the employed extraction methodologies. The optimal method for extracting protease was found to be through the use of triton treatments and ultrasound. Protein extraction agents are considered important against detergents. Nevertheless, this approach is generally only practical when utilizing substantial amounts. As an illustration, the utilization of sodium dodecyl sulfate (SDS) has the potential to extract a maximum of 3 mg of protein per milliliter from sludge samples that contain 5% solid. Nevertheless, SDS necessitates a concentration of more than 2% weight per volume and is known to pose challenges in terms of eliminating it. The combination of sodium palmitate and alkali, specifically at a concentration of 0.15 mM NaOH, has the potential to yield a protein output of 15 mg/mL from waste-activated sludge containing 5% solubilized solids. Nonetheless, the output of said production is limited due to the scarcity of alternative protein sources, i.e., horn, animal hair, leather, and hoof. Additionally, wastewater has been utilized in the creation of fertilizers containing amino acid–chelated trace elements (AACTE). A novel approach has been introduced by Liu *et al.* for the extraction of bacterial protein from sewage effluent with the aim of producing AACTE fertilizer. The purification procedure comprises five crucial phases, which mainly include acidic hydrolysis and decolorization of sludge. The most favorable results were obtained by subjecting the sludge to hydrolysis, using a pH of 0.5. The hydrolysis process involved the use of a 6 mol/L HCl solution in a ratio of 8-g dry sludge for the purpose of converting proteins into amino acids. The hydrolysis reaction was carried out for a period of 10 hours at a temperature of 121°C. Under ideal hydrolysis conditions, the protein recovery rate was found to be 78.5%. The application of amino acids demonstrates a unique potential as a sustainable corrosion inhibitor in the industrial pickling process [14–19]. Several research studies have provided evidence to support the effectiveness of amino acids in inhibiting corrosive reactions and controlling them in various metals under acidic conditions.

Su *et al.* reported on the production of a hydrolysate comprising 50 distinct amino acids through the utilization of a hydrothermal hydrochloric acid method that was aided by ultrasound. The adsorption of amino acids onto steel surfaces is facilitated by the presence of -COOH and $-NH_2$ (hydrophilic and hydrophobic groups, respectively), which leads to the formation of stable protective films. The corrosion behavior of steel surfaces was examined through a comparative analysis of specimens with and without prior application of a surface coating of amino acid. The results indicate a noteworthy decrease in the roughness of the surface and a sixfold reduction in the corrosion rate. The adhesive derived from wood is under discussion. Pervaiz and Sain utilized an alkaline cell-disruption method as employed by Hwang *et al.* (2008) to extract proteins. The protein acquired exhibited a shear strength of approximately 41%, which is comparable to that of conventional materials [20].

12.3 Fatty Acids Chains

The retrieval of the WAS is recognized as feasible through AD and thermal techniques, including the wet air oxidation method and hydrothermal treatment. Acetic, formic, and propionic acids are among the SCFAs that can be obtained through these processes. The acceleration of the initial two steps can lead to an increase in the accumulation of SCFAs. The primary factors that affect the initial two steps include temperature, sludge retention time, and pH. The presence of alkaline conditions has been observed to have advantageous effects such as enhanced sludge solubilization through chemical means and sludge hydrolysis. However, there is currently no agreement on which of these effects is the primary contributing factor [21–25]. According to the authors' report, it was observed that 35°C was the most suitable temperature under alkaline circumstances with a pH of 10. The authors have additionally noted that alterations in protease activity suggest that the enhancement of SCFAs under alkaline conditions. The phylum Firmicutes, specifically *Clostridium* and *Bacillus*, as well as the phylum Proteobacteria, specifically Gamma proteobacteria, were identified as the primary functional bacterial taxa responsible for the accumulation of high levels of fatty acids. In order to increase the production of SCFAs during AD of WAS, it is common practice to inhibit the final step of methanogenesis. This can be accomplished by introducing a methane analog, such as iodoform. The literature reports a dual-promoting effect of CaO_2 addition, namely, (1) an increase in alkalinity due to the formation of $Ca(OH)_2$ and (2) the release of additional oxygen through its gradual decomposition in aqueous settings. The introduction of CaO_2 was observed to result in a significant increase in the population of Firmicutes, a phylum of microorganisms [26–28].

The hydrothermal retrieval of SCFAs from WAS is typically a two-step process. First, hydrolysis occurs, leading to the creation and buildup of dissolved organic compounds. Second, oxidation (for example, using hydrogen peroxide) can aid in the removal of carbon and the addition of oxygen. The study conducted involved the utilization of an in situ transesterification technique on dried sludge which yielded 6.23% [29–32].

12.4 Enzymes

A significant proportion of components present in WAS is composed of particles exceeding 0.1 μm in size, which are not readily amenable to microbial assimilation. This fraction accounts for 30 to 85% of the total organic matter in WAS. Currently, ultrasonication is recognized as one of the most extensively documented techniques for the retrieval of the aforementioned enzymes. Consequently, there is an enhancement in the efficacy of the degradation of unstable organic components. In their study, Nabarlatz *et al.* (2010) utilized a complex extraction technique involving ultrasonication, dialysis, ammonium sulfate precipitation, and lyophilization to retrieve hydrolytic enzymes (specifically, protease and lipase) from WAS [33].

The literature reports that the hydrolytic enzymatic activities in WAS are affected by salinity, which is attributed to the presence of salts. The enzymatic activities exhibit a negative correlation with increased salinity. The acquisition of enzymes that are appropriate for specific industrial uses, such as serving as a detergent additive, necessitates a significant level of transparency. Consequently, additional procedures for purification and concentration are typically necessary. Developing a methodology for the purification and concentration of all enzymatic fractions found in WAS (waste-activated sludge) is a significant obstacle that must be addressed. This challenge is compounded by the natural fluctuations in the composition of activated sludge, which occur on a daily and seasonal basis, as well as the relatively low concentration of enzymes present in the sludge.

12.5 Biopesticides

Due to its high concentration of essential nutrients like phosphorus, carbon, and nitrogen, WAS (waste-activated sludge) exhibits promising potential as a viable substitute for a medium of growth for industrial microorganisms in the production of valuable metabolic products. The production of δ-endotoxin is a well-known characteristic of organisms like the aerobic spore-former *Bacillus thuringiensis* (Bt) during sporulation. This biopesticide is widely used in various sectors such as agronomy, forestry, and public health and is considered to be the most popular one globally. The cost of these components constitutes a significant proportion of the overall production cost, ranging from 35 to 59% depending on the capacity of plant production. The utilization of WAS sludge as an inexpensive and environmentally friendly medium for Bt production, and subsequently using it for pest control in agricultural crops and forests is a financially viable and compatible approach with current sludge disposal practices.

Adequate aeration during the process of fermentation is imperative in order to optimize the growth of cells, sporulation, and the production of δ-endotoxin by Bt. However, excessive aeration may produce significant foaming, which is dependent on the solid contents. Hence, it is imperative to conduct inquiries into the impact of foaming on the production of Bt through WAS fermentation and its subsequent management in order to enhance Bt yield. In their study, Vidyarthi and colleagues (2000) examined the impact of polypropylene glycol (synthetic) and natural antifoaming agents on the yield of Bt. Although all agents exhibited efficient foaming control, the utilization of synthetic antifoaming agents had a negative impact on cellular respiratory activity and growth due to their interference with the transportation of nutrients, including oxygen, through the cell walls. Conversely, the utilization of natural oils for foaming control exhibited a more advantageous outcome. The utilization of WAS has the tendency to be expanded to yield additional advantageous metabolic products that exhibit insecticidal properties and demonstrate entomotoxicity. Scientists have illustrated the utilization of WAS as a primary resource for the synthesis of bioherbicides/biopesticides based on Trichoderma sp. The bioherbicides/biopesticides exhibit a wider range of activity as compared to Bt. The study revealed that thermal alkaline pretreatment yielded favorable outcomes akin to Bt fermentation production [34].

12.6 Bioplastics

Polyhydroxyalkanoates (PHAs) are a class of biodegradable polyesters that have significant interest in petroleum-based plastic substitutes in various domains, including packaging films, single-use utensils, absorbing materials, cosmetic and liquid containers, vessels for beverages, and cutting-edge innovations, pharmaceutical treatment, PHB, and its copolymer, P(3HB-co-HV), are extensively utilized PHAs, with promising potential applications for other PHA variants in the future. The properties of PHB are known to exhibit heightened levels of crystallinity, stiffness, and brittleness, whereas PHV is recognized for its ability to enhance softness and flexibility. PHAs, which function as internal carbon and energy storage, are synthesized by more than 75 bacterial genera. They also serve as a reservoir for the reduction of electron flow under adverse ecological circumstances. PHAs of microbial origin can be synthesized through the process of bacterial fermentation. The utilization of organic carbon growth substrate, specifically carbohydrates, constitutes approximately 50% of the total cost involved in microbial PHA production. Consequently, any forthcoming modifications in sugar production and related fluctuations in commodity prices will have a direct effect on the cost-effectiveness and microbial PHA production feasibility. The dominant influence on feedstock competition is expected to be the bioethanol industry, which is projected to experience a 25% increase in global production and consumption of biofuel ethanol from 2009 to 2014. The implementation of competitive modifications has the potential to impact the forthcoming economic landscape of bioplastics manufacturing, rendering it more financially viable. Microorganisms that accumulate PHAs can be naturally found in WAS. These SCFAs can be easily obtained from various sources including agricultural and domestic WAS. The production of microbial PHAs from WAS (waste-activated sludge) is influenced by various process parameters, including temperature, carbon-to-nitrogen ratio, and dissolved oxygen (DO). Previous studies have indicated that elevated pH levels (approximately 8 to 9) and reduced SRT (3 days) are conducive to increased rates of PHA production. According to the research of Krishna and Van Loosdrecht (1999), a reduction in temperature resulted in an increase in PHA accumulation in WAS. The findings indicate that an elevated C/N ratio impedes cell growth while promoting PHA accumulation. Hence, the attainment of an ideal C/N ratio in a particular medium derived from WAS necessitates the contemplation of these conflicting factors. The potential of anaerobic-aerobic processes for the accumulation of PHAs has been studied, and it has been found that increasing the DO level in anaerobic zones significantly enhances this capacity. Notwithstanding the existence of conditions, it is infrequent to come across literature that reports full-scale or pilot-scale studies. The primary technical obstacle linked to the manufacturing of bioplastics utilizing WAS is the relatively low economic output [35].

12.7 Bio-Flocculants and Bio-Surfactants

Bio-flocculants and bio-surfactants, which are extracellular macromolecules, represent another set of significant metabolic by-products that are produced during the microbial conversion of organic substrates. When grown in synthetic growth media, microorganisms secrete numerous bio-flocculants and biopolymers. These bio-flocculants have found extensive applications. The efficacy of flocculation activities that are on par with synthetic flocculants has been exhibited, and they are anticipated to possess biodegradability and nontoxicity toward both the environment and human beings.

A comparative analysis was conducted by *Morgan et al.* (1990) to examine the characteristics of biopolymers. Significant variations were observed in the overall ECP yield per unit of suspended solids (SS) between the activated sludge and granular sludge samples. The former produced ECP at a rate of 70 to 90 mg/g SS, which was notably higher than the latter, which generated ECP at a rate of 10 to 20 mg/g SS. Henceforth, it was suggested that activated sludge, also known as WAS, could serve as a promising candidate for the generation of biopolymers, even when compared to alternative biological wastewater streams such as anaerobic granules. In contrast to activated sludge, where carbohydrates were dominant, the protein component of the isolated polymers was more abundant in anaerobic samples. Two separate wastewater treatment facilities produced the activated sludge. Extraction time and stirring intensity result in a significant improvement in EPS yield, surpassing that of conventional methods by a factor of two.

WAS (waste-activated sludge) is recognized as a highly effective environmental reservoir for the recovery of microorganisms capable of producing bio-flocculants. This is primarily due to the inherent bio-flocculation process that occurs in WAS during aerobic treatment. Studies comparing natural and synthetic flocculants have found that natural ones have shown that they typically perform equally or better than synthetic flocculants. The research of Sponza (2003) has indicated that the composition of EPS is significantly influenced by the type of wastewater, a finding that has been corroborated by subsequent studies. In contrast to their chemically synthesized counterparts, bio-surfactants exhibit superior efficacy and efficiency. It results in a lesser amount of surfactant mandatory to attain optimal reduction in surface tension. The utilization of WAS-derived media uses another substitute such as bio-surfactants. The feasibility of extracting bio-surfactants directly from domestic wastewater sludge through cell lysis and alkaline treatment has been established. However, it is important to acknowledge that the practicality of this approach may be affected by the pathogenic nature of certain microorganisms that are typically present in both domestic and industrial wastewater.

Following microbial cultivation in a medium derived from WAS, diverse microorganisms can produce bio-surfactants and bio-flocculants.

These techniques include alkaline treatment combined with heating, ion exchange combined with stirring, formaldehyde treatment combined with heating, and CER combined with sonication, among others. It is necessary to optimize cultivation processes by taking into account crucial process parameters in order to address potential challenges in scaling up production in the future. Furthermore, it is imperative to conduct a thorough characterization of the ionic properties and structure of every bio-flocculant and bio-surfactant acquired. This step is crucial in determining the appropriate operational parameters necessary for achieving optimal process outcomes [36].

12.8 Current Challenges

The valuable components of this waste stream are often overlooked. These advantages include offering an alternative treatment approach for an urban waste stream that is continuously expanding and posing challenges. Additionally, the process enables the retrieval of valuable resources and bio-products that have the potential to serve as sustainable green product substitutes. This aspect is particularly relevant in the current era where there is a growing interest. The evaluation of potential hazards, expenses, and advantages is imperative in determining the practicality of each biorefinery project that incorporates WAS. This topic has been extensively researched thus far. The future viability of WAS-derived processes is dependent upon the market values and quality of their resulting products. Currently, there exist various issues and challenges that are linked to biorefinery.

(1) The production of protein and enzymes through biorefinery methods is currently perceived as a more expensive approach when compared to the utilization of agro-industry residues and animal health. This is due to concerns regarding pathogenic and metal toxicity, which have impeded progress in this field.

The retrieval of metabolic waste products such as is still in its early stages. To advance this field of study, additional refinement is necessary. It is imperative to conduct further research and refinement on significant variables such as the ideal growth conditions and wastewater matrix.

The challenge of scaling up the production of biorefineries based on the *wet al*kaline oxidation and solubilization (WAS) process is a crucial area of focus for future endeavors, given that a significant portion of the current research has been conducted at the laboratory scale. In order to expand the implementation of the biorefinery method utilizing WAS, it is imperative to conduct additional investigations on a larger scale, including pilot and full-scale assessments, to enhance the efficiency of each biorefinery procedure. The complete techno-economic performance of said processes can only be comprehensively examined and comprehended under such circumstances.

The management of sewage sludge is becoming increasingly expensive and its constituents are often undervalued despite its growing status as a waste stream. It has been estimated that a significant proportion, ranging from 30 to 60%, of the expenses associated with the operation of sewage treatment plants can be attributed to the various activities. These benefits include offering an alternative treatment method for the continuously increasing and challenging urban waste stream, as well as the retrieval of valuable resources and bio-products that have the potential to serve as sustainable green product substitutes. The evaluation of technical feasibility, potential risks, associated costs, and benefits is crucial in determining the viability of each biorefinery development that involves WAS. This topic has been extensively researched thus far. The future viability of WAS-derived processes is contingent upon the value of their resulting goods and corresponding market values. Currently, there exist various challenges and obstacles that are linked to the production of biorefinery.

Further development is required for purification procedures that are both simple and efficient in order to achieve the necessary purity levels for specific applications of bio-products recovered from WAS. Such advancements would significantly enhance the economic viability.

The challenge of scaling up is a significant concern for future endeavors, given that a considerable portion of the existing research has been conducted at the laboratory level. In order to expand the implementation of the biorefinery strategy utilizing WAS, it is imperative to conduct additional investigations at greater magnitudes, including pilot and ultimately full scale, to enhance the efficiency of each biorefinery procedure. The comprehensive examination and comprehension of the techno-economic efficiency of these procedures can only be achieved under such circumstances.

12.8.1 Production of Food Wastes (FW)

The term can be accurately stated as the amount of food discarded within the segments of the food supply chain that are involved in producing edible products intended for human consumption. The treatment and valorization options for various industrial effluents vary due to their distinct characteristics. Therefore, this chapter solely concentrates on general and nonspecific FW, and discussing each industrial effluent separately is beyond its scope.

The issue of food waste (FW) has long been a topic of concern for societies, with various factors driving this concern. These factors include efficiency considerations, such as the waste of valuable resources, economic concerns related to financial losses, and moral dimensions that take into account the simultaneous presence of hunger and food wastage. The recognition among the global community regarding the escalating significance of the environmental and social issues linked to the rising production of FW. This increase is primarily attributed to the expanding population and the consequent elevation in living standards worldwide. The estimation of food loss along the food supply chain is a challenging task, and precise figures are difficult to obtain. However, current estimates suggest that approximately more than one billion tons of nutriment are gone annually. The failure to utilize food resources to their fullest potential not only results in economic losses but also entails squandering involved in the cultivation, processing, packaging, transportation, and promotion of food. This study examines the aggregate and individual levels of FW production across various nations. The escalation of inhabitants and enhancement of status has resulted in a significant surge in the quantity of FW. According to projections, if there are no significant changes in consumption patterns and food production practices, the global production of FW will exceed twice its current value by 2050. Thus, in order to decrease the generation of FW, it is imperative to modify wasteful behaviors, especially in high-income nations, and align them with contemporary practices.

The prioritization of prevention and, when unfeasible, the adoption of reuse and recycling methodologies must be emphasized, contrary to the prevailing notion among the masses. The stages of food acquisition constitute substantial portions of the overall FW, accounting for as much as 50, 25, and 20, respectively, totaling 95% collectively. However, the aforementioned ratio varies across different nations. For instance, in sub-Saharan Africa, the proportion of FW attributed to consumer losses is merely 5%. However, in Europe, this figure rises to 52%, and in North America and Oceania, it reaches 61%. This disparity is influenced by cultural and socioeconomic factors. Thus, in order to decrease the formation of FW, it is necessary to not only modify consumer behavior by altering consumption patterns. An instance of this can be observed by the producers of FW exceeding 10 tons FW $year^{-1}$. It is crucial to implement additional measures and impose penalties on conventional waste management methods, such as landfilling, in order to decelerate and ultimately diminish the present escalation in FW generation. A household-level illustration can be observed in the implementation of pay-as-you-throw strategies, wherein individuals are billed according to the quantity of waste they produce for disposal.

The implementation of this particular strategy has yielded positive results in various cities across Europe, North America, and Asia. It has demonstrated a significant decrease in the quantity of waste generated, particularly when complemented by efficient recycling practices and tax incentives to alleviate additional expenses for the general public. Currently, waste generation and management are considered to be economically inexpensive [37].

Sewage Sludge Characteristics

The disposal of excess biological sewage sludge (EBSS) has become a significant environmental concern on a global scale. The demand for ecologically sound and enduring disposal methods for EBSS is becoming pressing as a result of escalating volumes and apprehensions pertaining to its potential pollutants content, which dissuade or eliminate conventional approaches.

The effluent from biological wastewater treatment processes, commonly known as EBSS, has been found to harbor various hazardous substances such as pathogenic microorganisms. These substances are often incompletely degraded and remain embedded in the biomass.

Sewage Sludge Biorefinery

As per the preceding definition, the term biorefinery encompasses all technologies employed in the processing of energy biomass supply systems for the purpose of biofuel production and/or resource recovery. Effective integration of processes can facilitate the generation of a multitude of potential end products, including both energy and materials. The processes have the potential to enhance flexibility and optimize operations in anticipation of potential variations in the quantity, composition, and properties of sewage sludge over time. Ideally, it would be advantageous to introduce treatment processes or operational adjustments within WWTPs to enhance the characteristics of surplus biological sludge and improve its recovery options. The heterogeneity of sludge in terms of chemical composition and energy content suggests the potential for targeted processing of these streams to yield distinct end products. As previously stated, GS exhibits a greater EPS concentration in comparison to CAS.

Microbial electrochemical technology (MET) is a field of study that focuses on the use of microorganisms for the production of electrical energy mainly by the conversion process of chemical energy. This technology has the potential to revolutionize the way we generate and store energy, as it offers a sustainable and environmentally friendly alternative to traditional energy sources. Scientists are involved actively in exploring various applications of this technology in areas such as wastewater treatment, biofuel production, and bioremediation.

The electrochemical interplay between microorganisms and electrodes can now be harnessed as microbial electrochemical technologies. The primary application of MET involves the utilization of processes that are closely associated with microbial electrochemistry. The term bio-electrochemical system is frequently used to describe primary microbial electrochemical technologies.

One example of this interaction involves the utilization of electrochemical techniques to monitor or regulate the microbial response environment. It is noteworthy that the observed bonding is of the ionic type within an electrochemical system. In order to facilitate this monitoring or adaptation, a microbial system must be established. This implies that the proximity of microbial and electrochemical systems is imperative, and spatial separation between the two is not feasible.

Retrieval of Energy from Inland Wastewater

2014 is an important year, which has facilitated significant progress in human health, quality of life, and environmental enhancement over the past century. This occasion was marked by the recognition of notable advancements. Every cycle necessitated significant infrastructure investments, with a duration of roughly 50 years, which closely corresponds to the maximum life span of said infrastructure. The onset of a significant cycle is presently underway, which is attributed to the cessation of the current infrastructure's life span and the acknowledgment of the necessity to curtail the worldwide environmental footprint while promoting enduring societal sustainability. The objective of this endeavor is to mitigate the considerable utilization of resources such as energy, chemicals, and transportation in current wastewater treatment processes while simultaneously facilitating the retrieval of the inherent value present in wastewater. This section delineates the underlying factors driving the impetus for resource recovery, the various platforms that facilitate such recovery, and the pragmatic implementation of resource recovery within the context of a small urban community.

The utilization of domestic wastewater as a sole source of fertilizer is limited due to significant losses to both animal husbandry, which is typically not accounted for in urban treatment systems, and the surrounding environment. On a global scale, it has been estimated that roughly 20% of the nitrogen and phosphorous produced through manufacturing processes is present in domestic wastewater. The concentration of urban areas means that a significant portion of this waste has the potential to be recovered.

The situation pertaining to energy is weakened. According to estimates, the chemical energy that is contained within wastewater is approximately 1.3 MJ per individual on a daily basis, or 6.5 MJ per kiloliter. The conversion of this energy from a dispersed chemical form to a practical form necessitates a specific process. Nevertheless, it serves as a concentrated carbon source that could potentially be more efficiently employed as a resource.

In accordance with the principles of resource dissipation, it can be concluded that industrialized societies cannot rely solely on domestic wastewater to meet their elemental and energy requirements. Nevertheless, it constitutes a significant asset and ought to be maximally leveraged. Various process options are available to facilitate energy-efficient wastewater treatment. The present paper provides further discussion on the aforementioned topics. The majority of efforts are directed toward energy recovery. There exists a restricted range of procedures that strive to achieve complete or improved recovery of these resources [38].

Biohydrogen

Hydrogen has been identified as a significant energy carrier due to its advantageous characteristics, such as its lack of CO_2 emissions during combustion and its high energy density per unit mass. The primary methods utilized for hydrogen (H_2) production are characterized by their reliance on nonrenewable resources, high energy consumption, and significant carbon emissions. These methods include natural gas steam reforming, which accounts for approximately half of the world's H_2 production; oil reforming, which accounts for 30%; and coal gasification, which accounts for 18%. The production of hydrogen in a sustainable manner necessitates the utilization of technologies that are both environmentally benign and economically feasible. Biological mechanisms, encompassing autotrophic processes such as biophotolysis, as well as heterotrophic processes are considered to be sustainable means of producing H_2. Nevertheless, it is worth noting that DF represents the sole technology that effectively achieves the simultaneous objectives of waste management and energy generation, given its capacity to utilize organic waste and wastewater as viable sources of feedstock. The process of dark fermentation (DF) is a constituent of the complete AD process. During this process, anaerobic and facultative bacteria decompose substrates that are rich in carbohydrates into less complex organic compounds, primarily volatile fatty acids (VFA), while simultaneously generating hydrogen gas (H_2). The feasibility of DF is primarily constrained by its relatively low hydrogen production, which is at most 4 moles of H_2 per mole of glucose. This implies that it can only retrieve a maximum of 33% of the energy content of the biomass.

The potential of DF to obtain an effluent with a high concentration of unstable components renders it a highly desirable option for integration with other processes for the production of energy and chemicals, as well as for further enhancing the stability of the DF effluent. When compared to underdeveloped alternatives. The biohythane process, also referred to as DF followed by AD, generates a biogas that is rich in H_2. In Europe, it can be observed that the utilization of two-phase systems is limited to approximately 7% of the existing AD infrastructure, which is equivalent to approximately 250 plants. The benefits offered by two-stage systems are insufficient to offset the increased expenses associated with initial investment and operational intricacy.

Biodiesel

Biodiesel is an environmentally sustainable energy alternative that can serve as a partial substitute for conventional fossil fuels, particularly in the transportation industry, which accounts for approximately 23% of global greenhouse gas emissions, while H_2 and methane, following biogas upgrading, have the potential to serve as fuels. The oil sources are commonly referred to as first-generation biofuels and account for more than 95% of biodiesel production. The production of second-generation biofuels, specifically biodiesel derived from nonedible oils, has garnered interest.

Microbes with the ability to accumulate more than 20% of their dry weight as oil, commonly known as oleaginous microorganisms, present a promising alternative to vegetable oil. These microorganisms, which include microalgae, fungi, yeast, and bacteria, exhibit faster growth rates than plants. The cost of feedstock presents a significant obstacle in achieving profitability in the production of microbial biodiesel. As a result, there has been a thorough investigation into the integration of microbial lipid production with waste and wastewater treatment as a potential solution.

Nevertheless, this methodology is subject to certain limitations associated with the growth rates of microalgae, yield of lipids, and the process of lipid extraction. The extraction of lipids poses a common challenge for all microorganisms that are rich in oil. However, the growth rates and lipid yield of phototrophic microalgae are hindered by various limitations, including the penetration of light and mutual shading of cells. Heterotrophic microalgae have the ability to surmount the majority of constraints associated with the cultivation of phototrophic microalgae while exhibiting accelerated growth rates and increased lipid productivity. The co-production of bio-hydrogen and biodiesel through the utilization of heterotrophic microalgae presents a promising strategy for both waste management and energy generation. Nevertheless, the primary obstacles involve multiple factors to achieve a targeted VFA concentration and distribution while also preventing bacterial contamination. It is noteworthy to mention that bacteria, fungi, and yeast are currently being recognized as potential sources of microbial oil. Despite being in its initial phases, this methodology exhibits promising potential as a viable substitute for microbial lipids derived from microalgae, owing to its remarkable growth rates, productivities, and yields, which are achieved through the utilization of diverse carbon sources [39, 40].

References

1. Van Drecht, G., Bouwman, A.F., Harrison, J., Knoop, J.M., Global nitrogen and phosphate in urban wastewater for the period 1970 to 2050. *Global Biogeochem. Cycles.*, 23, n/a, 2009.
2. Peccia, J., Westerhoff, P., We Should Expect More out of Our Sewage Sludge. *Environ. Sci. Technol.*, 49, 8271, 2015.
3. Lachhab, K., Tyagi, R.D., Valéro, J.R., Production of Bacillus thuringiensis biopesticides using wastewater sludge as a raw material: Effect of inoculum and sludge solids concentration. *Process Biochem.*, 37, 197, 2001.
4. Vidyarthi, A.S., Tyagi, R.D., Valero, J.R., Surampalli, R.Y., Studies on the production of B. thuringiensis based biopesticides using wastewater sludge as a raw material. *Water Res.*, 36, 4850, 2002.
5. Egan, M., Biosolids management strategies: An evaluation of energy production as an alternative to land application. *Environ. Sci. Pollut. Res.*, 20, 4299, 2013.
6. Lundin, M., Olofsson, M., Pettersson, G.J., Zetterlund, H., Environmental and economic assessment of sewage sludge handling options. *Resour. Conserv. Recycl.*, 41, 255, 2004.
7. Campbell, H.W., Sludge management - Future issues and trends. *Water Sci. Technol.*, 41, 1, 2000.
8. Rulkens, W., Sewage sludge as a biomass resource for the production of energy: Overview and assessment of the various options. *Energy Fuels*, 22, 9, 2008.
9. Wang, H., Brown, S.L., Magesan, G.N., Slade, A.H., Quintern, M., Clinton, P.W., Payn, T.W., Technological options for the management of biosolids. *Environ. Sci. Pollut. Res.*, 15, 308, 2008.
10. Hadi, P., Xu, M., Ning, C., Sze Ki Lin, C., McKay, G., A critical review on preparation, characterization and utilization of sludge-derived activated carbons for wastewater treatment. *Chem. Eng. J.*, 260, 895, 2015.
11. Ukwatta, A., Mohajerani, A., Setunge, S., Eshtiaghi, N., Possible use of biosolids in fired-clay bricks. *Constr. Build. Mater.*, 91, 86, 2015.

12. Raunkjær, K., Hvitved-Jacobsen, T., Nielsen, P.H., Measurement of pools of protein, carbohydrate and lipid in domestic wastewater. *Water Res.*, 28, 251, 1994.
13. Adebayo, O.T., Fagbenro, O.A., Jegede, T., Evaluation of Cassia fistula meal as a replacement for soybean meal in practical diets of Oreochromis niloticus fingerlings. *Aquac. Nutr.*, 10, 99, 2004.
14. Shier, W.T., Purwono, S.K., Extraction of single-cell protein from activated sewage sludge: thermal solubilization of protein. *Bioresour. Technol.*, 49, 157, 1994.
15. Vriens, L., Nihoul, R., Verachtert, H., Activated sludges as animal feed: A review. *Biol. Wastes.*, 27, 161, 1989.
16. Chishti, S.S., Nazrul Hasnain, S., Altaf Khan, M., Studies on the recovery of sludge protein. *Water Res.*, 26, 241, 1992.
17. Hwang, J., Zhang, L., Seo, S., Lee, Y.W., Jahng, D., Protein recovery from excess sludge for its use as animal feed. *Bioresour. Technol.*, 99, 8949, 2008.
18. Zhang, W., Su, R., Guo, G., Li, D., Experimental research on protein extraction from excess activated sludge by papain hydrolysis. *Adv. Mater. Res.*, 518– 523, 3367, 2012.
19. Ras, M., Girbal-Neuhauser, E., Paul, E., Spérandio, M., Lefebvre, D., Protein extraction from activated sludge: An analytical approach. *Water Res.*, 42, 1867, 2008.
20. Kroiss, H., What is the potential for utilizing the resources in sludge? *Water Sci. Technol.*, 49, 1, 2004.
21. Gorbovskiy, K., Kazakov, A., Norov, A., Malyavin, A., Mikhaylichenko, A., Properties of complex ammonium nitrate-based fertilizers depending on the degree of phosphoric acid ammoniation. *Int. J. Ind. Chem.*, 8, 315, 2017.
22. Liu, Y., Kong, S., Li, Y., Zeng, H., Novel technology for sewage sludge utilization: Preparation of amino acids chelated trace elements (AACTE) fertilizer. *J. Hazard. Mater.*, 171, 1159, 2009.
23. Zhang, D.Q., Cai, Q.R., He, X.M., Gao, L.X., Zhou, G.D., Inhibition effect of some amino acids on copper corrosion in HCl solution. *Mater. Chem. Phys.*, 112, 353, 2008.
24. Gece, G., Bilgiç, S., A theoretical study on the inhibition efficiencies of some amino acids as corrosion inhibitors of nickel. *Corros. Sci.*, 52, 3435, 2010.
25. Khaled, K.F., Corrosion control of copper in nitric acid solutions using some amino acids - A combined experimental and theoretical study. *Corros. Sci.*, 52, 3225, 2010.
26. Eddy, N.O., Experimental and theoretical studies on some amino acids and their potential activity as inhibitors for the corrosion of mild steel, part 2. *J. Adv. Res.*, 2, 35, 2011.
27. Su, W., Tang, B., Fu, F., Huang, S., Zhao, S., Bin, L., Ding, J., Chen, C., A new insight into resource recovery of excess sewage sludge: Feasibility of extracting mixed amino acids as an environment-friendly corrosion inhibitor for industrial pickling. *J. Hazard. Mater.*, 279, 38, 2014.

28. Pervaiz, M., Sain, M., Protein extraction from secondary sludge of paper mill wastewater and its utilization as a wood adhesive. *BioResources*, 6, 961, 2011.
29. Rulkens, W.H., Sustainable sludge management - What are the challenges for the future? *Water Sci. Technol.*, 49, 11, 2004.
30. Yuan, H., Chen, Y., Zhang, H., Jiang, S., Zhou, Q., Gu, G., Improved bioproduction of short-chain fatty acids (SCFAs) from excess sludge under alkaline conditions. *Environ. Sci. Technol.*, 40, 2025, 2006.
31. Chen, Y., Jiang, S., Yuan, H., Zhou, Q., Gu, G., Hydrolysis and acidification of waste activated sludge at different pHs. *Water Res.*, 41, 683, 2007.
32. Yu, G.H., He, P.J., Shao, L.M., He, P.P., Toward understanding the mechanism of improving the production of volatile fatty acids from activated sludge at pH 10.0. *Water Res.*, 42, 4637, 2008.
33. Liu, Y., Li, X., Kang, X., Yuan, Y., Du, M., Short chain fatty acids accumulation and microbial community succession during ultrasonic-pretreated sludge anaerobic fermentation process: Effect of alkaline adjustment. *Int. Biodeterior. Biodegrad.*, 94, 128, 2014.
34. Rughoonundun, H., Granda, C., Mohee, R., Holtzapple, M.T., Effect of thermochemical pretreatment on sewage sludge and its impact on carboxylic acids production. *Waste Manage.*, 30, 1614, 2010.
35. Li, Y., Wang, J., Zhang, A., Wang, L., Enhancing the quantity and quality of short-chain fatty acids production from waste activated sludge using CaO2 as an additive. *Water Res.*, 83, 84, 2015.
36. Nagtode, V.S., Cardoza, C., Yasin, H.K.A., Mali, S.N., Tambe, S.M., Roy, P., Singh, K., Goel, A., Amin, P.D., Thorat, B.R., Cruz, J.N., Pratap, A.P., Green Surfactants (Biosurfactants): A Petroleum-Free Substitute for Sustainability-Comparison, Applications, Market, and Future Prospects. *ACS Omega*, 8, 11674, 2023.
37. Jiang, S., Chen, Y., Zhou, Q., Gu, G., Biological short-chain fatty acids (SCFAs) production from waste-activated sludge affected by surfactant. *Water Res.*, 41, 3112, 2007.
38. Shanableh, A., Jones, S., Production and transformation of volatile fatty acids from sludge subjected to hydrothermal treatment. *Water Sci. Technol.*, 44, 129, 2001.
39. Liao, P.H., Lo, K.V., Chan, W.I., Wong, W.T., Sludge reduction and volatile fatty acid recovery using microwave advanced oxidation process. *J. Environ. Sci. Health. - Part A Toxic/Hazard. Subst. Environ. Eng.*, 42, 633, 2007.
40. Tang, B., Yu, L., Huang, S., Luo, J., Zhuo, Y., Energy efficiency of pre-treating excess sewage sludge with microwave irradiation. *Bioresour. Technol.*, 101, 5092, 2010.

13

Wastewater Treatment Processes and Resource Recovery, Effectiveness, and Challenges

Urvashi Tomar[1] **and Pallavi Jain**[2*]

[1]*Department of Botany, Hans Raj College, University of Delhi, Delhi, India*
[2]*Department of Chemistry, SRM Institute of Science and Technology, Delhi NCR Campus, Modinagar, Uttar Pradesh, India*

Abstract
During the past few decades, environmental problems, specifically water pollution, have become a major concern for industries and public authorities. Water scarcity is one of the great issues and many people lack access to freshwater for household purposes. Wastewater is produced by domestic and industrial activities and mostly contributes to pollution by adding undesirable solid and liquid contaminates into the receiving water bodies. Wastewater treatment removes the contaminants (both solid and liquid) from wastewater; this water can be converted into a reusable form. This treated water can be used for domestic, irrigation, and industrial purposes. Wastewater treatment comprises three stages of treatment, namely, primary, secondary, and tertiary treatment. Primary and secondary treatments remove suspended solids and the majority of the BOD in the wastewater. Tertiary treatment is an advanced treatment that is employed to remove specific contaminants that cannot be removed after primary and secondary treatment. The wastewater treatment processes are a combination of chemical, physical, and biological methods with the purpose of removing all the contaminants from the effluent water. This chapter describes the overview of the treatments with the purpose of making wastewater reusable. It also provides insights into technology that can be used to harvest resources (water, energy, and materials) from this wastewater. Besides removing the contaminants from the wastewater, the new paradigm is to focus on recovery from the wastewater. Resource recovery systems are one such approach that contributes to global

**Corresponding author*: palli24@gmail.com

Inamuddin, Tariq Altalhi, Mohammad Luqman and Joseph Kapuku (eds.) Sewage and Biomass from Wastewater to Energy, (347–364) © 2024 Scrivener Publishing LLC

sustainability goals. This chapter discusses all the relevant factors that should be included while developing strategies for resource recovery from wastewater. Global environmental and ecological concerns cannot be solved without addressing public and social issues.

Keywords: Wastewater, resource recovery system, sustainability, environment, pollution

13.1 Introduction

The Earth's surface is covered with 71% of water, which makes up approximately 65% of human bodies. Although only 3% of the water is appropriate for consumption and other household work by humans, the rest 97% is salty and cannot be used for human consumption [1]. Water is required for human consumption, agriculture, industrial, and recreation purposes [2]. Water is a life-sustaining asset to us, but it is eventually being contaminated by the continuous use of chemicals that are toxic for industrial and agricultural progress; moreover, continuous growth in population leads to the depletion of suitable water in water bodies [3]. Water loses its value after getting polluted both economically as well as aesthetically and can be a threat to the survival of all the living organisms inside it [4]. Water is one such environmental factor that has been most affected by the continuous degradation of the environment [5]. Apparently, this creates a matter of concern due to two basic reasons. First, the quality of water that is available to us is an essential requirement to maintain the social and sustainable balance of the country [6]. Second, the unavailability of quality water supplies puts the risk of health in the population [7]. Hence, it has become a vital condition to religiously care for various water sources available [8]. Pollution of the rivers and streams occurring due to chemical contaminants is one major environmental problem [4]. Humans use water to carry out different activities domestic and economic, the same is being contaminated and become unsuitable for consumption by human beings [9]. This used water is called wastewater. Thus, the treatment of wastewater should be efficient before returning it to the environment in order to avoid its contribution to water pollution in the water bodies [10]. The untreated wastewater is generally discharged into rivers, oceans, or lakes, causing a tremendous amount of destruction due to the presence of pollutants [11]. As per data given by the United Nations [12], this wastewater is one of serious concern in both developing as well as developed countries, due to this approximately 297 thousand children die every year suffering from diseases which are caused by low-quality water, as about 80% of wastewater is still discharged in water bodies without treatment [13]. We can define wastewater as a combination of water-carried waste that is added as a result of usage from mining, industrial waste, sewage, pesticides, and agricultural fertilizers, combined with water from different water bodies [14]. This wastewater generally consists of a large amount of pathogen/disease-causing agents, oxygen-demanding wastes, organic and inorganic chemicals, nutrients that increase plant growth, minerals, sediments, and toxic chemicals. The key contaminants of wastewater effluents are heavy metals, dyes, halogenated hydrocarbons, dyes, salts, organic compounds, soluble bases, etc. [15, 16].

Wastewater can be categorized into four categories [4]:

- Domestic wastewater: it is the wastewater that is discharged from households and institutions.
- Industrial wastewater: it is the wastewater that contains effluents from industries and manufacturing units.
- Infiltration water that enters the sewer system through direct or indirect ways (like cracks, porous walls, or leaked joints).
- Stormwater runoff which is the result of flooding occurs due to rainfall.

However, wastewater recycling from all sources has received much attention in the view of sustainable development, and improvement in wastewater is one of the main concerns [17–20]. Hence, the treatment of wastewater has now become one of the priorities [14]. In addition to this, efforts have been made to design programs that could adequately manage and treat wastewater as untreated wastewater is a primary trigger of water pollution in water bodies [21]. In the past decades, several technologies have been developed including physical, chemical, and biological techniques. The techniques established for wastewater treatment include precipitation, evaporation, membrane filtration, solvent extraction, carbon adsorption, electrochemistry, flotation, oxidation, phytoremediation, and biodegradation [22–24]. Wastewater treatment techniques are however high-cost demanding as it is essential to remove all the contaminants present in the wastewater efficiently so that we can get water that is clean and reusable, although the techniques are designed for the purification of the effluents in wastewater without considering their impact on the environment [25].

The main aim of the treatment of wastewater was generally to decrease the number of suspended solids, dissolved inorganic compounds, oxygen-demanding substances, and disease-causing pathogens/bacteria. However, nowadays, improving the means of waste disposal methods from the treatment processes is being focused on more. There are three stages of municipal/domestic wastewater treatment: first, primary treatment which includes removal of grit, screening, and sedimentation; second, secondary treatment in which dissolved organic matter is oxidized using biologically active sludge; third, tertiary treatment, consisting of advanced biological, physical, and chemical methods such as carbon adsorption and granular filtration [4]. Industrial discharges have characteristics of various organic and inorganic substances, BOD, and suspended solids. To control industrial wastewater, the focus should be on the place of generation of wastewater in the plant itself; the treatment of wastewater must be done before its discharge into the treatment sources which can later be reused or discharged into the receiving water directly. Presently, there is no single technique available catering to the desired treatment of water because of the complex nature of effluents/contaminants in industrial wastewater [14].

An increase in awareness in society regarding wastewater as a valuable energy resource rather than just waste has changed our way of managing wastewater and creating the advancement in technologies that are energy-efficient as well as resource recovering. Furthermore, wastewater treatment can be considered a sustainable process if suitable technologies are adopted [26]. This chapter discusses the general processes involved in wastewater treatment and the limitations underlying the processes. It also deals with possible methods of resource recovery from wastewater and various tools and guiding principles to be considered for the development of resource recovery systems that can contribute to global sustainability.

13.2 Wastewater Treatment: Processes

Wastewater should be decontaminated using the best possible approach for purification which should be chosen such that the objectives established by legislation can also be achieved. Wastewater treatment includes five successive steps (Figure 13.1): (i) pre-treatment/preliminary treatment consists of physical and mechanical methods, (ii) primary treatment consisting of various physicochemical and chemical methods, (iii) secondary treatment consists of chemical and biological methods, (iv) tertiary treatment consisting of physical and chemical methods, and (v) unpolluted water [14].

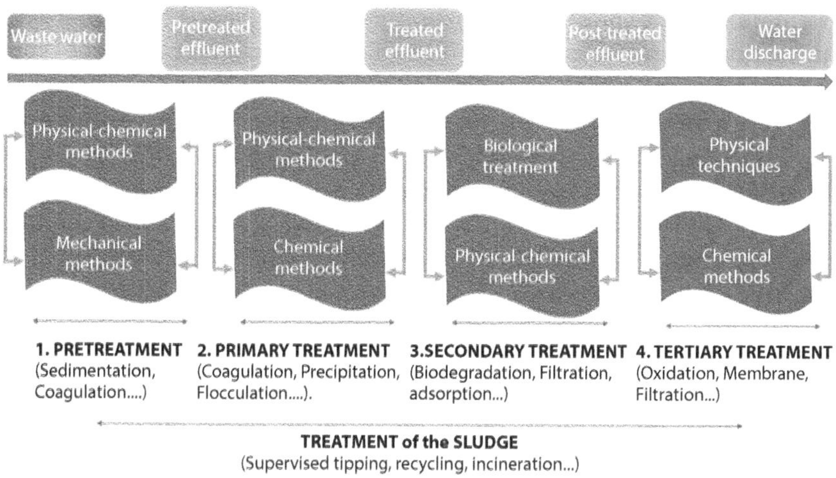

Figure 13.1 Wastewater treatment processes [14].

13.2.1 Primary Treatment

Primary treatment is the initial step that is developed in order to remove floating and suspended solids from raw wastewater. Solid objects and sedimentations are removed by gravity. This physical separation is done via a mechanical process but sometimes chemicals can also be used to accelerate the sedimentation process. Primary treatment reduces the total suspended solids approximately by 50 to 60% and BOD by 20 to 30% [27]. Some heavy metals, organic phosphorus, and organic nitrogen which are associated with solids are also removed in primary treatment. However, it will not affect colloidal and dissolved constituents present in the wastewater [4].

13.2.2 Secondary Treatment

The dissolved organic matter which was not removed by primary treatment is removed by secondary treatment. Organic matter is consumed by microorganisms on their own by converting it into water, carbon dioxide, and energy. Furthermore, more suspended solids can be removed by using additional settling methods, and later on, the biological process is followed. About 85% of the suspended solids and BOD are removed by this stage of treatment. The biological sludge is also removed by this process as carbonaceous pollutants are settled down in the settling tank. This sludge can later be used with other wastes for biogas production in a biogas plant. The leftover water is later processed to remove carbon and nitrogen.

Furthermore, for chlorine treatment, this water is passed through a sedimentation basin. Water at this stage will further contain different types of chemicals, microbial, and metallic contaminations. In order to make this water reusable, it is further filtered and then passed into a disinfectant tank. The disinfectant used is generally sodium hypochlorite. Now, this water treated by secondary treatment can be considered safe for irrigation purposes. The solid waste that is obtained after primary and secondary treatment can be used for many purposes like landfilling, buildings, and fertilizers but only after treatment in a gravity thickening tank, later put inside a dewatering tank, and lastly taken into a lime stabilization tank [27].

For wastewater treatment in addition to this ASP, several methods can be used in reactors like trickling filters, some anaerobic treatments like UASB, methanogenic reactors, MFC, and artificial wetlands. These reactors were used for the treatment of wastewater for a long time. UASB reactor performance was examined and it was observed that 85% of the total COD was removed in an experiment conducted for 200 days as a trial in which the maintained COD was at a 1.21 kg COD/m^3/day loading rate [28]. A combined model of a UASB-AS system was used for domestic wastewater of low strength with a BOD5 that amounted to 340 mg/L [29]. The result of this experiment showed a reduction in sludge construction by 60% and aeration energy consumption by 40%. A reduction in TSS, COD, and total sulfate by 75, 81, and 76%, respectively, was observed using a seeded UASB reactor that used cow manure for wastewater treatment [30].

13.2.3 Tertiary Treatment

It is also known as an advanced treatment. It is designed to achieve high-quality effluent which cannot be achieved by the secondary treatment processes [4]. The tertiary treatment is generally used to remove a specific contaminant that cannot be removed after primary as well as secondary treatment. Hence, approximately 99% of the impurities can be removed by this tertiary treatment from the wastewater. Water is individually treated or may be treated in combination with advanced methods which include UV light treatment, ultrasonication (US), and exposure to ozone. This helps in the removal of heavy metal and bacterial contaminants that might remain in the treated water. The water after secondary treatment undergoes US, which is later exposed to UV light and then finally entered into the ozone chamber for the effective removal of the contaminants. Cell membranes are physically disrupted and attacked by free radicals such that the inviable cells are rendered during the US method [31, 32]. The above treatment combining US+UV+O_3 results in the development of free radicals attached to the membrane of the contaminant (Biological). When this cell membrane gets ruptured, chemical oxidants can enter the cell to attack the internal structures. Therefore, the combined methods are used to promote the aggregation of microorganisms ultimately increasing chemical disinfectant efficiency [31–36]. This combined treatment process is considered highly efficient in wastewater from textile industries [37]. This process of combining UV and US is also employed as a pre-treatment step [38, 39] and can also be employed by combining it with other treatments like ozone [40] and TiO_2 photocatalysis [41] to optimize the process of disinfection. The wastewater treatment model is depicted in Figure 13.2, indicating the

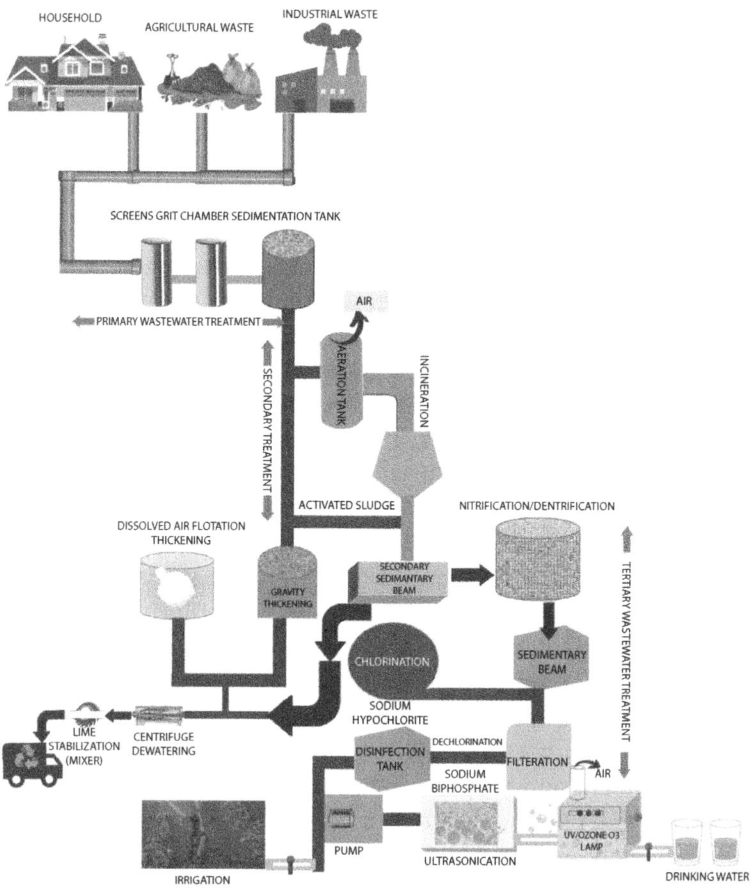

Figure 13.2 Wastewater treatment indicating various methods involved in the process and water after treatment can be reused for different purposes [27].

steps involved in wastewater treatment after ensuring proper purification of wastewater and quality testing, the treated water can be available for reuse (drinking, irrigation, and domestic use).

New technologies for wastewater treatment and fulfilling the requirement of clean water are also being developed progressively [42], although the discharged water treatment is challenging using available methods [43]. The method to be selected for the treatment depends on various factors including the composition of the sewage, process cost, dye concentration, or the other impurities that might be found in wastewater [44]. Each treatment has a few distinctive features which might be beneficial in some way but might restrict in others [45]. Techniques of wastewater treatment that require high running costs and high installation, low output, and increased processing time still produce toxic byproducts that are less considered for industrial applications [46]. Therefore, an alternative treatment should be developed that can degrade or remove the contaminants present in wastewater without any side effects on the environment [47].

13.3 Limitations of Wastewater Treatment Techniques

Even though the methods of wastewater treatment are beneficial as they are responsible for treating the wastewater and converting it into reusable water, each method has certain limitations. Besides being rapid, simple, efficient, and having multiple approaches when oxidants are considered, chemical methods are usually large-scale laboratory-based techniques that might not be economical for small industries such that they can achieve energy requirements. Similarly, many physical methods like membrane filtration are simple and quick techniques at high concentrations; these are suitable for mostly all types of pollutants like mineral ions, dyes, etc., but in contrast, high maintenance, energy, and operation cost are required for them hence cannot be considered for small industries. Physiochemical processes like coagulation and flocculation are simple but become very challenging to handle the generation of sludge volume; this will ultimately increase the cost of this process. Biological treatment is also considered simple and economical but it is a process having less biodegradability, requiring an optimum environment and proper management of microorganisms [45]. In other words, water and wastewater treatments are traditionally driven by considering safety, management, function, and cost-benefit analysis. These water and wastewater treatment systems are mainly focused on water stress [48, 49]. Water stress is considered a hazard with other global problems like land use changes, climate change, demographics, urbanization, and increasing population [49]. Water stress also contributes to an imbalance in nutrients and many other environmental detriments like loss of biodiversity in receiving water bodies, discharge of different pharmaceutical/industrial contaminants [50], and eutrophication [51]. If all the other relevant factors are included in the treatment process, then only the emphasis on cost and benefits can be considered, which unfortunately is not the case [52].

13.4 Resource Recovery Systems (RRS)

Resource recovery systems that can recover water, materials, and energy from wastewater can be developed by using technologies already available or developing new approaches [53]. In today's era of sustainable development, we must acknowledge water as a renewable resource that can be used to recover water, energy [54], fertilizers [55], and bioplastics [56]. Moreover, instead of focusing on what we should remove from the wastewater, now we are more focused on what we can recover from it, the processes that are applied for the same can be described as RRS. It is a conceptual change that allows us to achieve a net positive impact of wastewater on communities [52].

13.4.1 Water Recovery

Water recycling can be considered as one technique that can provide water resources in areas facing water stresses. Examples that have been successful in providing such indirect potable water for reuse for over 30 years include Orange County Water District (California) and the Upper Occoquan Sewage Authority (Virginia) [52]. Singapore is among the best examples of system-level reuse of water using the "four national taps" strategy. The diverse water network of Singapore includes (i) water imported from Malaysia, (ii) water supply from locals, (iii) desalination, and (iv) NEWater (indirect potable and direct non-potable reuse of reclaimed water). This NEWater is now meeting 30% of the water demand of the country after the Changi plant became operational in 2010 [57].

13.4.2 Energy and Material Recovery

Energy can also be recovered from wastewater and basically in the form of methane gas which contains biogas produced after the treatment of wastewater anaerobically and solid contaminant digestion. The Anaerobic reactors are used by the world throughout and are very helpful in the production of CH_4 which later can be (i) used for heat generation by combustion on-site, (ii) used by natural gas dealers locally, or (iii) used as fuel for automobiles [22]. MFCs are another example of wastewater energy recovery [58] which can also be used for buildings' heating and cooling by extraction of latent heat [59]. In the United States and the United Kingdom, biosolids are very well used as fertilizer is increasing nowadays [52]. According to recent developments, the harvesting of $MgNH_4PO_4{\cdot}6H_2O$ from solids treatment processes is also reported [60]. In the United States and the United Kingdom, biosolids are increasingly being used as fertilizer nowadays [52].

Despite all these mentioned advances, the negative impact of wastewater systems on local water cycles is much greater than on the consumption of materials and energy. Recently, much focus has been given to energy sustainability rather than water sustainability. There is a need for the reorientation of this idea such that the impact of different technologies used for wastewater treatment should be addressed first on hydrological cycles and later assessed to find any negative impact of the technologies on global energy or climate sustainability [52]. With this, we can not only reduce the negative impact on other aspects of sustainability but can also evolve our planning and designing of these processes so that the regional and global aim of water resource availability and quality will be enhanced.

13.5 Challenges to the Implementation of RRS Successfully

Due to the unavailability of an overall methodology that could cater to both social factors as well as include an agreement on sustainability goals and targets hence, we do not use the available technologies so frequently to recover resources from wastewater [61]. A report by the Urban Water Management Plan in 2005 [62] illustrated the importance of social factors. San Diego is a coastal city with semi-arid climatic conditions and the population of the city is approximately >1.3 million. The water sources in the city are the water imported from the Colorado River and Sacramento-San Joaquin Delta, distancing 390 and 715 km, respectively. The water imported contained discharge from about >200 wastewater treatment plants which ultimately contributed up to 90% water supply of San Diego. Aiming to enhance the water supply, two reclamation plants were also constructed resulting in the enhancement of the water supply by up to >25% from local sources. The U.S. EPA took a major step to enhance the reuse by mandating the capacity of one plant by 75%. However, the public rejected the water produced by this plant and about 73% of the water was thereby wasted in the sewer. This has made it quite clear that besides having proper technology to recover wastewater, factors that involve social sustainability in the planning and design process must also be considered [52]. Restrictions on the use of reclaimed water exist in several locations, including San Diego, Toowoomba (Australia) [63], Los Angeles [64], and San Ramon Dublin [64], due to public pressure. This reusable water challenges cultural notions, thus resulting in many risks [63]. Therefore, it has become necessary that decision-makers and engineers consider sociopolitical views [65] which include relations with public knowledge concerns and relevant experiences [66]. There are a few additional barriers to the path of wastewater treatment systems including the absence of a conducive environment (legislative, political, modes of public discourse, and financing) and the lack of political will [65, 67, 68]. To date, the factors outside the engineering scope have not yet been addressed. However, these factors can be traced by narrowing the defined approaches that are used to educate the professionals of water industries [52]. Therefore, these shortcomings and integrated resource management strategies should be focused on to make progress in emerging sustainable water management techniques [69].

13.6 Conclusion and Future Prospects

The crisis of water is one of the major growing concerns of the world in this century. Due to the increasing population, water resources are under threat. With the increasing population, the generation of wastewater which may be domestic, industrial, or agricultural is also increasing especially in the developed countries. However, these countries have the potential to serve themselves as an alternative to freshwater sources so that water can be reused after appropriate treatment and proper distribution. A large amount of wastewater is generated globally. This fact gives a clear vision of why it is now necessary to reuse wastewater in order to solve the problem of the water crisis. Hence, it is quite evident that treated industrial and domestic wastewater can be used to fulfill various water demands. Moreover, the development of low-cost, effective, and new methods of wastewater treatment is an active field of research. The focus is particularly on solving this problem of water such that the methods employed should not possess any negative impact on the environment, although wastewater is currently used for irrigation in some developing countries without treatment, causing various health problems at both the supplier as well as receiver end. The top priority of pollutant-free water in the receiving sources has put a lot of pressure on the decontamination of wastewater from the industries even though it is a quite challenging task. Moreover, it is quite evident that choosing a particular technology for the removal of all the contaminants/pollutants from the wastewater is very difficult. There are multiple techniques that can be employed for wastewater treatment and resource recovery. Eventually, when we proceed forward to make water a sustainable resource, the management strategies should focus more on the challenges discussed. Recourse recovery systems should be included in our water systems. RRS must be such that firstly, it can fulfill the demand for water supply; secondly, it should ensure resource recovery efficiently; and finally, it should provide sufficient acknowledgment of the importance of environmental as well as social aspects of using water as a sustainable resource. Hence, for the implementation of sustainable solutions to wastewater treatment and resource recovery besides assessing the techniques, equal importance is to be given to the planning and social dimension of the technique. Significant importance should be given to promoting positive social learning, enhancing and using local knowledge, resolving conflicts, and gaining higher public interest in accepting water management decisions. These approaches should also be included in RRS so that we can achieve global sustainability. Eventually, wastewater, if used as a renewable resource, can provide great openings for water management systems which can provide water, energy, and material recovery thus contributing to global sustainable development. However, it will require cooperation between various water management sectors so that a proper strategy can be set for the water industry to achieve the sustainability target. First, a proper research and implementation protocol must be identified that caters to the fact that resource recovery from wastewater is a very important strategy toward sustainability. Second, planning on different sociotechnical aspects should also be considered. Last, industry-based and locality-based initiatives should also be considered so that we can identify the best possible strategy which can be best as a resource recovery system. Furthermore, the discoveries in water industries are now contributing much to global sustainability, public health, and environmental protection; hence, the strategies should be implemented after taking care of sustainable planning and design.

References

1. Ahmad, M., Yousaf, M., Nasir, A., Bhatti, I. A., Mahmood, A., Fang, X., Jian, X., Zadeh, K., Mahmood, N., Porous eleocharis@ MnPE layered hybrid for synergistic adsorption and catalytic biodegradation of toxic azo dyes from industrial wastewater. *Environ. Sci. Technol.*, 53, 2161, 2019.

2. Hogeboom, R.J., Knook, L., Hoekstra, A.Y., The blue water footprint of the world's artificial reservoirs for hydroelectricity, irrigation, residential and industrial water supply, flood protection, fishing and recreation. *Adv. Water Resour.,* 113, 285-294, 2018.
3. Foster, J.E., Plasma-based water purification: Challenges and prospects for the future. *Phys. Plasmas*, 24, 055501, 2017.
4. Sonune, A., Ghate, R., Developments in wastewater treatment methods. *Desalination*, 167, 55, 2004.
5. Loucks, D.P., Managing water as a critical component of a changing world. *Water Resour. Manage.*, 31, 2905, 2017.
6. Zhang, S., Xiang, M., Xu, Z., Wang, L., Zhang, C., Evaluation of water cycle health status based on a cloud model. *J. Clean. Prod.*, 245, 118850, 2020.
7. Zuccarello, P., Manganelli, M., Conti, G.O., Copat, C., Grasso, A., Cristaldi, A., De Angelis, G., Testai, E., Stefanelli, M., Vichi, S., Fiore, M., Ferrante, M., Water quality and human health: A simple monitoring model of toxic cyanobacteria growth in highly variable Mediterranean hot dry environments. *Environ. Res.*, 192, 110291, 2021.
8. Chen, H., Chen, A., Xu, L., Xie, H., Qiao, H., Lin, Q., Cai, K., A deep learning CNN architecture applied in smart near-infrared analysis of water pollution for agricultural irrigation resources. *Agric. Water Manage.*, 240, 106303 2020.
9. Nong, X., Shao, D., Zhong, H., Liang, J., Evaluation of water quality in the South-to-North Water Diversion Project of China using the water quality index (WQI) method. *Water Res.*, 178, 115781, 2020.
10. Finlay, N.C., Peacock, C.L., Edwards, K.A., Johnson, K.L., Characteristics and mechanisms of Pb (II) sorption onto Fe-rich waste water treatment residue (WTR): A potential sustainable Pb immobilisation technology for soils. *J. Hazard. Mater.*, 402, 123433, 2021.
11. Wu, Z., Ye, Q., Water pollution loads and shifting within China's inter-province trade. *J. Clean. Prod.*, 259, 120879, 2020.
12. The United Nations World Water Development Report, Leaving No One behind. Paris, France, https://www.un.org/, 2019.
13. Khan, S.A.R., Ponce, P., Yu, Z., Golpîra, H., Mathew, M., Environmental technology and wastewater treatment: Strategies to achieve environmental sustainability. *Chemosphere*, 286, 131532, 2022.
14. Crini, G., Lichtfouse, E., Advantages and disadvantages of techniques used for wastewater treatment. *Environ. Chem. Lett.*, 17, 145, 2019.
15. Chauhan, P.S., Kant, R., Rai, A., Gupta, A., Bhattacharya, S., Facile synthesis of ZnO/GO nanoflowers over Si substrate for improved photocatalytic decolorization of MB dye and industrial wastewater under solar irradiation. *Mater. Sci. Semicond. Process.*, 89, 6, 2019.
16. Nagtode, V.S., Cardoza, C., Yasin, H.K.A., Mali, S.N., Tambe, S.M., Roy, P., Singh, K., Goel, A., Amin, P.D., Thorat, B.R., Cruz, J.N., Pratap, A.P., Green Surfactants (Biosurfactants): A Petroleum-Free Substitute for

Sustainability- Comparison, Applications, Market, and Future Prospects. *ACS Omega*, 8, 11674, 2023.

17. Kentish, S.E., Stevens, G.W., Innovations in separations technology for the recycling and re-use of liquid waste streams. *Chem. Eng. J.*, 84, 149, 2001.
18. Sharma, S.K., Sanghi, R., *Advances in water treatment and pollution prevention*, Springer Science & Business Media, 2012.
19. Khalaf, M.N., *Green polymers and environmental pollution control*, CRC Press, 2016.
20. Rathoure, A.K., Dhatwalia, V.K., *Toxicity and waste management using bioremediation*, IGI Global, Hershey, 2016.
21. Jiang, J., Okvitasari, R.A., Huang, F., Tsai, C., Characteristics, pollution patterns and risks of per fluoroalkyl substances in drinking water sources of Taiwan. *Chemosphere,* 264, 128579, 2021.
22. Liu, D. H., and Lipták, B. G., *Hazardous waste and solid*, CRC Press, 1999.
23. Harvey, P. J., Campanella, B. F., Castro, P. M., Harms, H., Lichtfouse, E., Schäffner, A. R., Reichhart, D., Phytoremediation of polyaromatic hydrocarbons, anilines and phenols. *Environ. Sci. Pollut. Res.*, 9, 29-47, 2002.
24. Forgacs, E., Cserháti, T., Oros, G., Removal of synthetic dyes from wastewaters: A review. *Environ. Int.*, 30, 953, 2004.
25. Khan, H., Khan, I., Tien, T., The heterogeneity of renewable energy consumption, carbon emission and financial development in the globe: A panel quantile regression approach. *Energy Rep.*, 6, 859, 2020.
26. Li, J., Lin, J., Xu, X., Zhang, X., Xue, Y., Mi, J., Mo, Z., Fan, Y., Hu, L., Yang, X., Porous boron nitride with a high surface area: Hydrogen storage and water treatment. *Nanotechnology*, 24, 155603, 2013.
27. Kesari, K. K., Soni, R., Jamal, Q.M.S., Tripathi, P., Lal, J.A., Jha, N.K., Siddiqui, M,H., Kumar, P., Tripathi, J., Ruokolainen, J., Wastewater treatment and reuse: A review of its applications and health implications. *Water Air Soil Pollut.*, 232, 1, 2021.
28. Behling, E., Diaz, A., Colina, G., Herrera, M., Gutierrez, E., Chacin, E., Fernandez, N., Forster, C.F., Domestic wastewater treatment using a UASB reactor. *Bioresour. Technol.*, 61, 239, 1996.
29. Sperling, M., Chernicharo, C.A.L., *Biological Wastewater treatment in warm climate regions*, pp. 405-407, IWA Publishing, 2005.
30. Rizvi, H., Ahmad, N., Abbas, F., Bukhari, I. H., Yasar, A., Ali, S., Yasmeen, T., Riaz, M., Start-up of UASB reactors treating municipal wastewater and effect of temperature/sludge age and hydraulic retention time (HRT) on its performance. *Arab. J. Chem.*, 8, 780, 2015.
31. Phull, S. S., Newman, A.P., Lorimer, J.P., Pollet, T.J., Mason, T.J., The development and evaluation of ultrasound in the biocidal treatment of water. *Ultrason. Sonochem.*, 4, 157, 1997.
32. Scherba, G., Weigel, R.M., Obrien, W.D., Quantitative assessment of the germicidal efficacy of ultrasonic energy. *Appl. Environ. Microbiol.*, 57, 2079, 1991.

33. Hua, I., Thompson, J.E., Inactivation of E. coli by sonication at discrete ultrasonic frequencies. *Water Res.*, 34, 3888–3893, 2000.
34. Kesari, K.K., Kumar, S., Verma, H.N., Behari, J., Influence of ultrasonic treatment in sewage sludge. *Hydrl. Curr. Res.*, 2, 115, 2011a.
35. Kesari, K.K., Verma, H.N., Behari, J., Physical methods in wastewater treatment. *Int. J. Environ. Technol. Manag.*, 14, 43, 2011b.
36. Petrier, C., Jeunet, A., Luche, J.L., Reverdy, G., Unexpected frequency effects on the rate of oxidative processes induced by ultrasound. *J. Am. Chem. Soc.*, 25, 148, 1992.
37. Pesoutova, R., Hlavinek, P., Matysikova, J., Use of advanced oxidation processes for textile wastewater treatment- A review. *Food Environ. Saf.*, 10, 59, 2011.
38. Blume, T., Neis, U., Improved wastewater disinfection by ultrasonic pre-treatment. *Ultrason. Sonochem.*, 11, 333, 2004.
39. Naddeo, V., Belgiorno, V., Ricco, D., Kassinos, D., Degradation of diclofenac by sonolysis, ozonation and their simultaneous application. *Ultrason. Sonochem.*, 16, 790, 2009.
40. Jyoti, K. K., Pandit, A. B., Ozone and cavitation for water disinfection. *Biochem. Eng. J.*, 38, 2249, 2004.
41. Paleologou, A., Marakas, H., Xekoukoulotakis, N.P., Moya, A., Vergara, Y., Kalogerakis, N., Gikas, P., Mantzavinos, D., Disinfection of water and wastewater by TiO_2 photocatalysis, sonolysis and UV-C irradiation. *Catal. Today.*, 129, 136, 2007.
42. Ahmed, S.N., Haider, W., Heterogeneous photocatalysis and its potential applications in water and wastewater treatment: A review. *Nanotechnology*, 29, 342001, 2018.
43. Nguyen, C.H., Juang, R.S., Efficient removal of cationic dyes from water by a combined adsorption-photocatalysis process using platinum-doped titanate nanomaterials. *J. Taiwan Inst. Chem. Eng.*, 99, 166, 2019.
44. Wawrzkiewicz, M., Adach, E., Hubicki, Z., Application of titania based adsorbent for removal of acid, reactive and direct dyes from textile effluents. *Adsorption*, 25, 621, 2019.
45. Rashid, R., Shafiq, I., Akhter, P., Iqbal, M.J., Hussain, M., A state-of-the-art review on wastewater treatment techniques: The effectiveness of adsorption method. *Environ. Sci. Pollut. Res.*, 28, 9050, 2021.
46. Wong, J.K.H., Tan, H.K., Lau, S.Y., Yap, P.S., Danquah, M.K., Potential and challenges of enzyme incorporated nanotechnology in dye wastewater treatment: A review. *J. Environ. Chem. Eng.*, 7, 103261, 2019.
47. Yahya, N., Aziz, F., Jamaludin, N.A., Mutalib, M.A., Ismail, A.F., Salleh, W.N.W., Jaafer, J., Yusof, N., Ludin, N.A., A review of integrated photocatalyst adsorbents for wastewater treatment. *J. Environ. Chem. Eng.*, 6, 7411, 2018.
48. Oki, T., Kanae, S., Global hydrological cycle and world water resources. *Science*, 313, 1068, 2006.

49. Zimmerman, J.B., Mihelcic, J.R., Smith, J., Global stressors on water quality and quantity. *Environ. Sci. Technol.*, 42, 4247, 2008.
50. Kolpin, D.W., Furlong, E.T., Meyer, M.T., Thurman, E.M., Zaugg, S.D., Barber, L.B., Buxton, H.T. Pharmaceuticals, hormones, and other organic wastewater contaminants in U.S. streams, 1999-2000: A national reconnaissance. *Environ. Sci. Technol.*, 36, 1202, 2002.
51. Conley, D.J., Paerl, H.W., Howarth, R.W., Boesch, D.F., Seitzinger, S.P., Havens, K.E., Lancelot, C., Likens, G.E., Controlling eutrophication: Nitrogen and phosphorus. *Science*, 323, 1014, 2009.
52. Guest, J.S., Skerlos, S.J., Barnard, J.L., Beck, M.B., Daigger, G.T., Hilger, H., Jackson, S.J., Karvazy, K., Kelly, L., Macpherson, L., Mihelcic, J.R., Pramanik, A., Raskin, L., Loosdrecht, M.C.M.V., Yeh, D., Love, N.G., A new planning and design paradigm to achieve sustainable resource recovery from wastewater. *Environ. Sci. Technol.*, 43, 6126, 2009.
53. Daigger, G.T., Evolving urban water and residuals management paradigms: Water reclamation and reuse, decentralization, and resource recovery. *Water Environ. Res.*, 81, 809, 2009.
54. Daigger, G.T., New approaches and technologies for wastewater management. *Bridge*, 38, 38, 2008.
55. Larsen, T.V., Alder, A.C., Eggen, R.I.L., Maurer, M., Lienert, J., Source separation: Will we see a paradigm shift in wastewater handling? *Environ. Sci. Technol.*, 43, 2009.
56. Kleerebezem, R., van Loosdrecht, M.C.M., Mixed culture biotechnology for bioenergy production. *Curr. Opin. Biotechnol.*, 18, 207, 2007.
57. Public Utilities Board of Singapore. Press Release: Sembcorp NEWater to start building Changi NEWater Plant, 2008.
58. Logan, B.E., Hamelers, B., Rozendal, R., Schröder, U., Keller, J., Freguia, S., Aelterman, P., Verstraete,W., Rabaey, K. Microbial fuel cells: Methodology and technology. *Environ. Sci. Technol.*, 40, 5181, 2006.
59. Funamizu, N., Iida, M., Sakakura, Y., Takakuwa, T., Reuse of heat energy in wastewater: Implementation examples in Japan. *Water Sci. Technol.*, 43, 277, 2001.
60. de-Bashan, L.E., Bashan, Y., Recent advances in removing phosphorus from wastewater and its future use as a fertilizer. *Water Res.*, 38, 4222, 2004.
61. Parris, T. M., Kates, R.W., Characterizing a sustainability transition: Goals, targets, trends, and driving forces. *Proc. Natl. Acad. Sci. U.S.A.*, 100, 8068, 2003.
62. Urban Water Management Plan; San Diego County Water Authority Water Resources Department: San Diego, CA, www.sdcwa.org/manage/2005U-WMP. phtml, 2005.
63. Marks, J., Martin, B., Zadoroznyj, M., How Australians order acceptance of recycled water-national baseline data. *J. Sociol.*, 44, 83, 2008.

64. Literature Review of Factors Influencing Public Perceptions of Water Reuse; Technical Report 54/03; CSIRO Land and Water: Clayton South, Australia, www.clw.csiro.au/publications/ technical2003/tr54-03.pdf, 2003.
65. Saravanan, V.S., McDonald, G.T., Mollinga, P.P., Critical review of integrated water resources management: Moving beyond polarised discourse. *Nat. Resour. Forum*, 33, 76, 2009.
66. Jasanoff, S., Designs on Nature: Science and democracy in Europe and the United States; Princeton., *Sci. Technol. Human Values*, 33, 134, 2005.
67. Fry, L.M., Mihelcic, J.R., Watkins, D.W., Water and nonwatery related challenges of achieving global sanitation coverage. *Environ. Sci. Technol.*, 42, 4298, 2008.
68. Starkl, M., Brunner, N., Flogl, W., Wimmer, J., Design of an institutional decision-making process: The case of urban water management. *J. Environ. Manage.*, 90, 1030, 2009.
69. Cheng, H., Hu, Y., Zhao, J., Meeting China's water shortage crisis: Current practices and challenges. *Environ. Sci. Technol.*, 43 (2), 240–244, 2009.

14

Circular Bioeconomy in the Recovery of Polymers from Sewage Sludge

Merry Meryam Martgrita[1]* and Siti Khodijah Chaerun[2,3]†

[1]Bioprocess Engineering Study Program, Faculty of Biotechnology, Institut Teknologi Del, Jl. Sisingamangaraja, Toba, Sumatera Utara, Indonesia
[2]Department of Metallurgical Engineering, Faculty of Mining and Petroleum Engineering, Institut Teknologi Bandung, Ganesha 10, Bandung, West Java, Indonesia
[3]Geomicrobiology-Biomining & Biocorrosion Laboratory, Microbial Culture Collection Laboratory, Biosciences and Biotechnology Research Center (BBRC), Institut Teknologi Bandung, Ganesha 10, Bandung, West Java, Indonesia

Abstract
This chapter explores the integration of circular bioeconomy principles in the palm oil and pulp and paper industries, highlighting innovative techniques for recovering and repurposing cellulose waste into valuable materials and products. Addressing environmental challenges associated with global population growth and industrialization, the circular bioeconomy is crucial for sustainable development. The recovery of cellulose is essential in managing urban wastewater and paper sludge due to its significant contribution to chemical oxygen demand (COD). Efficient degradation and recovery methods are necessary to transform cellulose into a variety of materials. Additionally, the extraction of extracellular polymeric substances (EPS) from biofilms plays a vital role in cell adhesion and sludge cohesiveness, offering potential applications as bioflocculants and in toxin removal. Research worldwide demonstrates various methods for cellulose recovery from sewage sludge, with yields ranging from 30% to 100%, suitable for use in energy, construction, and other industries. Direct utilization of cellulose in sewage sludge is pivotal for biofuel production, with microorganisms such as *A. niger* and *E. coli* effectively converting waste into cellulase enzymes. The palm oil industry's implementation of circular bioeconomy principles involves transforming waste into biofuels, enzymes, compost,

**Corresponding author*: merrymartgrita@gmail.com; merry.martgrita@del.ac.id
† *Corresponding author*: skchaerun@gmail.com; skchaerun@itb.ac.id

Inamuddin, Tariq Altalhi, Mohammad Luqman and Joseph Kapuku (eds.) Sewage and Biomass from Wastewater to Energy, (365–390) © 2024 Scrivener Publishing LLC

and electricity, thereby enhancing sustainability and waste utilization. Similarly, the pulp and paper industry applies these principles by using sludge for biofuel production, generating cellulose nanofibrils and biochar, leading to the development of valuable polymer materials. Although various cellulose recovery methods have been researched, many remain at the experimental stage and lack comprehensive economic analysis. Future developments may include innovative methods for optimal cellulose recovery. Implementing circular bioeconomy principles in these industries supports Sustainable Development Goals 11 and 12, promoting sustainable practices in cellulose recovery and product creation.

Keywords: Circular bioeconomy, cellulose waste recovery, sustainable development, biofuel production, waste utilization

14.1 Introduction

The escalating global population growth has brought about increased awareness and concern surrounding environmental issues. These issues encompass pollution, climate change, global warming, waste management, and the exhaustion of natural resources, all of which necessitate urgent attention. The extensive exploitation of fossil fuels for energy production, chemical synthesis, and the fabrication of synthetic materials primarily drives the depletion of renewable natural resources and the surge in greenhouse gas emissions. Both these phenomena have profound environmental repercussions. In response to these global challenges, solutions that can positively impact the environment are imperative. One such solution is the circular bioeconomy, which integrates the principles and frameworks of both the circular economy and the bioeconomy. The circular bioeconomy offers a sustainable approach to resource management and waste reduction, thereby mitigating environmental degradation and promoting sustainable development [1].

The circular economy is an economic model aimed at minimizing resource consumption and waste generation, thereby fostering sustainable economic growth [2]. It is underpinned by several key concepts:

1. The concept referred to as "10 R" encompasses a series of actions, including refuse, rethink, reduce, reuse, repair, refurbish, remanufacture, repurpose, recycle, and recover [3].
2. The cleaner production concept advocates for energy savings, emission reductions, and enhanced production efficiency, optimizing environmental management processes [4].

3. The industrial symbiosis concept, derived from industrial ecology, encourages interorganizational collaboration and sharing of resources such as water, energy, byproducts, and residual materials. This approach optimizes material and energy use, reduces waste production, and repurposes effluents for other processes [5].

In the context of the circular bioeconomy, these circular economy concepts are applied specifically to the efficient, sustainable use of biomass resources. The goal is to produce valuable bioproducts and promote a systemic approach to economic development that is both environmentally and economically sustainable [1].

The implementation of a circular economy, which yields positive impacts on the economy, health, and environmental quality, aligns with the 12th Sustainable Development Goal (SDG). This goal emphasizes the necessity of resource reuse to address the challenges of economic sustainability. Furthermore, the circular economy also supports the 11th SDG, as waste recycling initiatives can contribute to the sustainability of healthy and comfortable settlements and communities [6]. One strategy within the biobased circular economy is biorefining, a process that efficiently utilizes biomass resources to produce a variety of marketable products and metabolites, such as biomaterials and bioactive compounds. Waste biorefinery, a bioprocess that transforms waste into biomaterials and biofuels, has also garnered significant attention. The products generated through this process can reduce reliance on fossil resources and mitigate potential resource scarcity. This approach not only maintains the sustainability of the energy-environment nexus but also protects the environment by reducing the carbon footprint [7, 8].

The core tenets of a circular economy, which include waste and pollution reduction, product and material reuse, and the regeneration of natural systems, are designed to bolster economic growth, enhance human well-being, and foster social equity. The application of circular economy principles extends beyond the product level, permeating the entire supply chain and product lifecycle. This encompasses product design, production systems, raw material and energy consumption, waste collection, and disposal systems. The disposal stage is often identified as the most problematic [9]. The incorporation of circular economy principles at each stage of the supply chain can result in efficiencies in raw material procurement and usage, availability of raw materials from recycled products, energy utilization, and waste management. Consequently, the circular economy can contribute to more efficient industrial financial operations and increased profitability. For instance, in the food industry, waste can be generated at the points of sale and consumption, but waste reduction can also be implemented during the production process and throughout the supply chain [10]. Waste is generated at various points within the food supply chain, including the initial stages of production, during food distribution, and post-consumption. Therefore, the implementation of circular waste management principles at different stages in the supply chain would be beneficial [11, 12].

Implementing the circular concept to reduce food waste production yields significant benefits across economic, societal, and environmental dimensions. Economically, circular waste management can decrease expenditures due to enhanced production efficiency, which in turn can lower food prices and improve food quality and security [10, 13, 14]. Societally, the reduction of food waste production, particularly through the utilization of surplus food, can contribute to alleviating hunger [15]. Environmentally, the transition from a linear to a circular economy can mitigate waste production, thereby reducing greenhouse gas emissions that contribute to climate change, and conserving energy and materials [10, 13].

For recyclable materials, the primary challenge lies in creating products with extended life spans, or those that are reusable or bio-recyclable. However, not all materials can undergo biological recycling, be utilized as secondary raw materials for subsequent production, or be repurposed to create other products. Consequently, landfilling becomes the only viable solution when products cannot be reused. Organic waste, encompassing food and agricultural waste, is often disposed of in landfills or incinerated, despite its recyclability. This practice poses environmental risks as it can provide a substrate for pathogenic microorganisms, and the resulting leachate can contaminate surface and groundwater [16]. Moreover, the incineration of organic waste not only contributes to air pollution but also leads to the demise of soil microorganisms, thereby diminishing soil fertility.

The implementation of a circular economy across various sectors, including manufacturing, the food industry, agriculture, and households, yields not only financial advantages but also positive environmental outcomes. However, the circular economy also presents challenges that necessitate continuous technological innovation and environmental conservation. Alonso-Muñoz *et al.* [17] provided an analysis of the implementation of a circular economy in the food industry, concluding several key points:

1. Anaerobic Waste Treatment: The biological processing of food waste sludge using anaerobic microbes can generate

volatile fatty acids, which can subsequently be used to produce biomethane gas, thereby mitigating the emission of global warming gasses [18, 19].

2. Circular Food Supply Chain: This concept modifies the traditional linear supply chain, impacting trade frameworks' sustainability to enhance performance and management. This transformation presents challenges in consumption patterns and food recovery to prevent food surplus and achieve closed-loop supply chains [20, 21].
3. Value Addition to Food Waste: The extraction process plays a crucial role in enhancing the value of food waste and in refinery technology, while the fermentation process is associated with the conversion of food waste into biogas. Refinery technology facilitates the production of high-value bio-based products through organic waste recycling. This technology enables the implementation of a circular bioeconomy for sustainable food waste processing [22, 23].
4. Novel Technology: The technology for energy conversion from urban solid waste adheres to the principles of waste segregation and recycling [24]. Volatile fatty acids derived from the fermentation of food waste can be utilized to produce biogas in the form of green chemical-based methane gas [25].

14.2 Polymers Recovery in Waste Management

Cellulose, an organic polymer that is insoluble in water and common solvents, constitutes a significant component of plant cells along with hemicellulose and lignin. Toilet paper is a primary source of cellulose in urban wastewater, with usage rates varying significantly across regions [26]. Asian countries typically use up to 2 kg/person/year, while Europe and the United States use 11 and 18 kg/person/year, respectively [27, 28]. It is estimated that cellulose accounts for 13 to 35% w/w of the total COD component in urban wastewater treatment facilities in various countries [29–31]. In the paper and cardboard industry, paper sludge, a byproduct of the production process, contains both organic and inorganic material. The composition of paper sludge depends on the source material and the production process. The cellulose content in paper sludge, primarily derived from wood material, can reach up to 40% of the mass of the produced paper, with cellulose concentrations ranging from 28 to 80% [32–35].

The management and application of sewage are typically limited to biodegradation under anaerobic or aerobic conditions. Anaerobic biodegradation is more prevalent due to its ability to produce energy in the form of methane gas. Moreover, many sludge biodegradation processes that generate energy, such as hydrogen gas and ethanol, can be conducted anaerobically [36, 37].

Cellulose fiber, an organic polymer, presents a significant challenge for degradation. The initial stages of cellulose degradation under both aerobic and anaerobic conditions typically involve hydrolysis via cellulase or through the action of cellulose-digesting microorganisms [29, 38]. To effectively degrade cellulose in sewage sludge, various optimization methods are required, including physical methods such as temperature optimization, physicochemical methods such as steam treatment, and the use of acidic or basic chemical solutions. However, these methods are not suitable for cellulose recovery, as the goal of recovery is to separate cellulose fibers from the sludge. For cellulose recovery from sewage sludge, a viable method involves replacing primary sedimentation in the tank with a filter for enhanced separation efficiency. Although the use of filters is technically more costly than sedimentation, the employment of highly porous filters can reduce energy consumption in the sludge degradation process within the aeration tank [28]. Hydrolysis of sewage sludge devoid of cellulose fiber results in significantly less solid waste compared to sludge that still contains cellulose fiber [29]. The cellulose fibers obtained from cellulose recovery can be further utilized to produce a variety of materials, including construction materials for bicycle paths, building construction materials, binders or additives for asphalt, insulation materials, bioplastics, flocculants, and materials for bioethanol production [39, 40].

Extracellular polymeric substances (EPS) represent another class of biopolymers that constitute a significant component of bacterial biofilms. Biofilms are estimated to contain 50 to 90% of organic carbon in the form of biopolymers, which include polysaccharides such as hexoses, hexosamines, and ketoses [41]. EPS, generally hydrophobic in nature, contains polyhydroxyalkanoates (PHA) that play a crucial role in cell adhesion and sludge cohesiveness. They also influence sedimentation, strength, drying capacity, and environmental safety. The presence of EPS in biofilms within sludge provides a protective layer against external environmental disturbances, serves as a carbon source during biological processes, and forms bonds with other pollutants through hydrogen bonds, ionic bonds, or biochemical reactions [41–45]. EPS extraction can be achieved through several methods that do not induce cell death or damage the EPS structure. These methods include physical processes such as sonication, centrifugation, and heating, chemical processes involving acidic and basic compounds, and biological processes using enzymes. The recovered EPS can be utilized as a bioflocculant and as a compound for binding and removing toxic substances, including heavy metals and dyes [46].

14.3 Mechanisms for Cellulose Recovery from Sewage Sludge

Numerous investigations have been conducted to explore the mechanisms of cellulose recovery from a variety of waste sources (refer to Table 14.1). Research focusing on cellulose recovery from waste sources in municipal wastewater treatment plants (MWTP) has been undertaken in countries such as the Netherlands and Italy. The Netherlands' MWTP employs a filtration method for cellulose recovery, yielding approximately 30% of recovered cellulose, which is subsequently utilized as raw material for power plants [47]. Similarly, Italy's MWTP, which also uses filtration, produces about 38% of recovered cellulose, which finds application in building construction materials [27].

In Japan's MWTP, a physicochemical method involving the use of sulfuric acid and heating at 130°C with an autoclave is employed for cellulose recovery, yielding up to 70% of recovered cellulose [29]. Waste derived from *Quercus* sp. wood processing in Mexico [48], bamboo processing factories in China [49], and agricultural areas in China [50] also utilize this chemical method for cellulose recovery. The chemical compounds employed include ethanol as an organic solvent, and catalysts such as $AlCl_3$, H_2SO_4, and ethylene glycol (EG). The recovered cellulose produced is successively as much as 86.25% [48–50].

MWTPs in Spain [51, 52] effectively utilize a chemical compound, namely, ionic liquid (IL) tetrakis(hydroxymethyl)phosphonium chloride, yielding nearly 100% recovered cellulose. Additionally, sewage sludge from the MWTP is also utilized with the aid of an IL catalyst for the production of levulinic acid [53]. The IL is also employed for the cellulose recovery process from paper mill waste in Spain and pine wood processing factories in China. The cellulose recovery process in Spain's paper mill uses IL to produce 105.5% recovered cellulose [51], while the pine wood processing factory in China, which uses a mixed IL of [EMIM]DEP and DMSO, only yields 3% cellulose [54].

From the studies conducted in various countries across Europe, Asia, and South America, it is evident that several methods can be employed to

Table 14.1 Mechanisms of cellulose recovery from sewage sludge.

Types of waste source	Mechanisms	Utilization of recovery result	References
Municipal wastewater treatment plants, Netherlands	The main source is latrine tissue. Cellulose recovery was carried out by mechanical technique using a sieve with a size of <0.35 mm with a filtering rate of 30 $m^3/m^2/h$. As much as 30% of cellulose can be recovered using this method.	Biomass-based power plant raw materials	[47]
Municipal sewage processing plants, Italy	The cellulose recovery system uses a rotating belt filter (RBF) and a filter size of 90 μm to produce 38% pure cellulose.	Building construction material	[27]
Municipal wastewater treatment plants, Spain	Waste contains about 31.3% carbohydrates which are generally cellulose. The use of ionic liquid (IL) produces cellulose of about 25.6% at 60°C for 13 hours with a ratio of 1:12 by weight of solid to the volume of IL.	-	[53]
Municipal wastewater treatment plants and paper mill, Spain	Solid and liquid waste from wastewater treatment is used in this research. Use of tetrakis IL with a ratio of 1 gds:10 mL(IL) heated at 100°C for 24 hours.	Production of levulinic acid with the help of a *Bronsted Acidic Ionic Liquid* catalyst (as much as 31% from paper mill sludge and 25% from municipal sludge)	[52]
The paper mill, Spain	Paper processing waste contains 42.3% carbohydrates which are dominated by cellulose. Cellulose was recovered by physical and chemical methods using two types of ionic liquid (IL), namely, 1-butyl-3-methylimidazolium methyl sulfate and tetrakis(hydroxymethyl) phosphonium chloride, and heating at 100°C for 24 hours. The recovery results using both types of IL produced an average yield of 105.5% cellulose.	-	[52]

(*Continued*)

Table 14.1 Mechanisms of cellulose recovery from sewage sludge. (*Continued*)

Types of waste source	Mechanisms	Utilization of recovery result	References
Pinewood processing factory, Shijiazhuang, China	The cellulose recovery was carried out using two types of ionic liquid (IL), namely, [EMIM]DEP and DMSO with a ratio of 1:0.5 for 8 min at 105°C. As much as 3% cellulose can be recovered by this method.	-	[54]
Quercus sp. wood processing factory, Mexico	The recovery of cellulose in Oak sawdust is used with the Ethanosolv process. This process uses ethanol as an organic solvent and acid catalysts such as $AlCl_3$ and H_2SO_4. The optimum condition for this process is 160°C, 0.5% H_2SO_4, dissolved for 10 min in 50% ethanol:water solution. As much as 86.25% of cellulose was successfully recovered using this method.	-	[48]
Bamboo processing factory, Guangxi Zhuang, China	The main ingredient used is dry bamboo which is processed into powder form. Recovery of cellulose was carried out by chemical and physical methods as pre-treatment with 20% ammonium sulfate for 6 hours at 150°C and followed by enzymatic hydrolysis for 48 hours. This method can recover cellulose as much as 59.87%.	-	[49]
Agriculture, China	The waste used is rice straw. Enzymatic hydrolysis was carried out chemically and the yield can be up to 100% or 76% (w/w) in solid residue.	-	[50]
Palm oil factory, Malaysia	The method used is low-transition temperature mixtures (LTTMs) with an L-malic acid-sucrose-water molar ratio of 2:4:2 (w/w/w). Empty bunches were pre-treated with LTTM solution at 80°C for 6 hours.	Paper pulp	[42]

yield substantial quantities of recovered cellulose. These include chemical methods using various types of chemical catalysts, particularly chemical methods employing ILs such as tetrakis(hydroxymethyl)phosphonium chloride and 1-butyl-3-methylimidazolium methyl sulfate.

14.4 Mechanism of Cellulose Content Utilization in Sewage Sludge

The direct harnessing of cellulose content in sewage sludge, bypassing the necessity for a cellulose recovery process, is a viable approach. This direct utilization predominantly targets the production of cellulase enzymes and biofuels such as bioethanol, bio-hydrogen, and biomethane, facilitated through a fermentation process (as depicted in Table 14.2). Numerous studies have been undertaken to generate cellulase via microbial fermentation, exploiting the cellulose content in waste as a substrate. *Aspergillus niger* emerges as the most frequently employed microorganism in these studies. For instance, Abdullah *et al.* [55] utilized sugarcane plant waste to produce cellulase using *A. niger* strain ITBCC L74 through a solid-state fermentation (SSF) method, achieving an enzyme activity of 0.63 U/g. In another study by the same group [56], rice straw, water hyacinth, and corncob waste were used as substrates for *A. niger*, resulting in the highest cellulase activity of 2.569 U/mL with rice straw, followed by 1.631 U/mL with water hyacinth, and 1.302 U/mL with corncobs. Furthermore, Bansal *et al.* [57] reported the production of CMCase enzymes with an activity of 310.6 U/gds, FPase enzymes with an activity of 16.8 U/gds, and β-glucosidase enzymes with an activity of 33 U/gds, using wheat bran waste as a substrate for *A. niger* via SSF.

Several studies have demonstrated the production of cellulase by other microorganisms such as *Escherichia coli*, *Trichoderma* sp., *Bacillus pseudomycoides*, and *Cellulomonas uda* strain NCIM 2353, which exhibited higher enzymatic activity than *Aspergillus niger*. For instance, Amraini *et al.* [58] reported that *E. coli* fermentation of empty oil palm fruit bunches yielded cellulase with an activity of 2.665 U/mL. Gautam *et al.* [59] demonstrated that *Trichoderma* sp., grown on domestic waste, produced exoglucanase, endoglucanase, and β-glucosidase with activities of 1.77, 1.95, and 1.66 U/mL, respectively. *Bacillus pseudomycoides*, used in the fermentation of sugarcane processing waste, generated cellulase with a peak activity of 3.5 U/mL [60]. Moreover, *Cellulomonas uda* strain NCIM 2353, grown on date seed substrates with high cellulose and hemicellulose content (32.77% and 30.20%, respectively), exhibited cellulase activity of 175.96 U/mL [61].

Table 14.2 Mechanistic overview of cellulose content utilization in industrial waste streams.

Types of waste source	Mechanisms	References
Sugar factory	Sugar cane processing waste is used as a substrate to optimize cellulase production. The substrate was fermented using the SSF method using *Aspergillus niger* strain ITBCC L74 which was incubated at 35°C for 3 days. The optimal working conditions for *A. niger* were 4.5% w/w urea at pH 3.5 and 1 mM $MgCl_2$ concentration. This condition produces cellulase with an activity of 0.63 U/g. The cellulase produced from fermentation was extracted using sodium citrate buffer with a pH of 4.5 with a ratio to dry waste of 10:1 w/v.	[55]
Agricultural waste	Water hyacinths, corncobs, and rice straws are used as raw materials to produce cellulase. Solid fermentation was using *Aspergillus niger* strain ITBCC L74. The optimum condition for the fermentation time is 6 days. The highest enzyme activity (2.569 U/mL) was obtained from rice straw fermentation with 80% water content, followed by water hyacinth fermentation (70% water content) yielding 1.631 U/mL, while from corncobs, it was 1.302 U/mL with 60% water content.	[56]
Agricultural and food processing waste	Agricultural wastes such as corn cobs, grass, leaves, sugar cane processing waste, wheat straw, sawdust, and food processing wastes such as kitchen waste, carrot skins, pineapple skins, potato skins, rice bran, wheat bran, and orange peels are used as substrates to produce cellulase. The initial process was carried out using 1.0% sulfuric acid and 1% sodium hydroxide separately, and then 20 g of the substrate was added to 100 ml of acid/alkali and left for 2 hours at 24°C.	[57]

(*Continued*)

Table 14.2 Mechanistic overview of cellulose content utilization in industrial waste streams. (*Continued*)

Types of waste source	Mechanisms	References
	Cellulase production lasted for 96 hours at 30°C without the addition of other nutrients by the SSF method using *A. niger* strain NS-2 fungus. Wheat bran substrate without preprocessing produced the highest enzyme production (310.6 U/gds CMCase, 16.8 U/gds FPase, and 33 U/gds β-glucosidase) compared to other substrate sources without preprocessing, adding H_2SO_4 or adding NaOH.	
Palm oil factory	The empty palm fruit bunches biomass consists of cellulose about 44% (w/w), which is used as a substrate to produce cellulase. The study compared the use of *E. coli* strain EgRK2 and *Bacillus sp.* strain RK2. The optimum condition is 24 hours of fermentation time at pH 7 and a temperature of 40°C. The cellulase produced using *E.coli* strain EgRK2 was 2.665 U/mL while using *Bacillus sp.* obtained 2.498 U/mL.	[58]
Domestic trash	The optimum conditions for *A. niger* were at 40°C and pH 6–7, while the optimum conditions for *Trichoderma sp.* are 45°C and a pH of 6.5. Domestic trash is a decent carbon source to produce 1.189, 1.504, and 1.39 U/mL of exoglucanase, endoglucanase, and β-glucosidase, respectively, using *A. niger*, while *Trichoderma sp.* produced 1.77, 1.95, and 1.66 U/mL of exoglucanase, endoglucanase, and β-glucosidase, respectively.	[59]

(*Continued*)

Table 14.2 Mechanistic overview of cellulose content utilization in industrial waste streams. (*Continued*)

Types of waste source	Mechanisms	References
Sugar cane processing waste	Fermentation of waste from sugar cane processing using *Bacillus pseudomycoides* at pH 7, temperature 40°C, fermentation time of 72 hours, and conditions of adding peptone and magnesium sulfate. The best cellulase concentration was 4.91 mg/mL and activity was 2.84 U/mL with the addition of magnesium sulfate. When 1 and 1.5% of the waste were added to the yeast extract, cellulose was produced with a concentration of 5.6 mg/mL and an activity of 3.5 U/mL.	[60]
Dates processing industry	Dates seeds are a waste of date syrup processing and contain 37.03% lignin, 32.77% cellulose, and 30.20% hemicellulose, which have the potential to be used as substrates in producing cellulase. *Cellulomonas uda* strain NCIM 2353 was used to produce cellulase with a date seed concentration of 30.65 g/L, xylose concentration of 0.55 g/L, casein 7 g/L, and pH 7.4. The cellulase produced from this fermentation was 175.96 U/mL. Maximum enzyme activity was obtained at pH 7 and a temperature of 50°C. The resulting cellulase enzyme is used in the enzymatic hydrolysis process to produce bio-hydrogen from sugarcane processing waste raw materials. The maximum bio-hydrogen concentration produced was 187.44 mmol/L for 24 hours.	[61]
Agricultural waste	Wheat bran is an agricultural waste that can be used as a substrate to produce cellulase. The combination of *Trichoderma reesei* and *Penicillium citrinum* cultures was used for the production of cellulase using the SSF (solid state fermentation) method	[65]

(*Continued*)

Table 14.2 Mechanistic overview of cellulose content utilization in industrial waste streams. (*Continued*)

Types of waste source	Mechanisms	References
	at conditions of 70% water content, 10^6 inoculum spores suspended at 30°C and pH 5. The resulting cellulase was 6.71 FPU/gds on the 6th day of fermentation.	
Agricultural waste	Cellulase enzyme production by *Aspergillus japonicus* strain URM5620 resulted in the activity of filter paper cellulase and carboxy-methyl cellulase with a substrate concentration of 3% passion fruit peel and 1% glucose under 30°C SmF (submerged fermentation) conditions for 72 hours. Cellulase was used in the pre-treatment of the enzymatic phase in the hydrolysis process of passion fruit peel substrate to produce biogas at a bioreactor temperature of 37°C, with an agitation speed of 60 rpm for 30 days. The amount of methane produced in hydrolysate I was 97.02 mL (60% higher than hydrolysate II) because hydrolysate I produced enzymes with higher activity compared to hydrolysate II.	[62]
Microalgae	Mixed culture of microalgae with the same ratio as pure cellulose was co-cultured with *Trichoderma reesei* and *Aspergillus niger* and produced cellulase with the highest activity of 3.3 U/mL. Mixed culture samples of the same microalgae and cellulose were fermented by *Saccharomyces cerevisiae* and produced 0.46 g-bioethanol/g-glucose or a theoretical 92% recovery.	[64]

These cellulase enzymes, produced by microorganisms cultured on cellulose-containing substrates, can be further utilized in the biofuel production process. For example, Swathy *et al.* [61] employed cellulase in the enzymatic hydrolysis of sugarcane waste to generate bio-hydrogen, achieving concentrations up to 187.44 mmol/L within 24 hours. Silva *et al.* [62] used cellulase derived from *Aspergillus japonicus* strain URM5620 in the enzymatic hydrolysis of passion fruit peel substrates to produce biomethane, yielding a volume of 97.02 mL over 30 days. Direct utilization of cellulose content in waste for bioethanol production has also been explored. Chairul *et al.* [63] reported a 68.255% conversion of rejected pulp into bioethanol using a simultaneous saccharification and co-fermentation (SSCF) method with a mixed culture of *Saccharomyces cerevisiae* and *Pichia stipitis*. Furthermore, Shokrkar *et al.* [64] demonstrated that fermentation of microalgae mixed culture supplemented with pure cellulose by *Saccharomyces cerevisiae* could convert 92% of the substrate into bioethanol.

14.5 Implementing Circular Bioeconomy Principles in the Palm Oil Industry

The palm oil sector represents an industry capable of incorporating circular economy principles within its production processes, particularly in relation to waste utilization, which is typically accumulated at factory sites (Figure 14.1). The current waste management practices can lead to water body contamination, air emissions, landscape alterations, soil acidification, and leachate production, all of which have potential implications for human health [66, 67]. The palm oil industry generates both solid organic waste, including fiber and husk, and liquid waste in the form of palm oil mill effluent (POME). These waste products can be processed to produce

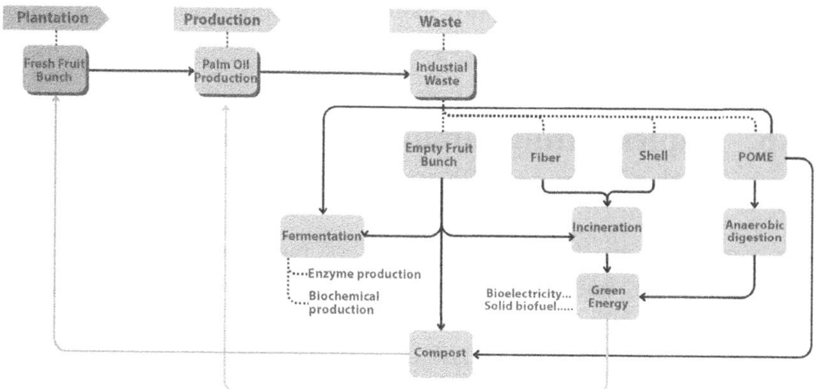

Figure 14.1 Diagram illustrating the implementation of circular bioeconomy principles within the palm oil industry.

biofuels, which serve multiple purposes within the industry, such as providing heat for boilers, generating electricity, and serving as raw materials for biodiesel production. The production of these biofuels can be achieved through various physical processes, including co-combustion, carbonization, gasification, pyrolysis, and thermolysis, as well as biological processes under anaerobic conditions [68–70]. The potential electricity generation from the utilization of palm oil industry waste, specifically empty fruit bunches, fiber, and husk, varies considerably depending on the efficiency of the conversion process. With a range of 74 to 119 MWh annually, this potential is contingent upon the productivity of various transformation processes. Specifically, pyrolysis, with an efficiency of 65%, gasification at 50% efficiency, and direct combustion, which boasts an efficiency of 80%, all contribute to this potential energy yield [71].

The solid and liquid waste generated by the palm oil industry has the potential to be effectively repurposed through various biological processes. One such process involves submerged fermentation using microorganisms such as *Aspergillus niger*, which results in the production of lipase enzymes [72]. Another approach employs *Streptomyces* to ferment cellulose extract from empty fruit bunches (EFB), yielding xylanase enzymes [73]. Alkaline hydrolysis of lignocellulose-rich EFB can produce significant amounts of vanillic acid and biovanilla [74]. Moreover, the waste from palm oil processing can be utilized for soil enrichment, serving as a substitute for nutrient-rich phosphorus, nitrogen, potassium, and magnesium. These nutrients can be derived from composting, bioreactor waste, and pyrolysis. For instance, Lim and Wu [75] used POME, which is rich in N, P, K, and Mg, as a raw material for composting to produce fertilizer. Similarly, Hau *et al.* [76] used a combination of POME, EFB, ashes, and residue from various production processes as raw materials for composting, resulting in a fertilizer rich in carbon and nitrogen.

14.6 Implementing Circular Bioeconomy Principles in the Pulp and Paper Industry

The comprehensive production process of pulp and paper generates a variety of wastes and byproducts, with approximately 87% constituting pulp and paper mill sludge and around 13% comprising impurities such as chemical waste and gas emissions [77]. The disposal or landfilling of paper mill sludge can have environmental implications, including the production of gas and liquid leachate emissions, which can contaminate soil, aquifers, and terrestrial water bodies. Furthermore, the incineration of paper mill sludge can lead to air, water, and soil contamination [34]. The principles of the circular economy could be applied in pulp and paper mills to utilize their sludge effectively. Paper mill sludge can serve as a raw material for biomass energy. Biofuels, such as biodiesel and bioethanol, can be derived from the biological and chemical conversion of industrial and agricultural waste. Primary resources for bioethanol, such as sugar, can be obtained from the processing of paper mill sludge [34]. Paper mill sludge contains 28 to 68% cellulosic fiber and 8.4 to 10.7% hemicellulose. Sludge hydrolysis using cellulase enzymes can transform sludge into glucose with a yield ranging from 10.5 to 98%. Industrial yeast strains can ferment this sugar into bioethanol, achieving a conversion productivity of 0.18 to 1.64 g/L/h. Alkasrawi *et al.* [78] demonstrated that the simultaneous saccharification and fermentation (SSF) process for converting paper mill slurry into bioethanol reduces energy use by 1.2 MJ/L of bioethanol. The efficiency of the SSF method is further enhanced by the selection of an effective fiber recovery method combined with the addition of an accelerant. The chemical fiber recovery method and the addition of an accelerant resulted in an increase in bioethanol yield of up to 10%.

Cellulose nanofibrils (CNFs) are emerging nanomaterials that are garnering significant industrial interest due to their exceptional properties [79]. Notably, CNFs possess superior mechanical characteristics, are biocompatible, and exhibit biodegradability [80], rendering them potential candidates for applications in nanotechnology [81], biomedicine [82, 83], and energy storage devices [83]. Du *et al.* [84] implemented an efficient methodology to produce nanofibrils and nanopaper from paper mill wastewater pretreated with formic acid. The resultant CNFs exhibited high thermal stability, a significant crystallinity index, and a high yield exceeding 70%. This research further revealed that over 90% of the formic acid could be recovered, and the decomposed sugar could be repurposed for the production of various chemicals. The cellulose nanopaper, fabricated using the derived CNF as a raw material, demonstrated excellent mechanical properties, including high tensile strength and durability.

Paper mill sludge is also characterized by the presence of a primary byproduct known as biosolids, which are solid organic materials derived from liquid waste. Studies conducted by McNamara *et al.* [85] and Carey *et al.* [86] have demonstrated that dry biosolids when subjected to pyrolysis, can generate energy, and the biochar resulting from biosolids can enhance the growth rate of grass, surpassing the effects of commercial fertilizers. Further research by Liu *et al.* [87] has revealed that biochar, derived from the pyrolysis of paper mill sludge, possesses a significant calcium content that acts as a catalyst. The application of this biochar can yield up to 40% of the total pyrolysis product in the form of pyrolysis gas and generate energy up to 8400 kJ, a value that surpasses the energy generated by biochar catalysts obtained from wood or corncobs and mineral catalysts such as dolomite.

The sector of wood-based industries can also adopt the principles of a circular economy by efficiently recovering and reutilizing lignocellulosic materials from wood waste to procure cellulose byproducts. These cellulose byproducts can be further processed to generate cellulose plastic composites, which are anticipated to substitute wood plastic composites that are constrained by limited wood raw materials due to competition with primary needs. Wood-plastic composites are materials that are increasingly being utilized in building and construction products, automotive components, and consumer goods [88]. In a study conducted by Zhou *et al.* [89], it was demonstrated that a maleic anhydride coupling agent could enhance the interfacial bonding in cellulose plastic composites and wood plastic composites. The findings revealed that while wood plastic composites exhibited superior thermal stability compared to cellulose plastic composites, the latter were more flexible, environmentally friendly, and economically beneficial. The collective results from these studies suggest that the principles of a circular economy can be implemented through the processing of paper mill sludge, resulting in the production of high-value polymer materials. The implementation of circular bioeconomy principles within the pulp and paper industry is clearly demonstrated in Figure 14.2.

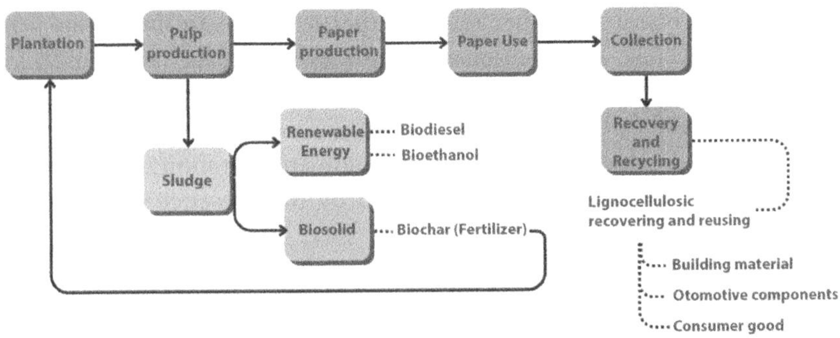

Figure 14.2 Diagram illustrating the implementation of circular bioeconomy principles within the pulp and paper industry.

14.7 Future Perspectives

The preceding discussion elucidates that numerous mechanisms have been extensively researched and developed globally to facilitate cellulose recovery from a variety of cellulose-rich waste types. Concurrently, significant strides have been made in the development of waste treatment mechanisms that directly harness the cellulose content to produce cellulase enzymes or biofuels such as bioethanol, bio-hydrogen, or biomethane. While some of these researched methods have yielded high returns, others have demonstrated relatively lower yields. Most of these investigations remain confined to the laboratory scale and have yet to incorporate an economic analysis pertaining to the optimal performance of the mechanism. Consequently, the pursuit of developing methods for cellulose recovery and processing cellulose-containing waste warrants further enhancement. This could potentially involve the utilization of chemical compounds, such as ILs, or microorganisms that can yield optimal cellulose recovery results.

14.8 Conclusions

Upon reviewing the existing literature, it is evident that the principle of a circular bioeconomy has begun to be integrated into various industries, including palm oil processing, pulp and paper production, agriculture, and municipal wastewater treatment across numerous countries. Numerous research endeavors have been undertaken to develop methods for cellulose recovery, with the recovered cellulose being repurposed as a raw material for the production of more valuable and socially beneficial products. These methods encompass both physical and chemical processes.

Furthermore, researchers have devised a range of waste treatment methodologies aimed at harnessing the cellulose content within waste to generate valuable products such as cellulase enzymes and biofuels. These methodologies predominantly employ biological processes, leveraging a variety of potent microorganisms. These research efforts, both directly and indirectly, bolster the implementation of circular bioeconomy principles, particularly within the industrial sector and government-run municipal wastewater treatment facilities. Given the availability of diverse recovery and treatment alternatives, industries and governments have a plethora of options to choose from, depending on the nature of the waste and the associated costs. The availability of these alternative methods, coupled with their economic analysis, is expected to encourage more industries to adopt the principles of a circular bioeconomy in their operations. By doing so, these industries acknowledge the importance of waste recycling into products of higher economic value, thereby supporting the sustainability of health and environmental comfort for society. This aligns with the 11th and 12th SDGs, emphasizing the importance of sustainable cities and communities, and responsible consumption and production, respectively.

References

1. Venkata Mohan, S., Modestra, J.A., Amulya, K., Butti, S.K., Velvizhi, G., A Circular Bioeconomy with Biobased Products from CO2 Sequestration. *Trends Biotechnol.*, 34, 506, 2016.
2. Kouhizadeh, M., Zhu, Q., Sarkis, J., Blockchain and the circular economy: potential tensions and critical reflections from practice. *Prod. Plan. Control.*, 31, 950, 2020.
3. Bag, S., Gupta, S., Kumar, S., Industry 4.0 adoption and 10R advance manufacturing capabilities for sustainable development. *Int. J. Prod. Econ.*, 231, 107844, 2021.
4. Ma, S., Zhang, Y., Liu, Y., Yang, H., Lv, J., Ren, S., Data-driven sustainable intelligent manufacturing based on demand response for energy-intensive industries. *J. Clean. Prod.*, 274, 123155, 2020.
5. Sehnem, S., Vazquez-Brust, D., Pereira, S.C.F., Campos, L.M.S., Circular economy: benefits, impacts and overlapping. *Supply Chain Manage.*, 24, 784, 2019.
6. Sharma, H.B., Vanapalli, K.R., Samal, B., Cheela, V.R.S., Dubey, B.K., Bhattacharya, J., Circular economy approach in solid waste management system to achieve UN-SDGs: Solutions for post-COVID recovery. *Sci. Total Environ.*, 800, 149605, 2021.
7. Dahiya, S., Kumar, A.N., Shanthi Sravan, J., Chatterjee, S., Sarkar, O., Mohan, S.V., Food waste biorefinery: Sustainable strategy for circular bioeconomy. *Bioresour. Technol.*, 248, 2, 2018.
8. Moraes, E.S., Reis, G.F., Cruz, J., Cozzolino, K., Neto, A.F.G., Andrade-Filho, T., Neto, A.M.J.C., Thermodynamics study of biokerosene from coconut and palm kernel oils and JP-8 aircraft fuels in the gas phase by the DFT method. *J. Mol. Model.*, 26, 79, 2020.
9. Neto, G.C. de O., Teixeira, M.M., Souza, G.L.V., Arns, V.D., Tucci, H.N.P., Amorim, M., Assessment of the Eco-Efficiency of the Circular Economy in the Recovery of Cellulose from the Shredding of Textile Waste. *Polymers (Basel)*, 14, 1317, 2022.
10. Narvanen, E., Mesiranta, N., Mattila, M., Heikkinen, A. (Ed.), *Food waste management: solving the wicked problem*, Palgrave Macmillan, Cham, 2020.

11. Parfitt, J., Barthel, M., MacNaughton, S., Food waste within food supply chains: Quantification and potential for change to 2050. *Philos. Trans. R. Soc. B Biol. Sci.*, 365, 3065, 2010.
12. Lipinski, B., Hanson, C., Lomax, J., Kitinoja, L., Waite, R., Searchinger, T., Toward a sustainable food system Reducing food loss and waste. International Food Policy Research Institute 2016.
13. Krishnan, R., Agarwal, R., Bajada, C., Arshinder, K., Redesigning a food supply chain for environmental sustainability – An analysis of resource use and recovery. *J. Clean. Prod.*, 242, 118374, 2020.
14. Despoudi, S., Bucatariu, C., Otles, S., Karta, C., Food waste management, valorization, and sustainability in the food industry. *Food Waste Recovery*, 1, 3-19, 2021.
15. Chauhan, A., Debnath, R.M., Singh, S.P., Modelling the drivers for sustainable agri-food waste management. *Benchmarking*, 25, 981, 2018.
16. Sikorska, W., Musioł, M., Zawidlak-Węgrzyńska, B., Rydz, J., End-of-Life Options for (Bio)degradable Polymers in the Circular Economy. *Adv. Polym. Technol.*, 2021, 1, 2021.
17. Alonso-Muñoz, S., García-Muiña, F.E., Medina-Salgado, M.S., González-Sánchez, R., Towards circular economy practices in food waste management: a retrospective overview and a research agenda. *Br. Food J.*, 124, 478, 2022.
18. Battista, F., Frison, N., Pavan, P., Cavinato, C., Gottardo, M., Fatone, F., Eusebi, A.L., Majone, M., Zeppilli, M., Valentino, F., Fino, D., Tommasi, T., Bolzonella, D., Food wastes and sewage sludge as feedstock for an urban biorefinery producing biofuels and added-value bioproducts. *J. Chem. Technol. Biotechnol.*, 95, 328, 2020.
19. Paul, S., Dutta, A., Defersha, F., Dubey, B., Municipal Food Waste to Biomethane and Biofertilizer: A Circular Economy Concept. *Waste Biomass Valorization*, 9, 601, 2018.
20. Fogarassy, C., Nagy-Pércsi, K., Ajibade, S., Gyuricza, C., Ymeri, P., Relations between circular economic "principles" and organic food purchasing behavior in Hungary. *Agronomy*, 10, 616, 2020.
21. Teigiserova, D.A., Hamelin, L., Thomsen, M., Towards transparent valorization of food surplus, waste and loss: Clarifying definitions, food waste hierarchy, and role in the circular economy. *Sci. Total Environ.*, 706, 136033, 2020.
22. Moretto, G., Russo, I., Bolzonella, D., Pavan, P., Majone, M., Valentino, F., An urban biorefinery for food waste and biological sludge conversion into polyhydroxyalkanoates and biogas. *Water Res.*, 170, 115371, 2020.
23. Mak, T.M.W., Xiong, X., Tsang, D.C.W., Yu, I.K.M., Poon, C.S., Sustainable food waste management towards circular bioeconomy: Policy review, limitations and opportunities. *Bioresour. Technol.*, 297, 122497, 2020.
24. Istrate, I.R., Medina-Martos, E., Galvez-Martos, J.L., Dufour, J., Assessment of the energy recovery potential of municipal solid waste under future scenarios. *Appl. Energy*, 293, 116915, 2021.

25. Tampio, E.A., Blasco, L., Vainio, M.M., Kahala, M.M., Rasi, S.E., Volatile fatty acids (VFAs) and methane from food waste and cow slurry: Comparison of biogas and VFA fermentation processes. *GCB Bioenergy*. 11, 72, 2019.
26. Gupta, M., Ho, D., Santoro, D., Torfs, E., Doucet, J., Vanrolleghem, P.A., Nakhla, G., Experimental assessment and validation of quantification methods for cellulose content in municipal wastewater and sludge. *Environ. Sci. Pollut. Res.*, 25, 16743, 2018.
27. Palmieri, S., Cipolletta, G., Pastore, C., Giosuè, C., Akyol, Ç., Eusebi, A.L., Frison, N., Tittarelli, F., Fatone, F., Pilot scale cellulose recovery from sewage sludge and reuse in building and construction material. *Waste Manage.*, 100, 208, 2019.
28. Li, S., Wu, Z., Wu, Z., Liu, G., Enhancing fiber recovery from wastewater may require toilet paper redesign. *J. Clean. Prod.*, 261, 121138, 2020.
29. Lee, K.W., Lee, W.C., Lee, H.J., Dong, J.I., Gasification characteristics of sewage sludge combined with wood biomass. *J. Mater. Cycles Waste Manage.*, 16, 642, 2014.
30. Champagne, P., Li, C., Enzymatic hydrolysis of cellulosic municipal wastewater treatment process residuals as feedstocks for the recovery of simple sugars. *Bioresour. Technol.*, 100, 5700, 2009.
31. Ahmed, A.S., Bahreini, G., Ho, D., Sridhar, G., Gupta, M., Wessels, C., Marcelis, P., Elbeshbishy, E., Rosso, D., Santoro, D., Nakhla, G., Fate of cellulose in primary and secondary treatment at municipal water resource recovery facilities. *Water Environ. Res.*, 91, 1479, 2019.
32. Jele, T.B., Sithole, B., Lekha, P., Andrew, J., Characterisation of pulp and paper mill sludge for beneficiation. *Cellulose*, 29, 4629, 2022.
33. Nair, A.S., Al-Battashi, H., Al-Akzawi, A., Annamalai, N., Gujarathi, A., Al-Bahry, S., Dhillon, G.S., Sivakumar, N., Waste office paper: A potential feedstock for cellulase production by a novel strain Bacillus velezensis ASN1. *Waste Manage.*, 79, 491, 2018.
34. Duncan, S.M., Alkasrawi, M., Gurram, R., Almomani, F., Wiberley-Bradford, A.E., Singsaas, E., Paper mill sludge as a source of sugars for use in the production of bioethanol and isoprene. *Energies*, 13, 4662, 2020.
35. Peretz, R., Mamane, H., Wissotzky, E., Sterenzon, E., Gerchman, Y., Making Cardboard and Paper Recycling More Sustainable: Recycled Paper Sludge For Energy Production and Water-Treatment Applications. *Waste Biomass Valorization*, 12, 1599, 2021.
36. Rosiek, K., Directions and challenges in the management of municipal sewage sludge in Poland in the context of the circular economy. *Sustainability*, 12, 3686, 2020.
37. Micolucci, F., Gottardo, M., Cavinato, C., Pavan, P., Bolzonella, D., Mesophilic and thermophilic anaerobic digestion of the liquid fraction of pressed biowaste for high energy yields recovery. *Waste Manage.*, 48, 227, 2016.
38. Lynd, L.R., Weimer, P.J., van Zyl, W.H., Pretorius, I.S., Microbial Cellulose Utilization: Fundamentals and Biotechnology. *Microbiol. Mol. Biol. Rev.*, 66, 506, 2002.

39. Akyol, C., Eusebi, A.L *et al*, Cellulosic materials recovery from municipal wastewater: from treatment plants to the market, in: *Clean Energy and Resource Recovery: Wastewater Treatment Plants as Biorefineries*, K.V. Tyagi, K.J. An, Z. Cetecioglu (Ed.), pp. 125-136, Elsevier, Amsterdam, 2022.
40. Yuan, J.S., Tiller, K.H., Al-Ahmad, H., Stewart, N.R., Stewart, C.N., Plants to power: bioenergy to fuel the future. *Trends Plant Sci.*, 13, 421, 2008.
41. Tsuneda, S., Aikawa, H., Hayashi, H., Yuasa, A., Hirata, A., Extracellular polymeric substances responsible for bacterial adhesion onto solid surface. *FEMS Microbiol. Lett.*, 223, 287, 2003.
42. Yiin, C.L., Ho, S., Yusup, S., Quitain, A.T., Chan, Y.H., Loy, A.C.M., Gwee, Y.L., Recovery of cellulose fibers from oil palm empty fruit bunch for pulp and paper using green delignification approach. *Bioresour. Technol.*, 290, 121797, 2019.
43. Lee, K.S., Kwon, T.H., Park, T., Jeong, M.S., Microbiology and microbial products for enhanced oil recovery, in: *Theory and Practice in Microbial Enhanced Oil Recovery*, K.S. Lee, T-H. Kwon, T. Park, M.S. Jeong (Ed.), pp. 27-65, Elsevier, Houston, 2020.
44. Guangyin, Z., Youcai, Z., Harvest of bioenergy from sewage sludge by anaerobic digestion, in: *Pollution Control and Resource Recovery for Sewage Sludge*, Z. Guangyin (Ed.), pp. 181-273, Elsevier, Amsterdam, 2017.
45. Nagtode, V.S., Cardoza, C., Yasin, H.K.A., Mali, S.N., Tambe, S.M., Roy, P., Singh, K., Goel, A., Amin, P.D., Thorat, B.R., Cruz, J.N., Pratap, A.P., Green Surfactants (Biosurfactants): A Petroleum-Free Substitute for Sustainability-Comparison, Applications, Market, and Future Prospects. *ACS Omega*, 8, 11674, 2023.
46. Bohra, V., Ahamad, K.U *et al*, Energy and resources recovery from wastewater treatment systems, in: *Clean Energy and Resource Recovery: Wastewater Treatment Plants as Bio-refineries*, A. An, V. Tyagi, M. Kumar, Z. Cetecioglu (Ed.), pp. 17-36, Elsevier, Amsterdam, 2022.
47. Ruiken, C.J., Breuer, G., Klaversma, E., Santiago, T., van Loosdrecht, M.C.M., Sieving wastewater - Cellulose recovery, economic and energy evaluation. *Water Res.*, 47, 43, 2013.
48. Ruiz Cuilty, K., Ballinas-Casarrubias, L., Rodríguez de San Miguel, E., de Gyves, J., Robles-Venzor, J.C., González-Sánchez, G., Cellulose recovery from Quercus sp. sawdust using Ethanosolv pretreatment. *Biomass Bioenergy*, 111, 114, 2018.
49. Zhang, L., Liu, Y., Li, Z., Effects of reduced severity of ammonium sulfite pretreatment on bamboo for high cellulose recovery. *RSC Adv.*, 9, 30489, 2019.
50. Tang, S., Dong, Q., Fang, Z., Miao, Z. diao, Complete recovery of cellulose from rice straw pretreated with ethylene glycol and aluminum chloride for enzymatic hydrolysis. *Bioresour. Technol.*, 284, 98, 2019.
51. Glińska, K., Ismail, M.S. Bin, Goma-Camps, J., Valencia, P., Stüber, F., Giralt, J., Fabregat, A., Torrens, E., Olkiewicz, M., Bengoa, C., Recovery and characterisation of cellulose from industrial paper mill sludge using tetrakis and imidazolium based ionic liquids. *Ind. Crops Prod.*, 139, 111556, 2019.

52. Glińska, K., Lerigoleur, C., Giralt, J., Torrens, E., Bengoa, C., Valorization of cellulose recovered from wwtp sludge to added value levulinic acid with a brønsted acidic ionic liquid. *Catalysts*, 10, 1, 2020.
53. Glińska, K., Stüber, F., Fabregat, A., Giralt, J., Font, J., Mateo-Sanz, J.M., Torrens, E., Bengoa, C., Moving municipal WWTP towards circular economy:Cellulose recovery from primary sludge with ionic liquid. *Resour. Conserv. Recycl.*, 154, 104626, 2020.
54. Liu, R., Zhang, J., Sun, S., Bian, Y., Hu, Y., Dissolution and recovery of cellulose from pine wood bits in ionic liquids and a co-solvent component mixed system. *J. Eng. Fiber. Fabr.*, 14, 155892501983844, 2019.
55. Abdullah, Hamid, Christwardana, M., Hadiyanto, H., Optimization of cellulase production by aspergillus NigerITBCC L74 with bagasse as substrate using response surface methodology. *HAYATI J. Biosci.*, 25, 115, 2018.
56. Abdullah, B., Maftukhah, S., Listyaningrum, E., Faradhiba, F., Effect of some variable in cellulase production by Aspergillus niger ITBCC L74 using solid state fermentation. *IOP Conf. Ser. Mater. Sci. Eng.*, 316, 12066, 2018.
57. Bansal, N., Tewari, R., Soni, R., Soni, S.K., Production of cellulases from Aspergillus niger NS-2 in solid state fermentation on agricultural and kitchen waste residues. *Waste Manage.*, 32, 1341, 2012.
58. Amraini, S.Z., Ariyani, L.P., Hermansyah, H., Setyahadi, S., Rahman, S.F., Park, D.H., Gozan, M., Production and characterization of cellulase from E. coli EgRK2 recombinant based oil palm empty fruit bunch. *Biotechnol. Bioprocess Eng.*, 22, 287, 2017.
59. Gautam, S.P., Bundela, P.S., Pandey, A.K., Khan, J., Awasthi, M.K., Sarsaiya, S., Optimization for the Production of Cellulase Enzyme from Municipal Solid Waste Residue by Two Novel Cellulolytic Fungi. *Biotechnol. Res. Int.*, 2011, 1, 2011.
60. Pramanik, S.K., Mahmud, S., Paul, G.K., Jabin, T., Naher, K., Uddin, M.S., Zaman, S., Saleh, M.A., Fermentation optimization of cellulase production from sugarcane bagasse by Bacillus pseudomycoides and molecular modeling study of cellulase. *Curr. Res. Microb. Sci.*, 2, 100013, 2021.
61. Swathy, R., Rambabu, K., Banat, F., Ho, S.H., Chu, D.T., Show, P.L., Production and optimization of high grade cellulase from waste date seeds by Cellulomonas uda NCIM 2353 for biohydrogen production. *Int. J. Hydrogen Energy*, 45, 22260, 2020.
62. Silva, A.F.V., Santos, L.A., Valença, R.B., Porto, T.S., Da Motta Sobrinho, M.A., Gomes, G.J.C., Jucá, J.F.T., Santos, A.F.M.S., Cellulase production to obtain biogas from passion fruit (Passiflora edulis) peel waste hydrolysate. *J. Environ. Chem. Eng.*, 7, 103510, 2019.
63. Rosyida, V.T., Hayati, S.N., Indrianingsih, A.W., Maryana, R., Purwesti, Y.A., Ayesda, C.S., Enzim Selulase Kasar Aspergillus niger FNCC 6018 untuk Produksi Bioetanol melalui Proses Sakarifikasi dan Fermentasi Serentak. *J. Mikol. Indones.*, 2, 77, 2018.

64. Shokrkar, H., Zamani, M., Ebrahimi, S., Exploring strategies for the use of mixed microalgae in cellulase production and its application for bioethanol production. *Biofuels, Bioprod. Biorefin.*, 16, 816, 2022.
65. Lodha, A., Pawar, S., Rathod, V., Optimised cellulase production from fungal co-culture of Trichoderma reesei NCIM 1186 and Penicillium citrinum NCIM 768 under solid state fermentation. *J. Environ. Chem. Eng.*, 8, 103958, 2020.
66. Elias, M.A., Hadibarata, T., Sathishkumar, P., Modified oil palm industry solid waste as a potential adsorbent for lead removal. *Environ. Chem. Ecotoxicol.*, 3, 1, 2021.
67. Fan, Y. Van, Lee, C.T., Lim, J.S., Klemeš, J.J., Le, P.T.K., Cross-disciplinary approaches towards smart, resilient and sustainable circular economy. *J. Clean. Prod.*, 232, 1482, 2019.
68. Sukiran, M.A., Abnisa, F., Wan Daud, W.M.A., Abu Bakar, N., Loh, S.K., A review of torrefaction of oil palm solid wastes for biofuel production. *Energy Convers. Manage.*, 149, 101, 2017.
69. Lam, S.S., Tsang, Y.F., Yek, P.N.Y., Liew, R.K., Osman, M.S., Peng, W., Lee, W.H., Park, Y.K., Co-processing of oil palm waste and waste oil via microwave co-torrefaction: A waste reduction approach for producing solid fuel product with improved properties. *Process Saf. Environ. Prot.*, 128, 30, 2019.
70. Mekhilef, S., Saidur, R., Safari, A., Mustaffa, W.E.S.B., Biomass energy in Malaysia: Current state and prospects. *Renew. Sustain. Energy Rev.*, 15, 3360, 2011.
71. Nyakuma, B.B., Bioelectricity potential of oil palm waste in Malaysia, *Proceeding of the 3rd International Conference Research & Education in Natural Sciences (HERTSPO 2015)*, Shkodra BENA, Shkoder, Albania, 2015.
72. Silveira, E.A., Tardioli, P.W., Farinas, C.S., Valorization of Palm Oil Industrial Waste as Feedstock for Lipase Production. *Appl. Biochem. Biotechnol.*, 179, 558, 2016.
73. Sinjaroonsak, S., Chaiyaso, T., H-Kittikun, A., Optimization of Cellulase and Xylanase Productions by Streptomyces thermocoprophilus TC13W Using Low Cost Pretreated Oil Palm Empty Fruit Bunch. *Waste Biomass Valorization*, 11, 3925, 2020.
74. Zulkarnain, A., Bahrin, E.K., Ramli, N., Phang, L.Y., Abd-Aziz, S., Alkaline Hydrolysate of Oil Palm Empty Fruit Bunch as Potential Substrate for Biovanillin Production via Two-Step Bioconversion. *Waste Biomass Valorization*, 9, 13, 2018.
75. Lim, S.L., Wu, T.Y., Characterization of Matured Vermicompost Derived from Valorization of Palm Oil Mill Byproduct. *J. Agric. Food Chem.*, 64, 1761, 2016.
76. Hau, L.J., Shamsuddin, R., May, A.K.A., Saenong, A., Lazim, A.M., Narasimha, M., Low, A., Mixed Composting of Palm Oil Empty Fruit Bunch (EFB) and Palm Oil Mill Effluent (POME) with Various Organics: An Analysis on Final

Macronutrient Content and Physical Properties. *Waste Biomass Valorization*, 11, 5539, 2020.
77. Norrie, J and Fierro A., Paper sludge as soil conditioners, *Handbook of Soil Conditioners*, pp. 97-118, CRC Press, 2020.
78. Alkasrawi, M., Al-Othman, A., Tawalbeh, M., Doncan, S., Gurram, R., Singsaas, E., Almomani, F., Al-Asheh, S., A novel technique of paper mill sludge conversion to bioethanol toward sustainable energy production: Effect of fiber recovery on the saccharification hydrolysis and fermentation. *Energy*, 223, 120018, 2021.
79. Keplinger, T., Wang, X., Burgert, I., Nanofibrillated cellulose composites and wood derived scaffolds for functional materials. *J. Mater. Chem. A.*, 7, 2981, 2019.
80. Mokhena, T.C., John, M.J., Cellulose nanomaterials: new generation materials for solving global issues. *Cellulose*, 27, 1149, 2020.
81. Bian, H., Wei, L., Lin, C., Ma, Q., Dai, H., Zhu, J.Y., Lignin-Containing Cellulose Nanofibril-Reinforced Polyvinyl Alcohol Hydrogels. *ACS Sustain. Chem. Eng.*, 6, 4821, 2018.
82. Du, H., Liu, W., Zhang, M., Si, C., Zhang, X., Li, B., Cellulose nanocrystals and cellulose nanofibrils based hydrogels for biomedical applications. *Carbohydr. Polym.*, 209, 130, 2019.
83. Zhang, Y., Hao, N., Lin, X., Nie, S., Emerging challenges in the thermal management of cellulose nanofibril-based supercapacitors, lithium-ion batteries and solar cells: A review. *Carbohydr. Polym.*, 234, 115888, 2020.
84. Du, H., Parit, M., Wu, M., Che, X., Wang, Y., Zhang, M., Wang, R., Zhang, X., Jiang, Z., Li, B., Sustainable valorization of paper mill sludge into cellulose nanofibrils and cellulose nanopaper. *J. Hazard. Mater.*, 400, 123106, 2020.
85. McNamara, P.J., Koch, J.D., Liu, Z., Zitomer, D.H., Pyrolysis of Dried Wastewater Biosolids Can Be Energy Positive. *Water Environ. Res.*, 88, 804, 2016.
86. Carey, D.E., McNamara, P.J., Zitomer, D.H., Biochar from Pyrolysis of Biosolids for Nutrient Adsorption and Turfgrass Cultivation. *Water Environ. Res.*, 87, 2098, 2015.
87. Liu, Z., Hughes, M., Tong, Y., Zhou, J., Kreutter, W., Lopez, H.C., Singer, S., Zitomer, D., McNamara, P., Paper mill sludge biochar to enhance energy recovery from pyrolysis: A comprehensive evaluation and comparison. *Energy*, 239, 121925, 2022.
88. Keskisaari, A., Kärki, T., The use of waste materials in wood-plastic composites and their impact on the profitability of the product. *Resour. Conserv. Recycl.*, 134, 257, 2018.
89. Zhou, Y., Stanchev, P., Katsou, E., Awad, S., Fan, M., A circular economy use of recovered sludge cellulose in wood plastic composite production: Recycling and eco-efficiency assessment. *Waste Manage.*, 99, 42, 2019.

15

Recycled Wastewater from Sewage Treatment Plant for Sustainable Agriculture: An Indian Scenario

Aruna Jyothi Kora[1,2]

[1]National Centre for Compositional Characterisation of Materials (NCCCM), Bhabha Atomic Research Centre (BARC), ECIL PO, Hyderabad, Telangana, India
[2]Homi Bhabha National Institute (HBNI), Anushakti Nagar, Mumbai, India

Abstract

Global food security depends on sustainable water management, particularly in light of climate change and unpredictable weather. The agricultural industry uses around 70% of the world's freshwater, while water use efficiency is about 40% in countries like India. With more than 17% of the world's population, India has to live on 4% of the world's renewable water. It indicates tremendous pressure on using these limited resources and the need to focus on efficient use. In this context, an alternate option for water conservation is the utilization of recycled wastewater from sewage treatment plants (STPs) for sustainable agriculture. As the technology for STP is well established in India, if it is clubbed with water disinfection, the produced water with reduced microbial load can be effectively utilized for agriculture. The wastewater treatment at STPs involves primary, secondary, and tertiary treatments. The recycled/treated water obtained from the STPs is disinfected using a combination of chlorination, ozonation, ultraviolet, and gamma irradiation. Disinfection aims to reduce the number of microorganisms such as bacteria, fungi, viruses, and protozoa, so the disinfected water is used for irrigation and farming. The generated sludge after disinfection can be used as compost. Thus, the recycled water generated from STPs after disinfection treatment should meet the specifications required for irrigation water. The infrastructure needed for treating, storing, and delivering reclaimed water from the STP should be built in every municipality in the country. In addition, adopting measures such as installing better watering systems, selecting drought-tolerant crops, rainwater harvesting,

Email: koramaganti@gmail.com; koraaj@barc.gov.in

Inamuddin, Tariq Altalhi, Mohammad Luqman and Joseph Kapuku (eds.) Sewage and Biomass from Wastewater to Energy, (391–412) © 2024 Scrivener Publishing LLC

organic farming, and best agricultural practices such as crop rotation lead to sustainable agriculture in India.

Keywords: Agriculture, disinfection, irrigation, sewage treatment plant, sustainable, wastewater, irradiation, chlorination

15.1 Introduction

In India, agriculture farming is the main source of livelihood [1] for 60% of the country's population and generates 14% of the country's GDP and 11% of exports. Notably, India occupies first and second positions globally, in terms of arable and gross irrigated land areas, respectively [2]. Approximately, 70% of the global freshwater is exclusively used in the agricultural sector [3, 4]. Although 80% of current surface water is used for irrigation, India records only 38% of water use efficiency in agriculture [2, 5]. India, which has 18% of the global population, only has 4% of the world's freshwater resources [6]. Nearly half the world's population will reside in overstressed water supply countries during the next 10 to 15 years [7]. Therefore, tremendous pressure is exerted on efficiently utilizing these limited resources [8]. Therefore, effective water management is essential for ensuring global food security, particularly in light of climate change and the irregular nature of the weather [9, 10].

Gray water is the residential wastewater produced from bathing showers, hand-washing basins, kitchen sinks, kitchen sewage, laundry, and washing machines without toilet wastes [11]. Meanwhile, sewage is the wastewater generated from human activity in homes, agriculture, industries, hospitals, offices, and other settings [12, 13]. Wastewater is a significant, stable, reliable, and cost-effective source of water that is underutilized in many urban and peri-urban areas of India [14, 15]. The means of support and food security of numerous marginal farmers in Hyderabad, particularly those positioned around the Musi River, directly benefit from wastewater irrigation through the cultivation of para grass and leafy vegetables [16–18]. It was estimated that approximately 1.1 million hectares of land in urban and peri-urban areas could be irrigated based on wastewater availability from chosen Indian cities [17]. Also, the generated sludge from treated wastewater can serve as manure for 3 million hectares of land, thus reducing fertilizer dependency by 40% [15].

There are many concerns and issues about the utilization of untreated wastewater for irrigation and agriculture, such as high chemical oxygen demand (COD) and biological oxygen demand (BOD) [19]; soil salinization [16, 20]; pathogenic microbial, protozoan, and helminth load [12, 16, 17, 21]; heavy metal buildup and microbial contamination in soil and grown leafy greens and vegetables [4, 16, 17, 21, 22]; the stigma associated with wastewater cultivated vegetables; skin-related health problems; elevated hookworm; and Hepatitis E infections in wastewater irrigating farmers and workers [14, 16]. There are four important pathways through which human exposure to wastewater occurs: eating food that is produced on land cultivated on wastewater, eating animal products that graze on wastewater-irrigated land, drinking water from groundwater that is polluted with wastewater, and inhaling volatile compounds during irrigation [21].

Additionally, contaminated water is the primary cause of many waterborne infections in India [12]. For example, waterborne disease outbreaks and deaths were reported in Bolakpur, Hyderabad, due to the contamination of drinking water with sewage via its seepage into drinking water distribution pipes [23]. In the current scenario of the COVID-19 pandemic, wastewater was found to be the source of fecal shed virus in many countries, including India, and the spread of SARS-CoV-2 viral infection in a given locality was estimated and monitored with sewage surveillance [11, 24].

15.2 Sewage Treatment Plants

The sewage treatment is a multistep process for cleaning wastewater before discharge or reuse. The wastewater at sewage treatment plants (STPs) typically undergoes primary (sedimentation and primary filtration), secondary (trickling filter, aeration, and oxidation), and tertiary (alum injection, sand media filtration, and activated carbon filtration) treatments [14, 20, 25]. In India, an array of technologies such as activated sludge process (ASP), extended aeration (EA), fluidized aerobic bed reactor (FAB), moving bed biofilm reactor (MBBR), oxidation pond (OP), sequencing batch reactor (SBR), upflow anaerobic sludge blanket (UASB), and waste stabilization pond (WSP) is used for treating wastewater [13]. The image of sewage effluents undergoing secondary treatment in the aeration tank of an STP is given in Figure 15.1. In the year 2020–2021, in 35 states and union territories of India; there are 1631 STPs with a cumulative capacity of 36,668 million liters per day (MLD) for treating generated sewage of 72,368 MLD from urban centers [13]. Most of the operating STPs cannot meet the existing demands due to their limited capacity; only 28% are treated and the remaining 72% are untreated and disposed into water bodies

Figure 15.1 The image of sewage effluents undergoing secondary treatment in the aeration tank of an STP.

and groundwater [15]. A list containing some of the running STPs operated by the Hyderabad Metropolitan Water Supply and Sewerage Board (HMWSSB) at Hyderabad city of Telangana state is given in Table 15.1, in terms of their location, MLD capacity, technology adopted, and receiving water body [13, 26–28]. The HMWSSB is working on the abovementioned STPs and constructing additional STPs based on SBR technology across the city with a total capacity of 1259.5 MLD to achieve 100% treatment of generated sewage [29].

The treated wastewater from STPs can be utilized in an array of fields and avenues such as irrigation, farm forestry, horticulture, groundwater recharge, fish culture, toilet flushing, construction, and industrial reuse [14]. Therefore, recycled water offers numerous benefits, including easing the burden on available freshwater supplies, minimizing effluent dumping to surface or coastal waters, and providing a consistent water supply independent of rainfall sources [13, 14, 20]. Many countries like Israel, the United States, Kuwait, Italy, Spain, Malta, Germany, and Jordan are exploiting reclaimed gray water for irrigation [4, 11]. With improved treatment techniques and incentives for reclaimed sewage water to farmers, in terms of water recycling, Israel leads the globe by 87%, followed by Spain (20%) and the United States (5%). Fifty percent of the water in Israel's agriculture and farming originates from treated sewage [30].

The pathogens commonly found in wastewater are *Escherichia coli*, *Streptococcus*, *Salmonella*, *Shigella*, *Pseudomonas aeruginosa*, *Giardia lamblia*, enteroviruses, hookworms, etc. [4, 12, 16, 18, 21, 31, 32]. Of which, the total coliform bacteria (*Klebsiella*, *Enterobacter*, *Citrobacter*, *Salmonella*, *E. coli*, *Shigella*, *Proteus*, and *Serratia*) and fecal coliform (*E. coli*) serve as indicator bacteria for the hygienic quality of the water, as they are excreted by all warm-blooded animals, abundant in wastewater, outlive many pathogenic bacteria in the natural environment, and simple to detect and

Table 15.1 A list of running STPs operated at Hyderabad, in terms of their location, MLD capacity, technology adopted, and receiving water body.

STP name	Installed capacity (MLD)	Technology adopted	Receiving water body
Amberpet	339	UASB	Musi River
Attapur 1	23	SBR	Musi River
Attapur 2	51	SBR	Musi River
Durgam Cheruvu	5	ASP	Durgam Cheruvu
Khairatabad	20	SBR	Hussain Sagar Lake
Khaja Kunta	12	MBBR	Khaja Kunta Lake
Khajaguda	7	MBBR	Khajaguda Lake
KIMS	30	ASP	Hussain Sagar lake
Langer Houz	1.2	ASP + EA	Langer Houz Lake
Mir Alam Lake 1	10	ASP + EA	Mir Alam Cheruvu
Mir Alam Lake 2	5	ASP + EA	Mir Alam Cheruvu
Nacharam	10	EA	Pedda Cheruvu
Nagole	172	UASB	Musi River
Nalla Cheruvu	30	UASB	Musi River
Nanakramguda	4.5	MBBR	Pedda Cheruvu
Noor Mohammed Kunta	4	ASP + EA	Noor Mohammed Lake
Patel Cheruvu	2.5	EA	Patel Cheruvu
Pragati Nagar	2.5	ASP	Pragati Nagar Lake
Rangadhamuni	5	ASP	Rangadhamuni Cheruvu
Safilguda	0.6	ASP	Safilguda Cheruvu
Saroor Nagar	2.5	ASP	Saroor Nagar Cheruvu

enumerate [12, 33, 34]. The photograph of the growth of total coliforms and *E. coli* on differential, chromogenic culture media sampled from wastewater is shown in Figure 15.2, in terms of salmon red– (total coliforms) and blue-colored colonies (*E. coli*). Additionally, wastewater was found to be a possible hotspot for antibiotic-resistant bacteria [12].

Usually, the guidelines imposed for released effluents from STPs are limited to bulk parameters such as pH, temperature, conductivity, total suspended solids (TSS) (mg/L), total dissolved solids (TDS) (mg/L), alkalinity, hardness, dissolved oxygen (DO) (mg/L), COD (mg/L), biological oxygen demand (BOD) (mg/L), total nitrogen (mg/L), and phosphate (mg/L) and some extent coliforms [21, 25, 35]. Most of the STPs are made for effective organic matter and nutrient removal, but they are not planned and not effective for pathogenic microorganisms and coliform removal from wastewater. Although conventional STPs reduce the enteric microbe and pathogen levels, they depend on factors including influent sewage characteristics, treatment process type, pH, temperature, TSS removal, hydraulic retention time, and biological flora [14, 32, 34]. Thus, utilizing treated effluents from STPs requires mandatory incorporation of disinfection pre-treatment step at the tertiary level/stage before release for application, especially irrigation of farmlands [14]. The use of indicator bacteria such as total coliforms and fecal coliform assesses the efficacy of disinfection.

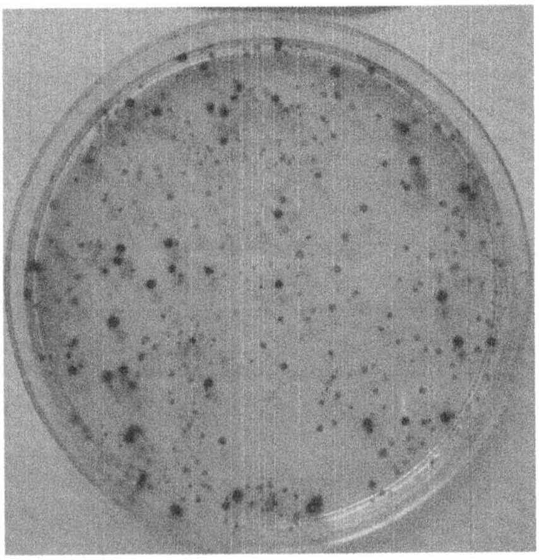

Figure 15.2 The photograph of the growth of salmon red–(total coliforms) and blue-colored colonies (*E. coli*) sampled from wastewater.

Based on the above facts, there is an urgent need to focus on reuse and it is evident that wastewater recycling is one of the possible solutions for attaining sustainable water management. Using reclaimed wastewater from STPs for sustainable agriculture also serves as an alternate choice for water conservation. As the technology for STPs is well established in India, if it is clubbed with water disinfection, the produced water with reduced microbial load can be effectively utilized for agriculture.

15.3 Water Disinfection

The Musi River–based farmers benefit from the high nutrient content of the wastewater in terms of faster growth and high yields and using any fertilizer/chemical inputs. However, they prefer further pre-treatment and are interested in paying a modest amount for good-quality irrigation water [16]. The common methods used to disinfect drinking water and reclaimed wastewater are chlorination, ozonation, UV irradiation, and gamma irradiation.

15.3.1 Chlorination

Among the different methods used, chlorination is widely utilized for the prevention of waterborne diseases, including cholera, typhoid, and dysentery, due to its low cost, well-established technology, ease of operation, availability of required materials, etc. [31, 33]. A typical chlorine dose of 5 to 20 mg/L for 20-min contact time is generally used for municipal wastewater disinfection dependent on effluent quality [33]. The process is also used for biofouling control and prevention of microbial corrosion in cooling water circuits of power plants [36–39]. The chlorination of drinking water and wastewater is achieved through sodium dichloroisocyanurate tablets, sodium hypochlorite solution, and chlorine gas cylinders (Figure 15.3). The parameters such as chlorine dose, contact time, and free residual chlorine levels should be monitored for efficient disinfection of pathogens [12]. The microbicidal action of the strong oxidizing agent, chlorine, is attributed to the inactivation of pathogens through cell wall destruction, alteration of cell permeability, oxidation of lipids, proteins, and enzymes, inhibition of enzyme activity, etc. [33]. It is important to study the formation of potential carcinogenic disinfection byproducts, organochlorine compounds, especially in waters of high ammonia and organic load and their continuous monitoring is required, in addition to safety aspects associated with handling chlorine gas [4, 21, 31, 33, 40].

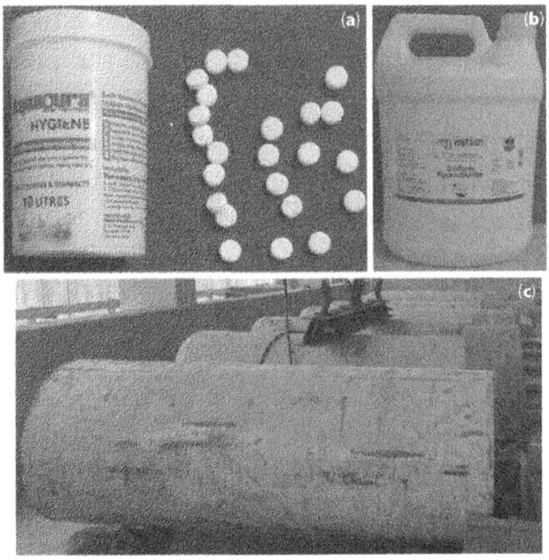

Figure 15.3 The photographs of (a) chlorine tablets, (b) sodium hypochlorite solution, and (c) chlorine gas cylinders used for chlorination.

15.3.2 Ozonation

Another important disinfection technology is ozonation which is mainly used for the disinfection of chlorine-resistant bacteria and the degradation of micropollutants such as antibiotics, biocides, pesticides, pharmaceuticals, and personal care products from wastewater [33, 41–43]. The colorless gas with strong germicidal and oxidizing properties disinfects bacteria, fungi, viruses, and protozoa, oxidizes metals (As, Fe, and Mn) to metal oxides, and eliminates odors, tastes, colors [43, 44]. The residue-free, environment-friendly ozone serves as a coagulant, deodorizer, detoxifier, potent disinfectant, parasiticide, germicide, and viricide [43–45]. In addition to water treatment, it is also exploited for sanitization, washing, and disinfection of equipment; removal of fungicides and pesticides; disinfection; spoilage prevention of foods; and processing of fruits, vegetables, meats, and seafood for retaining sensory, nutritional, physicochemical, and organoleptic characteristics [46]. It is generated onsite at home or in industrial settings based on corona discharge, UV radiation, and electrolysis techniques. Using commercially available *in situ* ozone generators or electrostatic air purifiers safety issues related to transport and gas storage can be avoided [44, 46]. The photograph of a kitchen ozonator used for the degradation of pesticides present on the fruit and vegetable surface is shown in Figure 15.4. The disinfection efficiency of ozone depends on ozone concentration, ambient temperature,

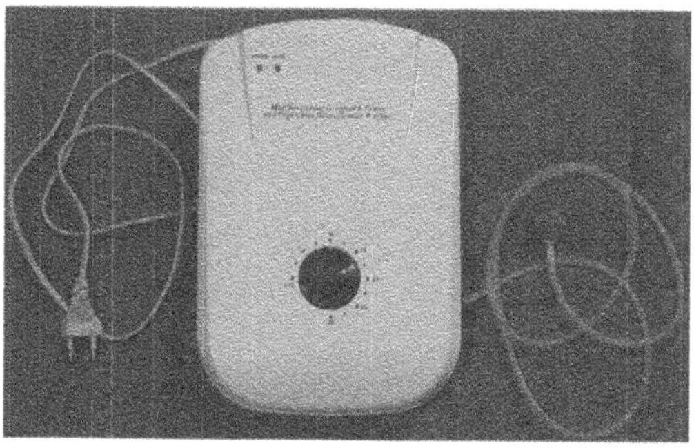

Figure 15.4 The photograph of the kitchen ozonator used for pesticide degradation on fruit and vegetable surfaces.

and humidity; exposure duration and a typical residual dose of 0.5 mg/L after a contact time of 20 min are used [44]. Reactive oxygen species generation, oxidation of proteins, enzymes, lipids, nucleic acids, cell wall, membrane, and nucleic acid damage, as well as double-stranded breaks, decreased transcription, cell lysis, etc., are accounted for the antimicrobial action of ozone [44, 46]. Important limitations such as poor solubility, toxicity, fire hazards, high investment costs, and untested ozonation byproducts should be considered while designing a disinfection strategy for wastewater [4, 45].

15.3.3 Ultraviolet (UV) Irradiation

The germicidal UV C (200 to 280 nm) light is commonly used for drinking water, wastewater, air disinfection and purification, and surface decontamination at typical doses in the range of 10 to 30 mJ/cm^2 [47, 48]. It serves as an alternative photochemical disinfection method in sanitizing open and public areas and decontaminating N95 respirators, surgical face masks, and personal protective equipment, especially during the COVID-19 pandemic [47, 49]. The nonionizing UV irradiation is also employed as a nonthermal method for treating liquid foods (fruit juices, alcoholic beverages, and soft drinks), disinfection of dairy products, processing of ready to eat meat products, and increasing the shelf life of fresh produce via inactivation prevention of foodborne pathogens [34]. UV light kills or inactivates microbes, including bacteria, fungi, and viruses, by damaging and destroying the nucleic acids (DNA and RNA) via pyrimidine dimer formation, preventing their replication and reproduction [40, 47, 48]. The photographs of the germicidal UV C lamp, UV-based air purifier, and UV lamp–bearing laminar flow cabinet that used surface sterilization

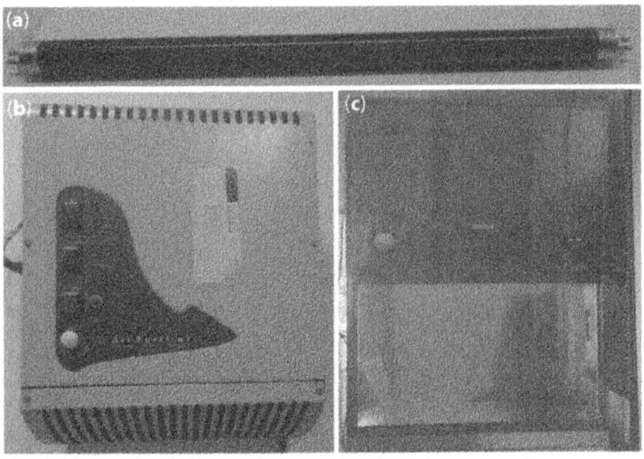

Figure 15.5 The photographs of (a) the germicidal UV C lamp, (b) UV-based air purifier, and (c) UV lamp–bearing laminar flow cabinet used surface sterilization.

are shown in Figure 15.5. The important limitations are low penetration depth, shadowing effects, environmentally hazardous mercury content in fragile quartz–based UV lamps, and associated human skin carcinogenicity [34, 47].

15.3.4 Gamma Irradiation

The ionizing gamma irradiation is used to disinfect drinking water, wastewater, and sewage sludge. In this scenario, a ^{60}Co source–based gamma irradiator, known as sludge hygenization research irradiator (SHRI), was commissioned by the Department of Atomic Energy, Government of India, to treat municipal raw sewage, treated effluents, and generated sludge at Vadodara, Gujarat. The radiation-disinfected, coliform load–reduced effluents are successfully utilized for irrigation [50]. If the microbial load is reduced, the sewage sludge generated from STPs is abundant in different nutrients and minerals which can be utilized as agriculture manure. The SHRI is employed commercially for radiation-based disinfection of sewage sludge of 110-m^3 volume, at a dose of 3 kGy [51]. The hygienized sludge can be used as a carrier for inoculation and enrichment of biofertilizers such as nitrogen-fixing (*Rhizobium*, *Azotobacter*, *Azospirillum*, and *Acetobacter*) and phosphate-, potassium-, and zinc-and-silicate–solubilizing and siderophore-producing bacteria (*Pseudomonas* sp, *Bacillus* sp) [52] and used as organic manure and fertilizer for agriculture [53, 54]. The eco-friendly, cold physical process of gamma irradiation is extensively exploited for food perseveration at low (<1 kGy), medium-low (1 to 10 kGy), and high-low (>0 kGy) doses. The low doses are applied for spouting inhibition in stems, tubers, bulbs, and rhizomes (onion, potato, ginger, and corm); delayed ripening and senescence in fruits and vegetables (mango); controlling insect and parasite infestation in stored products (cereals, pulses, oil seeds, and nuts); and quarantining. The medium doses are utilized for shelf-life extension of seafood, poultry, meat, meat products, and ready-to-eat products as well as inactivation of the virus on food packaging materials, and the high doses are applied for the hygenization of spices and sterilization of medical and healthcare products [34, 55–58]. The important aspects related to radiation safety and hazard issues and high initial capital cost should be considered [34].

The selection or combination of disinfection technologies (chlorination, ozonation, UV, and gamma irradiation) depends upon various factors such as wastewater volume, STP design, infrastructure, and operational costs, as well as micropollutant levels, microbial load, residual dose, disinfection by-products, etc. After disinfection of recycled/treated water from the STPs, the generated sludge can be commercialized as crop manure/compost with or without microbial inoculation [31, 59].

15.4 Government Policies, Initiatives, and Incentives

Different policies linked to water, sanitation, urban development and environment (national water policy, national urban sanitation policy, national policy on fecal sludge and septage management, national policy on safe reuse of treated water, and national environment policy); regulating bodies [National Green Tribunal, State, Central Pollution Control Board, Central Public Health and Environmental Engineering Organization (CPHEEO)]; programs [Atal Mission for Rejuvenation of Urban Transformation (AMRUT), *Swachh Bharat* Mission (SBM), Jal Jeevan Mission (JJM), National Mission for Clean Ganga (NMCG), and National Rural Drinking Water Programme (NRDWP)]; and monitoring authorities/ministries (Ministry of Jal Shakti, Ministry of Water Resources, River Development and Ganga Rejuvenation, Ministry of Environment, Forest and Climate Change, Ministry of Housing and Urban Affairs, and Ministry of Urban Development) under government of India are working in tandem toward the goal of sustainable water management [15]. The National Green Tribunal, an Indian statutory body dealing with environmental cases and speedy implementation of decisions related to it, passed an order for stringent implementation of norms for discharge of effluents from STPs in terms of TSS (10 mg/L), BOD (<10 mg/L), nitrogen (<10 mg/L), fecal coliform (<230 MPN/100 mL) [60].

The HMWSSB started a campaign to sell 5000 L of cleaned sewage water for ₹ 125 [61]. Under the *Swachh Bharat* Mission (SBM), the CPHEEO, the Ministry of Housing and Urban Affairs, the government of India gave guidelines for declaring a city/ward/circle/zone as SBM Water Plus. Under this directive, a minimum of 25% of treated wastewater from STPs should be reused for horticulture, agriculture, industrial use, construction, cleaning roads, etc. [62]. Hyderabad was one of the eight cities selected for SBM Water Plus status and was recycling 28% of treated wastewater. Most of the treated water is used to cultivate paddy, leafy greens and vegetables along the Musi River, irrigation of parks, and construction of flyovers, roads, footpaths, etc. [63]. Even small villages and Gram Panchayats such as Gowtampur in Telangana state stood first at the national level in Arogya (healthy) Panchayat theme in recognition of efforts toward sewage management through STP, utilization of reclaimed wastewater for farming and gardening, as well as better sanitation practices [64]. Also, housing and residential societies and associations are requesting subsidized or reduced electricity tariffs for the operation and maintenance of STPs constructed in gated communities, residential complexes, and group apartments [65].

15.5 Irrigation and Agriculture

For optimum utilization of reclaimed wastewater from STPs, the following criteria should be taken care of: groundwater contamination, deterioration of soil properties, the effect of salts on water uptake and plant metabolism, crop salt tolerance capability, and spread of infection among the farm workers and consumers devouring the produce. The crops grown on reclaimed wastewater can be categorized as nonedible, commercial crops (cotton, jute, fodder, and tobacco), crops that are cooked before consumption (cereals, pulses, and potatoes), and crops that are not consumed (seed multiplication) [66]. The norms recommended by CPHEEO for treated sewage used for horticulture and agriculture are given in Table 15.2, in terms of different parameters such as color, odor, pH, temperature, turbidity, TSS, TDS, oil and grease, COD, BOD, total Kjeldahl nitrogen as N, nitrate Nitrogen as N, dissolved phosphorous as P, fecal coliform, helminthic eggs, and minimum residual chlorine [13]. The recommended microbial standards in terms of fecal coliforms by the World Health Organization (WHO) for treated wastewater are 10^3 most probable number (MPN)/100 mL for irrigation and aquaculture [4, 14].

Table 15.2 The maximum permissible limits of various parameters for treated sewage used for horticulture and agriculture, as per Central Public Health and Environmental Engineering Organization.

Parameter	Maximum permissible limit			
	Horticulture crops	Nonedible crops	Edible crops	
			Raw	Cooked
Color	Colorless	As arising	Colorless	Colorless
Odor	Aseptic	Aseptic	Aseptic	Aseptic
pH	6.5–8.3	6.5–8.3	6.5–8.3	6.5–8.3
Temperature	Ambient	Ambient	Ambient	Ambient
Turbidity (NTU)	<2	As arising	<2	As arising
TSS (mg/L)	Nil	30	Nil	30
TDS (mg/L)	2100	2100	2100	2100
Oil and grease (mg/L)	10	10	Nil	Nil
COD (mg/L)	As arising	30	As arising	30
BOD (mg/L)	10	20	10	20
Total Kjeldahl nitrogen as N (mg/L)	10	10	10	10
Nitrate nitrogen as N (mg/L)	10	10	10	10
Dissolved phosphorous as P	2	5	2	5
Fecal coliform (MPN/100 mL)	Nil	230	Nil	230
Helminthic eggs (No/L)	As arising	<1	<1	<1
Minimum residual chlorine (mg/L)	1	Nil	Nil	Nil

In India, the Indian Standard (IS) IS 11624: 2019 adopted by the Bureau of Indian Standards (BIS) and approved by the Food and Agriculture Division Council regulates irrigation water quality. The suitability of irrigation water depends upon various factors, including water quality, soil type, crop characteristics, irrigation technique, drainage, and local climatic conditions [16, 17, 22, 66, 67]. The standard incorporated various aspects of irrigation water, including groundwater quality grouping, toxic, trace, and heavy metal concentration; irrigation techniquebased water quality; high saline and alkaline water suitability; and alkaline stress tolerance capabilities of crops [67]. Based on pH, electrical conductivity (EC) (dS/m), sodium absorption ratio (SAR) (mmol/L), and residual sodium carbonate (RSC) (me/L), the water quality is grouped as good, saline (marginal, saline, high), alkali (marginal, alkali, high), and toxic water (Table 15.3). Based on the boron (mg/L) content, the irrigation water quality is rated as low (<1 mg/L), medium (1 to 2 mg/L), high (2 to 4 mg/L), or very high (>4 mg/L). Depending on the irrigation technique (surface, sprinkler, and drip), the maximum permitted concentrations of ferrous iron (>1.5 mg/L), bicarbonate (8.5 me/L), and manganese (Mn) (>1.5 mg/L) are monitored for prevention of black or brown stains on foliage, as well as leaf burn and clogging of drip systems. The crops are categorized as sensitive, semi-tolerant, and tolerant concerning soil type (fine, moderately fine, moderately coarse, and coarse), rainfall, and salt tolerance [67].

The recommended upper permissible concentrations of trace elements such as aluminum (Al), arsenic (As), beryllium (Be), cadmium (Cd), cobalt (Co), chromium (Cr), copper (cu), fluoride (F), iron (Fe), lithium (Li), manganese (Mn), molybdenum (Mo), nickel (Ni), lead (Pb), selenium (Se), vanadium (V), and zinc (Zn) in irrigation water as per IS 11624: 2019 are given in Table 15.4. Based on the response to soil salinity, crops are grouped as highly sensitive (lentil, mash, chickpea, beans, peas, carrot, onion, lemon, orange, grape, peach, plum, pear, and apple), medium sensitive (radish, cowpea, broad bean, vetch, cabbage, cauliflower, cucumber, gourds, tomato, sweet potato, sorghum, minor millets, maize, clover, and berseem), moderately tolerant (spinach, sugarcane, Indian mustard, transplanted rice, wheat, pearl millet, oats, alfalfa, blue panic grass, para grass, Rhodes grass, Sudan grass, guava, pomegranate, and *Acacia*), and highly tolerant (barley, cotton, sugar beet, turnip, tobacco, safflower, rapeseed, Karnal grass, date palm, ber, *Mesquite*, *Casuarina*, *Tamarix*, and *Salvadora*).

Table 15.3 The quality grouping of irrigated water based on electrical conductivity (EC), sodium absorption ratio (SAR), and residual sodium carbonate (RSC), as per IS 11624: 2019.

	Water quality							
		Saline			**Alkali**			
	Good	**Marginal**	**Saline**	**High**	**Marginal**	**Alkali**	**High**	**Toxic**
EC (dS/m)	<0.7	0.7–3.0	>3.0	>3.0	<4	<4	Variable	Variable
SAR (mmol/L)	<10	<10	<10	>10	<10	<10	>10	Variable
RSC (me/L)	<2.5	<2.5	<2.5	<2.5	2.5–4.0	>4.0	>4.0	Variable

Table 15.4 The recommended upper permissible concentrations of trace and toxic elements in irrigation water, as per IS 11624: 2019.

Element	Maximum permissible concentration (mg/L)
Al	5.0
As	0.1
Be	0.1
Cd	0.01
Co	0.05
Cr	0.1
Cu	0.2
F	1.0
Fe	5.0
Li	2.5
Mn	0.2
Mo	0.01
Ni	0.2
Pb	5.0
Se	0.02
V	0.1
Zn	2.0

In addition to the effective utilization of recycled wastewater for agriculture, many sustainable agricultural practices, including rainwater harvesting, zero tillage, exploitation of micro irrigation systems, cultivation of drought-tolerant crops, organic farming, crop rotation, etc., should be followed and practiced in India [2, 66, 68]. A schematic diagram demonstrating the utilization of recycled wastewater generated from STPs toward sustainable agriculture, after microbial disinfection, is given in Figure 15.6.

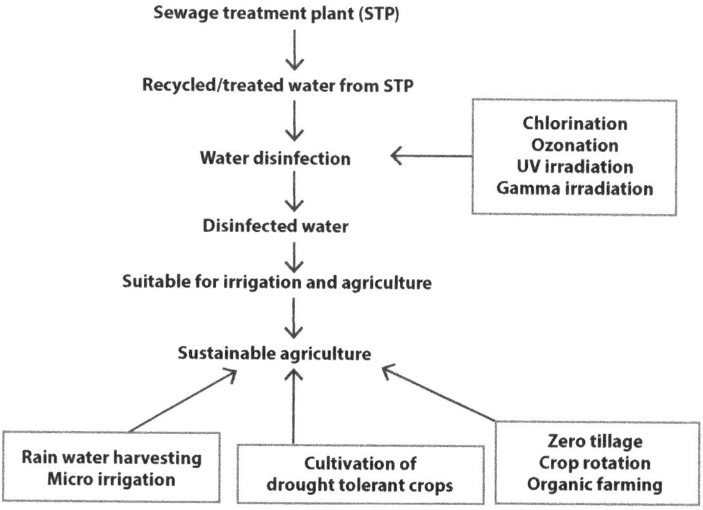

Figure 15.6 A schematic diagram demonstrating the utilization of disinfected recycled wastewater toward sustainable agriculture.

15.6 Conclusions

Wastewater serves as an undervalued/untapped water source and the technological improvement and economic affordability of STPs have made wastewater recycling a reality and expanded the most sustainable applications of reclaimed water. Thus, the recycled water generated from STPs after disinfection treatment meeting the specifications can be used for irrigation and agriculture. Sustainable agriculture improves water supply, food production, and, ultimately, food security, which is the need of the hour for a country like India. The infrastructure needed for treating, storing, and delivering reclaimed water from the STPs should be built in every municipality in the country. Issues such as chemical and radiation safety, inter-basin water transfer rights, and education of wastewater/sewage farm workers on sanitary rules, personal hygiene, sewage irrigation, etc., should be taken care of. Studies should be carried out on residual chlorine stress on crops and pilot and field-level demonstrations with different edible and nonedible crops using disinfected reclaimed wastewater should be carried out. Also, revenue generation through the sale of disinfected treated wastewater and sludge, biogas utilization for power generation, pisciculture, carbon credit, etc., should be evaluated and monitored [31, 59].

Acknowledgments

The author would like to thank Dr Mullapudi Venkata Balarama Krishna, Head, Environmental Science and Nanomaterials Section and Dr Athyala Christian Sahayam, Head, NCCCM/BARC, for their continuous support and encouragement throughout the work.

References

1. Namburi, K.R., Kora, A.J., Chetukuri, A., Kota, V.S.M.K., Biogenic silver nanoparticles as an antibacterial agent against bacterial leaf blight causing rice phytopathogen *Xanthomonas oryzae* pv. *oryzae*. *Bioprocess Biosyst. Eng.*, 44, 1975, 2021.
2. India Water Portal, Agriculture. https://www.indiawaterportal.org/topics/agriculture, 2023 (Accessed 7th April 2023).
3. Khokhar, T., Chart: Globally, 70% of freshwater is used for agriculture. https://blogs.worldbank.org/opendata/chart-globally-70-freshwater-used-agriculture#:~:text=In%20most%20regions%20of%20the,percent%20increase%20in%20water%20withdrawals., 2017 (Accessed 7th April 2023).
4. Norton-Brandão, D., Scherrenberg, S.M., van Lier, J.B., Reclamation of used urban waters for irrigation purposes – A review of treatment technologies. *J. Environ. Manage.*, 122, 85, 2013.
5. Vaidyanathan, A., Sivasubramaniyan, K., Efficiency of water use in Indian agriculture. Working Paper No. 183, 1-86, 2004.
6. The World Bank, How is India addressing its water needs? https://www.worldbank.org/en/country/india/brief/world-water-day-2022-how-india-is-addressing-its-water needs#:~:text=Communities%20lead%20India's%20groundwater%20conservation,highest%20rates%20of%20groundwater%20depletion., 2023 (Accessed 07th April 2023).
7. UNICEF, Water scarcity. https://www.unicef.org/wash/water-scarcity, 2023 (Accessed 07th April 2023).
8. Rana, M., Guleria, V., Water scarcity in India: A threat to sustainable management of water. *Essence-Int. J. Environ. Rehab. Conserv.*, 9, 35, 2018.
9. FAO, Sustainability, food security and climate change: Three intertwined challenges. https://www.fao.org/climate-smart-agriculture-sourcebook/concept/module-a1-introducing-csa/chapter-a1-1/fr/, 2023 (Accessed 07th Apr 2023).
10. FAO, Climate change and food security: risks and responses. I5188E/1/11.15, 1-122, 2015.
11. Filali, H., Barsan, N., Souguir, D., Nedeff, V., Tomozei, C., Hachicha, M., Greywater as an alternative solution for a sustainable management of water resources-A review. *Sustainability*, 14, 665, 2022.

12. Kora, A.J., Rastogi, L., Kumar, S.J., Jagatap, B.N., Physico-chemical and bacteriological screening of Hussain Sagar lake: An urban wetland. *Water Sci.*, 31, 24, 2017.
13. Meena, S.D., National inventory of sewage treatment plants, Central Pollution Control Board Delhi, India, pp. 1-187, 2021.
14. Jamwal, P., Mittal, A.K., Reuse of treated sewage in Delhi city: Microbial evaluation of STPs and reuse options. *Resources, Conserv. Recycycling*, 54, 211, 2010.
15. Indian Institute of Technology, Bombay, Niti Aayog, National Mission for Clean Ganga, Royal Danish Embassy, India, Urban wastewater scenario in India, NITI Aayog, New Delhi, pp. 1-108, 2022.
16. Mahesh, J., Amerasinghe, P., Pavelic, P., An integrated approach to assess the dynamics of a peri-urban watershed influenced by wastewater irrigation. *J. Hydrol.*, 523, 427, 2015.
17. Amerasinghe, P., Jampani, M., Drechsel, P., Cities as sources of irrigation water: An Indian scenario. *Water Policy Res. Highlight*, 53, 1, 2012.
18. Amerasinghe, P., Weckenbrock, P., Simmons, R., Acharya, S., Drescher, A., An atlas of water quality, health and agronomic risks and benefits associated with wastewater irrigated agriculture: A study from the banks of the Musi river, India, German Federal Ministry for Economic Cooperation and Development, 2009.
19. Deekshish, M., High contamination found 40 meters below river-Musi water unfit even for agri, Deccan Chronicle, Deccan Chronicle Holdings Limited, Hyderabad, pp. 1-2, 2021.
20. Rahman, M.M., Hagare, D., Maheshwari, B., Bayesian belief network analysis of soil salinity in a peri-urbanagricultural field irrigated with recycled water. *Agric. Water Manage.*, 176, 280, 2016.
21. Weber, S., Khan, S., Hollender, J., Human risk assessment of organic contaminants in reclaimed wastewater used for irrigation. *Desalination*, 187, 53, 2006.
22. Alghobar, M.A., Suresha, S., Evaluation of metal accumulation in soil and tomatoes irrigated with sewage water from Mysore city, Karnataka, India. *J. Saudi Soc. Agric. Sci.*, 16, 49, 2017.
23. Sudhir, U., Contaminated water kills 4 in Hyderabad. https://www.ndtv.com/india-news/contaminated-water-kills-4-in-hyderabad-393548, 2009 (Accessed 26th April 2023).
24. Mishra, R., COVID-19 surveillance using sewage samples in Hyderabad, in: CCMB (Ed.) CCMB, 2020.
25. Sagar, T.V., Sewage and effluents disposal facilities in Hyderabad metro development authority area, India. *Int. J. Emerging Technol. Innovative Eng.*, 1, 11, 2015.
26. Bommala, N., Hyderabad adopts smart tech for sewage treatment, Telangana Today, Telangana Publications Pvt Ltd, Hyderabad, 2022.

27. Dhareshwar, Y., Generation and treatment of sewage in Hyderabad. https://numerical.co.in/numerons/collection/5e8fd4f6f3c42b5803c09a3b, 2021 (Accessed 27th April 2023).
28. NEERI, Action plan for rejuvenation of river stretches (priority I and II) in Telangana state, Revised Report, CSIR-National Environmental Engineering Research Institute (NEERI), Hyderabad, 2019, pp. 1-202.
29. Singh, S.B.J., Hyderabad officials expedite works on Sewerage Treatment Plants as deadline nears, The New Indian Express, Indian Express Ltd, New Delhi, India, 2023.
30. Rinat, Z., Israel is undisputed world leader in using purified wastewater for crops. https://www.haaretz.com/life/2016-12-11/ty-article/.premium/israel-is-undisputed-champion-in-using-purified-wastewater-for-crops/0000017f-e191-d9aa-afff-f9d9d8f60000, 2016 (Accessed 2nd May 2023).
31. Norwalk Wastewater Equipment Company, Wastewater chlorination: An enduring public health practice. https://www.norweco.com/2019/08/wastewater-chlorination-an-enduring-public-health-practice/, 2019 (Accessed 04th November 2019).
32. Pant, A., Mittal, A.K., Monitoring of pathogenicity of effluents from the UASB based sewage treatment plant. *Environ. Monitor. Assess.*, 133, 43, 2007.
33. Verma, K., Gupta, A.B., Singh, A., Optimization of chlorination process and analysis of THMs to mitigate ill effects of sewage irrigation. *J. Environ. Chem. Eng.*, 5, 35409, 2017.
34. Singh, H., Bhardwaj, S.K., Khatri, M., Kim, K.-H., Bhardwaj, N., UVC radiation for food safety: An emerging technology for the microbial disinfection of food products. *Chem. Eng. J.*, 417, 128084, 2021.
35. CPHEEO, Compendium of recycle and reuse of wastewater in 54 Million Plus Cities. 1-131, 2021.
36. Rao, T.S., Kora, A.J., Chandramohan, P., Panigrahi, B.S., Narasimhan, S.V., Biofouling and microbial corrosion problem in the thermo-fluid heat exchanger and cooling water system of a nuclear test reactor. *Biofoul.*, 25, 581, 2009.
37. Kora, A.J., Toleti, S.R., Narasimhan, S.V., Evaluation of cooling water treatment programme at RAPS 3 & 4 with reference to chlorination and microbial control, *Symposium on Operational and Environmental Issues Concerning Use of Water as Coolant in Power Plants and Industries: Proceedings, Department of Atomic Energy*, Mumbai Kalpakkam, India, 2008, pp. 318-321.
38. Kora, A.J., Toleti, S.R., Microbiological investigations in the context of cooling water treatment programme in a test reactor, *Proceedings of the special workshop on water chemistry*, Department of Atomic Energy, India, Kalpakkam, India, 2001, pp. 74-80.
39. Satpathy, K.K., Anupkumar, B., Kora, A.J., Bio-fouling in the PFBR cooling system, *One day workshop on corrosion management in fast breeder reactors*, Kalpakkam, India, 2001, pp. 1-20.

40. Antonelli, M., Mezzanotte, V., Nurizzo, C., Wastewater disinfection by UV irradiation: Short and long-term efficiency. *Environ. Eng. Sci.*, 25, 363, 2008.
41. Ashauer, R., Post-ozonation in a municipal wastewater treatment plant improves water quality in the receiving stream. *Environ. Sci. Eur.*, 28, 1, 2016.
42. Kreuzinger, N., Schaar, H., Ozonation of tertiary treated wastewater – A solution for micro pollutant removal?, in: P. Hlavinek, I. Winkler, J. Marsalek, I. Mahrikova (Eds.) *Advanced Water Supply and Wastewater Treatment: A Road to Safer Society and Environment*, Springer Netherlands, Dordrecht, 2011, pp. 131-146.
43. Hussain, K., Khan, N.A., Vambol, V., Vambol, S., Yeremenko, S., Sydorenko, V., Advancement in Ozone base wastewater treatment technologies: Brief review. *Ecol. Questions*, 33, 7, 2022.
44. Cristiano, L., Could ozone be an effective disinfection measure against the novel coronavirus (SARS-CoV-2)? *J. Prevent. Med. Hygiene*, 61, E301, 2020.
45. Sensorex, Ozone disinfection in water treatment. https://sensorex.com/2021/07/19/ozone-disinfection-treatment/, 2021 (Accessed 27th April 2023).
46. Prabha, V., Barma, R.D., Singh, R., Madan, A., Ozone technology in food processing: A review. *Trends Biosci.*, 8, 4031, 2015.
47. Kumar, A., Raj, A., Gupta, A., Gautam, S., Kumar, M., Bherwani, H., Anshul, A., Pollution free UV-C radiation to mitigate COVID-19 transmission. *Gondwana Res.: Int. Geosci J.*, 114, 78, 2023.
48. Acher, A., Fischer, E., Turnheim, R., Manor, Y., Ecologically friendly wastewater disinfection techniques. *Environ. Sci. Eur.*, 31, 1398, 1997.
49. Banerjee, R., Roy, P., Das, S., Paul, M.K., A hybrid model integrating warm heat and ultraviolet germicidal irradiation might efficiently disinfect respirators and personal protective equipment. *Amer. J. infection Control*, 49, 309, 2021.
50. Rawat, K.P., Sharma, A., Rao, S.M., Microbiological and physicochemical analysis of radiation disinfected municipal sewage. *Water Res.*, 3 2, 737, 1998.
51. Benny, P.G., Shah, M.R., Sabharwal, S., Evaluation of the efficiency and throughput of a gamma irradiator treating municipal sewage sludge. *Appl. Radiat. Isot.*, 69, 604, 2011.
52. Kora, A.J., Commercial bacterial and fungal microbial biostimulants used for agriculture in India: An overview, in: Inamuddin, Charles Oluwaseun Adetunji, Mohd Imran Ahamed, T. Altalhi (Eds.), Microbial Biostimulants for Sustainable Agriculture and Environmental Bioremediation, CRC Press, USA, 2022, pp. 159-175.
53. Gautam, S., Shah, M.R., Sabharwal, S., Sharma, A., Gamma irradiation of municipal sludge for safe disposal and agricultural use. *Water Environ. Res.*, 77, 472, 2005.
54. Garg, N.K., Varshney, L., Gamma Radiation Technology for Hygienization of Municipal Dry Sewage Sludge. In: Singh, V.P., Yadav, S., and Yadava, R.N. Sewage and Biomass from Wastewater to Energy (eds.) *Energy and Environment*, pp. 103-106. Springer Singapore, Singapore, 2018.

55. Tripathi, J., Saxena, S., Gautam, S., Simulation study to assess the effectiveness of gamma radiation for inactivation of viruses on food packaging material. *Rad. Phys. Chem.*, 204, 110678, 2023.
56. Gautam, S., Ghanty, T.K., Radiation processing as a sustainable and green technology to ensure food security, safety and promote international trade, In: A.K. Tyagi , P.R.V. Rao (Eds.), *Atomic Energy in India: Achievements since Independence,* Homi Bhabha National Institute, Mumbai, 2022, pp. 211-217.
57. Jan, K., Bashir, K., Jan, S., Khan, A.L., Principles and Sources of Gamma Irradiation. In: Knoerzer, K.C.O.M.A., and Muthukumarappan, K.B.T.-I.F.P.T. (eds.) *Innovative Food Processing Technologies: A Comprehensive Review*, pp. 67–70. Elsevier, Oxford, 2020.
58. BARC, Radiation technology for food &agro commodities. https://www.barc.gov.in/pubaware/food.html, 2023 (Accessed 11th May 2023).
59. Kumar, A., Guidelines for preparation of project reports under national river conservation plan and national Ganga river basin authority in: N.R.C.D. Ministry of Environment & Forests (Ed.) Alternate Hydro Energy Centre, Indian Institute of Technology, Roorkee, New Delhi, 2010, pp. 1-176.
60. Press Trust of India, NGT orders stricter norms for effluent discharge from sewage treatment plants, Business Standard, Business Standard Private Ltd, New Delhi, India, 2019.
61. Staff Reporter, HMWSSB to sell treated sewage water. https://www.thehindu.com/news/cities/Hyderabad/HMWSSB-to-sell-treated-sewage-water/article17396706.ece, 2017 (Accessed 04th November 2019).
62. Ministry of Housing and Urban Affairs, Swachh Bharat Mission (Urban) SBM Water Plus Protocol. SBM Water Plus Protocol, 1-12, 2020.
63. Eenadu, Sewage treatment became plus, Eenadu, Vasundhara Publications Pvt Ltd, Hyderabad, p. 10, 21st Aug 2021.
64. Eenadu, Village with fast growth, Eenadu, Vasundhara Publications Pvt Ltd, Hyderabad, p. 14, 19th April 2023.
65. Eenadu, Special electricity tariff for STPs, Eenadu, Vasundhara Publications Pvt Ltd, Hyderabad, p. 9, 08th April 2023.
66. CPHEEO, Chapter 7: Recycling and reuse of sewage, in: P.A. Engineering (Ed.) 7.3.1 Agriculture, Ministry of Housing and Urban Affairs, Government of India, New Delhi, India, 2013.
67. Indian Standard IS 11624: 2019 Quality of Irrigation Water-Guidelines (First Revision), Bureau of Indian Standards, New Delhi, India, pp. 1-9, 2019.
68. Kora, A.J., Applications of Waste Decomposer in Plant Health Protection, Crop Productivity and Soil Health Management. In: Inamuddin, Ahamed, M.I., and Prasad, R. (eds.) *Application of Microbes in Environmental and Microbial Biotechnology*, pp. 609–624. Springer Nature Singapore, Singapore, 2022.

Index

Also of Interest

Check out these other forthcoming and published titles from Scrivener Publishing

Books by the same editor from Wiley-Scrivener

SOLAR ENERGY CONCENTRATORS, Edited by Inamuddin, ISBN: 9781394204328. Discover the latest techniques and applications for solar energy concentrators in this essential guide for academics, researchers, environmentalists, and professionals seeking to harness the power of solar energy while reducing environmental impact and costs.

WIND ENERGY STORAGE AND CONVERSION: From Basics to Utilities, Edited by Inamuddin, Tariq Altalhi, and Mohammad Luqman, ISBN: 9781394204335. This book provides a comprehensive guide to the benefits and developments of wind energy, including energy storage and conversion methods, making it a must-read for those interested in sustainable energy.

ADVANCED REDOX FLOW TECHNOLOGIES, Edited by Inamuddin and Tariq Altalhi, ISBN: 9781119904793. This book serves as a comprehensive guide to redox flow technologies, from their basic principles to their applications. As a result, this book provides a thorough review of craftsmanship in the subject, as well as the potential for future advances.

HANDBOOK OF WATER POLLUTION, Edited by Inamuddin, Tariq Altalhi, and Arwa Alrooqi, ISBN: 9781119904809. This handbook covers a wide range of water pollution issues and remedies. Researchers, students, and other governmental and non-governmental groups will all benefit from this informative guide. In detail, this book will include chapters in the areas of water pollution, analysis, and solutions.

ESSENTIAL OILS: Extraction Methods and Applications, Edited by Inamuddin, Tariq Altalhi, and Jorddy Neves Cruz, ISBN: 9781119829355. Written and edited by some of the world's top experts in the field, this exciting new volume provides up-to-date information on essential oils' chemical structure, therapeutic, and biological actions, and defines their functional uses and evaluates the advantages and disadvantages of their use in a variety of sectors.

PROTON EXCHANGE MEMBRANE FUEL CELLS: Electrochemical Methods and Computational Fluid Dynamics, Edited by Inamuddin, Omid Moradi, and Mohd Imran Ahamed, ISBN: 9781119829331. Edited by one of the most well-respected and prolific engineers in the world and his team, this book provides a comprehensive overview of hydrogen production, conversion, and storage, offering the scientific literature a comprehensive coverage of this important fuel.

MATERIALS FOR HYDROGEN PRODUCTION, CONVERSION, AND STORAGE, edited by Inamuddin, Tariq Altalhi, Sayed Mohammed Adnan, and Mohammed A. Amin, ISBN: 9781119829348. Edited by one of the most well-respected and prolific engineers in the world and his team, this book provides a comprehensive overview of hydrogen production, conversion, and storage, offering the scientific literature a comprehensive coverage of this important fuel.

FUNDAMENTALS OF SOLAR CELL DESIGN, edited by Inamuddin, Mohd Imran Ahamed, Rajender Boddula, and Mashallah Rezakazemi, ISBN: 9781119724704. Edited by one of the most well-respected and prolific engineers in the world and his team, this book provides a comprehensive overview of solar cells and explores the history of evolution and present scenarios of solar cell design, classification, properties, various semiconductor materials, thin films, wafer-scale, transparent solar cells, and other fundamentals of solar cell design.

Biofuel Cells, Edited by Inamuddin, Mohd Imran Ahamed, Rajender Boddula, and Mashallah Rezakazemi, ISBN: 9781119724698. This book covers the most recent developments and offers a detailed overview of fundamentals, principles, mechanisms, properties, optimizing parameters, analytical characterization tools, various types of biofuel cells, edited by one of the most well-respected and prolific engineers in the world and his team.

Biodiesel Technology and Applications, Edited by Inamuddin, Mohd Imran Ahamed, Rajender Boddula, and Mashallah Rezakazemi, ISBN: 9781119724643. This outstanding new volume provides a comprehensive overview on biodiesel technologies, covering a broad range of topics and practical applications, edited by one of the most well-respected and prolific engineers in the world and his team.

Applied Water Science Volume 1: Fundamentals and Applications, Edited by Inamuddin, Mohd Imran Ahamed, Rajender Boddula and Tauseef Ahmad Rangreez, ISBN: 9781119724766. Edited by one of the most well-respected and prolific engineers in the world and his team, this is the first volume in a two-volume set that is the most thorough, up-to-date, and comprehensive volume on applied water science available today.

Applied Water Science Volume 2: Remediation Technologies, Edited by Inamuddin, Mohd Imran Ahamed, Rajender Boddula and Tauseef Ahmad Rangreez, ISBN: 9781119724735. The second volume in a new two-volume set on applied water science, this book provides understanding, occurrence, identification, toxic effects and control of water pollutants in aquatic environment using green chemistry protocols.

Potassium-Ion Batteries: Materials and Applications, Edited by Inamuddin, Rajender Boddula, and Abdullah M. Asiri, ISBN: 9781119661399. Edited by one of the most well-respected and prolific engineers in the world and his team, this is the most thorough, up-to-date, and comprehensive volume on potassium-ion batteries available today.

Rechargeable Batteries: History, Progress, and Applications, edited by Rajender Boddula, Inamuddin, Ramyakrishna Pothu, and Abdullah M. Asiri, ISBN: 9781119661191. Edited by one of the most well-respected and prolific engineers in the world and his team, this is the most thorough, up-to-date, and comprehensive volume on rechargeable batteries available today.

Printed and bound by CPI Group (UK) Ltd, Croydon, CR0 4YY

17/09/2026

14964280-0006